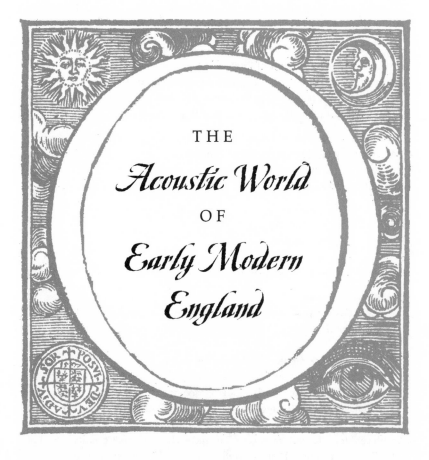

THE

Acoustic World

OF

Early Modern

England

ATTENDING TO THE O-FACTOR

Bruce R. Smith

THE UNIVERSITY OF CHICAGO PRESS

Chicago and London

Bruce R. Smith is professor of English at Georgetown University. He is the author of *Ancient Scripts and Modern Experience on the English Stage, 1500–1700* (1988) and *Homosexual Desire in Shakespeare's England: A Cultural Poetics* (1991), the latter published by the University of Chicago Press.

The University of Chicago Press, Chicago 60637
The University of Chicago Press, Ltd., London
© 1999 by The University of Chicago
All rights reserved. Published 1999

08 07 06 05 04 03 02 01 00 99   1 2 3 4 5
ISBN: 0-226-76376-5 (cloth)
ISBN: 0-226-76377-3 (paper)

Library of Congress Cataloging-in-Publication Data

Smith, Bruce R., 1946–
    The acoustic world of early modern England : attending to the O-factor /
    Bruce R. Smith.
        p.   cm.
    Includes bibliographical references and index.
    ISBN 0-226-76376-5 (alk. paper).—ISBN 0-226-76377-3 (pbk. : alk. paper)
        1. English literature—Early modern, 1500–1700—History and criticism.
    2. Popular literature—England—History and criticism.   3. Popular culture—
    England—History—16th century.   4. Popular culture—England—History—
    17th century.   5. Oral tradition—England—History—16th century.   6. Oral
    tradition—England—History—17th century.   7. Language and culture—
    England—History.   I. Title.
    PR428.P65S65   1999
    820.9'003—dc21   98-46445
                                                                                       CIP

FOR

Mae Reed Chunn,

Elias F. Mengel Jr.,

AND

Maggie Patten Smith,

THREE WHOSE VOICES

ABIDE.

Contents

Illustrations

$\mathcal{A}$ c k n o w l e d g m e n t s

"Huh?" was not quite everybody's reaction when I told them what I was about in this project. It is a pleasure now to thank some of those people and institutions. Research and writing were generously supported by a traveling fellowship from the Agecroft Association, Richmond, Virginia; a short-term research fellowship at the Folger Shakespeare Library, Washington, D.C.; three Summer Academic Research Grants, sabbatical leave, and a two-year reduced teaching schedule from Georgetown University; an International Globe Fellowship at Shakespeare's Globe, London; a short-term Mellon Fellowship at the Huntington Library, San Marino, California; and a residential fellowship at the Virginia Foundation for the Humanities and Public Policy, Charlottesville, Virginia. The most sustained and diverse support came from the Folger Shakespeare Library. I wish to thank in particular Werner Gundersheimer, Director, for his confidence in the project at a crucial time in its genesis; Barbara Mowat, Director of Academic Programs, and Lena Orlin, former Executive Director of the Folger Institute, for inviting me to offer a seminar on the subject of the book; Betsy Walsh, Harold Batie, LuEllen DeHaven, Rosalind Larry, Camille Seerattan, and other members of the reading room staff for their efficiency, patience, and friendliness; and Georgianna Ziegler, Reference Librarian, for her substantive suggestions and logistical support.

Several of the texts I consider here were first brought to my attention by scholars with areas of expertise beyond my own, including Linda Austern, Steve Buhler, Tom Cogswell, Mary Fuller, Andy Gurr, Kim Hall, Fred Hoxie, Karen Kupperman, David Linton, Rick Mallette, Rich McCoy, Jean Miller, Gail Paster, Peter Patrick, Francesca Santovetti, Mike Santulli, Michael Schoenfeld, Debora Shuger, Patrick Spottiswoode, and Stanley Wells. There are doubtless others who made such suggestions, and I beg their forgiveness if I have not included them here. Alan Dessen, Lynne Magnusson, Rick Mallette, Leah Marcus, Lois Potter, and Leslie Thomson generously shared work in advance of its publication. For help in construing Greek texts I am indebted to Victoria Pedrick, and for checking references to Martha Fay, Jennifer Margiotta, and Georgianna Ziegler. Wilson Pang of Honolulu executed the musical quotations as well as figures 8.7–8.9, and the artistry of Barbara Pope Book Design, also of Honolulu, turned my rough-and-ready sketches into the finished illustrations that appear as figures 3.3, 8.1–8.4, and 8.6 and the unnumbered figures in chapters 1 and 2.

Fittingly, many of the ideas recorded in these pages emerged during conversations in three seminars I convened on the subject of the book in 1994 and 1996. For these creative interchanges I am grateful to the following individuals: Catherine Fitzmaurice, Noam Flinker, Andy Gurr, Patricia Harris, Dakin Hart, Frederik Jonsson, Charles King, Sean Moore, Alan Nelson, Janet Starner, Jeffrey Weinstock, and Eric Wilson at the Folger Institute; Mark Bailey, Michael Blackie, Anne Flentgen, Sheila Gagen, Kimberly Miller, Mary Mullane, Rachel Owen, Arun Rath, Scott Rubenberg, Kendra Stitt, and Marcus Trompos at Georgetown University; and Kim Bannigan, Bobby Carey, Nicola Niedermair, and Todd Warner at the New Mexico campus of Middlebury College's Bread Loaf School of English. Many of these people will recognize their voices here.

The Acoustic World of Early Modern England had its origin in a paper titled "Towards an Acoustic Criticism of Shakespeare's Scripts" (original title "Lately I've Been Hearing Voices"), which I delivered at the annual meeting of the Shakespeare Association of America in Kansas City in 1992. Some of this material now figures in chapter 8. It was the response of the audience in Kansas City that inspired me to listen for voices and sounds outside the theater as well as within. I especially appreciate the encouragement of Jim Siemon, who helped judge the competition in which the paper was chosen for the program. The part of chapter 4 having to do with punctuation schemes originally appeared as an essay titled "Prickly Characters" in the anthology *Reading and Writing in Shakespeare,* edited by David M. Bergeron (1996). The University of Delaware Press has granted permission for this material to be included here.

For his faith in the project from the very beginning and for his unfailing support the whole way through I am grateful to my partner Gordon Davis.

—B.R.S.
Bread Loaf School of English
South San Ysidro, New Mexico
July 1998

PART
ONE
Around

1 Opening

Let us begin with a sound.

To begin with a sound, we must, in our present situation, begin with you. Make a sound. Any sound will do, but, in terms of what has come before and what is coming after, one sound in particular would be most apt:

[o:].

(That's the vowel sound in "oh.") Now, there are several ways to think about what you have just performed: (1) as a physical act, as something you have done with your body; (2) as a sensory experience, as something you have heard; (3) as an act of communication, as something you have projected into the world around you; and (4) as a political performance, as something you have done *because of* other people, if not in this particular case *with* other people, *for* other people, and *to* other people. Each of these ways of thinking about ○ makes its own demands on your attention, each comes equipped with its own terms of analysis, each bears its own intellectual history, each suffers its own limitations.

○ AS PHYSICAL ACT

To produce the [o:], you have performed, first of all, an extremely complex physical feat, and you haven't even had to think about it. For that, you can thank years of practice. What do you notice when you *do* think about it? Most likely, your first impulse is to try and visualize the process. John Hart does just that in *An Orthographie, conteyning the due order and reason, howe to write or paint th[']image of ● mannes voice, most like to the life or nature* (1569). The vowel *o* is made, he observes, "by taking away of all the tongue, cleane from the teeth or gummes, as is sayde for the a, and turning the lippes rounde as a ring, and thrusting forth of a sounding breath, which roundnesse to signifie the shape of the letter, was made (of the first inventor) in like sort, thus o" (H2ᵛ). That is to say, [o:] is O because the lips make an ○.

Instead of *seeing* the sound, however, try to *feel* the sound. Put a thumb and a finger on each side of your larynx or Adam's apple and make the sound again:

[o:].

Can you feel the vibration? The source is inside your larynx. Two gristly tissues, the vocal cords, are vibrating. The energy that drives them comes from your chest, from your lungs. The steady stream of air from your lungs has been transformed by your vocal cords into several different series of air waves. As these different air waves have reached your mouth, you have given the sound its distinctive quality—its [o:]-ness—by opening your throat, or pharynx, to a certain width and by positioning your tongue, teeth, and lips in precise ways: pharynx expanded, tongue relatively low and slightly cupped, teeth relatively far apart, lips pursed. (You won't be able to feel this, but you have also closed off the nasal cavity that you might have opened up if the sound in question had been [n] or [ng].) Pharynx, oral cavity, and nasal cavity together form a resonator for the sound waves generated by your larynx. The particular way in which you have shaped pharynx, oral cavity, and nasal cavity serves to accentuate certain of the sound waves you have produced in your larynx and to suppress others. The result is [o:]. To appreciate the range of possible positions for sounds in your mouth, you might want to pronounce this sequence of vowel sounds: [u] as in "hook," [a] as in "ah ha," [i] as in "hit." Can you feel the way the sounds move progressively forward in your mouth? Can you hear the shifts in pitch from high to low? Can you hear the changes in loudness? In physiological terms, the lungs, the larynx, and the "supralaryngeal" structure of pharynx, oral cavity, and nasal cavity form a three-part apparatus for making sounds. In acoustical terms, those three parts constitute an energy source (the lungs), a sound source (the larynx), and a mechanism for amplifying the sound and propagating it into the air (the supralaryngeal vocal tract). What you have, in effect, is a sound-producing device in the shape of a long, flexible cylinder, open at one end.

About this vocal apparatus early modern physiology was well informed (fig. 1.1). Helkiah Crooke in *A Description of the Body of Man* (1616) discusses its three parts separately: the lungs (384–388), the windpipe or "weazon" that leads to the larynx or "throttle" (388–391, 633–646), and the mouth, palate, tongue, and uvula (621–631). Pierre de La Primaudaye (1618) compares the whole thing to a portative organ, in which the lungs act as bellows and the larynx as the pipes. The big difference is that an organ needs many pipes to produce a variety of sounds, whereas the vocal tract uses only one, thanks to the capacity of muscles and cartilage to vary the size of the opening. The only confusion concerned which was the primary fissure for producing sound—the glottis (as Galen believed) or the recently anatomized vocal cords (as Arantius was claiming) (Crooke 1616: 641). Whichever fissure it might be, knowledge about how various sounds are produced was secure. When the space between the tissues is open

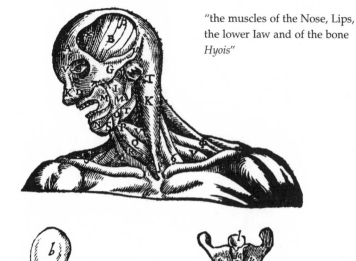

"the muscles of the Nose, Lips, the lower Iaw and of the bone *Hyois*"

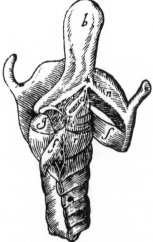

"all the proper Muscles [of the larynx], the Clefte, the Epiglottis or After-Tongue and the Gristles" N.B. especially *d,* "the *glottis* cleft or whistle"

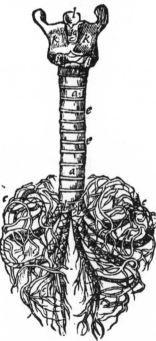

"the foreside of the throttle and the weazon, and the branches of the same disseminated through the substance of the lungs"

Figure 1.1. The three parts of the human vocal apparatus. From Helkiah Crooke, *Microcosmographia: A Description of the Body of Man* (1616). Reproduced by permission of the Folger Shakespeare Library, Washington, D.C.

to its fullest, La Primaudaye explains, the voice sounds "big and ob-scure"; when the tissues are pressed together, it sounds "small, cleere, and shril." The result is an instrument of extraordinary range and flexibility:

> For as euery one is disposed to lift vp or to depresse his voice, to in-large or restraine the pipes & instruments thereof, hee may speake ei-ther higher or lower, bigger or smaller, or clearer, and set what sound, tune, and accent he pleases vpon the speach, which hee will pro-nounce. Wherefore we may very well say, that euery one carrieth about with him and within him selfe very faire and strange Organes, vpon which hee may play at all houres at his pleasure, either in singing or speaking.

The slit between the tissues—now shorter, now longer—operates in effect like "a litle tongue." Once the sounds produced by the vocal cords have reached the mouth, they are shaped by the tongue proper, as well as by the palate, teeth, lips, and nose (La Primaudaye 1618: 377–382). In its contributions to the final sound, the oral cavity is compared by Ambroise Paré to the upper part of a lute: the uvula in particular acts "as a certaine quill to play withall" (1634: 143). From one point of view, then, ○ can be thought about in terms of speech physiology.

○ AS SENSORY EXPERIENCE

To see sound production is easy enough in an organ or a lute; to see it in your own vocal tract takes an extraordinary act of imagina-tion. It is not, after all, vision or even the feel of tongue against teeth that you most depend upon to monitor your making of sounds, but hearing. Or rather listening. You can *hear* sound waves within a cer-tain range of frequencies and a certain range of amplitudes, but you choose to *listen* with varying degrees of attention. About hearing you have no choice: you can shut off vision by closing your eyes, but from birth to death, in waking and in sleep, the coils of flesh, the tiny bones, the hair cells, the nerve fibers are always at the ready (Fry 1977: 62; Handel 1989: 461–545). To listen, however, is a choice. What's more, you can choose *how* to listen (Truax 1984: 19–24). At your most alert, you can listen *in search* of something you want to hear. A perfect example is speech. What we *hear* when someone speaks is a stream of constantly changing sounds in which consonant sounds merge with vowel sounds merge with consonant sounds merge with vowel sounds, according to a principle speech physiologists call "coarticula-tion." What we *listen for* when someone speaks is a series of discrete, recognizable sounds, or phonemes (Handel 1989: 134–162, 185–217). With less concentration, you can listen *in readiness* for sounds that may or may not occur. Walking in the park, you happen to hear some dogs barking nearby. "Ah!" you think to yourself. "Dogs." Finally, if your aural attention is focused elsewhere, you can listen *to the background.* Dogs may bark and you will not hear them at all. Hearing is a physiological constant; listening is a psychological variable. Make the [o:] again. Listen. What is there to listen *to*? If you think about it, you can probably distinguish the resonance your voice made inside your own body from the sound you projected into the space around you. A "person" like you is, after all, a "through-sounding," a "*persona*." Practically all of what you hear of yourself comes not from the air around you, but through vibrations in the bones and tissues of your skull. This pathway through the head modifies the sound waves, so that you hear a version of the sound that is higher and lighter than what I hear as an outsider. Hence the sense of estrangement we both feel when we listen to our voices on a tape recorder: "Who? That *can't* be me!" In acoustic terms, there are, then, two versions of my speech: the "private" speech I hear inside my body and the "public" speech that others hear outside (Fry 1977: 94).

Since you are reading this text, you are probably sitting all by yourself. If you can find someone else to occupy the position of "by" for a moment, ask that person to make the sound of [o:] back to you. (For the maximum effect, you might want to try this with your eyes closed.) What did you hear this time? Because the sound was coming from somewhere else, not from inside your body, you probably experienced the sound as an object outside yourself. But not totally. Visualized objects stay "out there"; heard sounds penetrate the body of the listener. They are out there and in here at the same time. This experience suggests that, from the listener's standpoint, there are two quite distinct ways of attending to sound: one that focuses on the *there*ness of the sound, on the sound-producer; and one that focuses on the *here*ness of the sound, on the physiological and psychological effects of sound on the listener. Both dimensions are present all the time, and we can readily shift focus from one to the other.

The *there*ness of sound, its "outside" qualilty, is a matter of time and space. What you are hearing, in fact, *is* time and space. Sound is a periodic displacement of molecules in the air. Now closer together, now farther apart, the molecules set up a sound wave that takes a certain number of microseconds to complete a cycle. The number of times per second a vibration pattern repeats itself constitutes its frequency. That much is a matter of time. But the wave also takes place in space: the vibration exerts pressure on the air molecules, and the greater that pressure happens to be, the greater the displacement of air molecules. The degree of the molecules' displacement constitutes that sound wave's amplitude (Fry 1977: 40–60; Fry 1979: 5–36; Lieberman and Blumstein 1988: 16–33; Handel 1989: 7–72). What you are hearing in [o:], then, is a certain set of frequencies (in this case, frequencies of 450 and 740 cycles per second) and, depending on how much air you put into the effort, a certain air pressure. No matter how reticent or forceful you were, [o:] happens to be, relatively speaking, the most intense phoneme in English: typically it strikes a listener's ear at a pressure about 1,000 times greater than the least pressure the human ear can detect (Fry 1979: 127). What you are listening to in [o:] is time (frequencies) and space (amplitude).

The *there*ness of sound becomes the *here*ness of sound in the ear of the receiver. The physical facts of time and space become the psychological experience of time and space. In the listener's ear, frequency becomes pitch. Pressure becomes loudness. Noise, timbre, inharmonics, and vibrato become roughness, brightness, warmth, and quaver. Rhythm becomes pulse. Patterns of radiation become distance and closeness, immediacy and delay. The context, no longer an outside entity, becomes something I belong to, something I am immersed in. It is just at this juncture, where *there*ness impinges on *here*ness, that Western ways of conceptualizing experience become as dogmatic as they are imprecise. Sensations versus perceptions, sense impressions versus intellectual discriminations, fallible evidence versus cerebral judgment: these dichotomies are as old as the Greeks. What seems to be missing are connections between the two. With listening, the science of "psychophysics" can take us only so far. To understand listening in its totality, we must take into account factors that are usually beyond the purview of science (Handel 1989: 547). First of all there is the intractable individual listener, with his distinctive knowledge and experience, her own particular goals and intentions. To understand these factors, we need a *psychology* of listening. Since knowledge and intentions are shaped by culture, we need to attend also to cultural differences in the construction of aural experience. The multiple cultures of early modern England may have shared with us the biological materiality of hearing, but their protocols of listening could be remarkably different from ours. We need a *cultural poetics* of listening. We must take into account, finally, the subjective experience of sound. We need a *phenomenology* of listening, which we can expect to be an amalgam of biological constants and cultural variables.

Through all the variables, cultural as well as individual, one thing is certain: sound is inescapable. It is as pervasive as the air that constitutes its primary medium. Barnaby Google hears, and sees, the situation exactly in his translation of Marcellus Palingenius's *The Zodiac of Life*:

> For ayre of slendrest substance is, and moueth by and by,
> Which beaten with ye noyse, doth shunne, and from ye stroke doth flye,
> And pearcing breakes into the eares, though close be kept the glasse,
> And close the doore, so fine it is that inward it will passe.
>
> (125)

Close your eyes for a moment, wherever you happen to be. Listen. Take stock of the sounds you can hear. Where are the sounds? How are you positioned vis-à-vis those sounds? Can you hear anything that you could not see? What are the limits of what you can hear? To ask such questions is to "bracket" listening as an experience, to attempt to understand listening on its own terms. Cognitively at least, you know by now that sound is the heard experience of two dimensions: space and time. How do those dimensions present themselves to your ears? It takes only a moment to discover that true silence does not exist. As human beings we are surrounded—and filled—by a continuous field of sound, by sounds outside our bodies as well as by metabolic sounds within. It is out of this continuous chaos of sounds, Michel Serres remarks, that meanings emerge: "Background noise is the ground of our perception, absolutely uninterrupted, it is our perennial sustenance, the element of the software of all our logic. It is the residue and the cesspool of our messages. No life without heat, no matter, neither; no warmth without air, no logos without noise" (1995: 7). The restless, unending noise of the sea, in Serres's version of *Genesis*, best manifests this existential condition.

Under the circumstances, silence is something I know about only by inference. Silence exists, not right here, but just beyond what I can hear. Silence defines the "horizon" of hearing. Where is that horizon? What lies within it? What shape does the horizon describe? Is it a straight line? No, because sounds can be "above" and "below," as well as "over there" and "right here." A cone directed into the space around you like a telescope? No, because a sound can be "behind" and "in front," as well as "above" and "below," as well as "over there" and "right here." A circle, then? A sphere? Geometrically a sphere seems right, but the positioning of your ears means that some parts of that sphere are more present than others. Extending to sound the phenomenological inquiry that Husserl and Heidegger brought to vision, Don Ihde isolates three characteristics of the auditory field: (1) surroundability, (2) directionality, and (3) continuity (1976: 76–81). Sound immerses me in the world: it is there and here, in front of me and behind me, above me and below me. Sound moves into presence and moves out of presence: it gives me reference points for situating myself in space and time. Sound subsumes me: it is continuously present, pulsing within my body, penetrating my body from without, filling my perceptual world to the very horizons of hearing. The shape of this auditory field approximates a sphere. ○ defines the perceptual limits of reverberating sound.

In several respects, the field of hearing differs from the field of seeing. Look around you for a moment. Take stock of what you can see, just as you took stock of what you could hear. Once again, keep yourself at the center: try to describe the seeing itself, not the objects that present themselves to sight. Prominent among the things that you can see, but could not hear, is this book. This is a book about sound, but it communicates through vision. Surely there are also things that you could hear but now cannot see: the radio in the next room, a barking dog down the street, your heart beating. What are the differences between seeing and hearing? Ihde speaks for many observers in noting how vision fixes objects "out there": I am connected to the objects I see, not by a sphere, but by a line. Objects that I see remain situated in geometric space; sounds that I hear reverberate inside me. In Stephen Handel's formulation, "Listening is centripetal; it pulls you into the world. Looking is centrifugal; it separates you from the world" (1989: xi). It is this sense of separation, perhaps, that explains why Western ways of knowing are grounded in metaphors of seeing. Descartes was only giving refinement to classical Greek assumptions when, in the *Discourse on Method* (1637), he looked to mathematics as a model for inscribing knowledge. Mathematics *proposes*: it positions ideas "in front," "out there." Knowledge is presumed to exist quite apart from the body of the knower (M. Johnson 1987: xxv–xxix). We find ourselves, once again, standing at that clueless, labyrinthine juncture between *there*ness and *here*ness.

Listening does not allow us the secure detachment that vision does. Hence the philosophers' distrust. What I am inviting you to negotiate in these pages is the difference between *ontology*, which assumes a detached, objective spectator who can see the whole, and *phenomenology*, which assumes a subject who is immersed in the experience she is trying to describe. Anthropologist Steven Feld has described the situation exactly: "By bringing a durative, motional world of time and space simultaneously to front and back, top and bottom, and left and right, an alignment suffuses the entire fixed or moving body. This is why hearing and voicing link the felt sensations of sound and balance to those of physical and emotional presence" (1996: 97). Listening is accepting presence—and not being apologetic about it.

Altogether reasonably, Derrida has argued against the logical primacy of voice over writing. Voice *seems*, of course, to be self-present.

When I speak, it belongs to the phenomenological essence of this operation that *I hear myself* [je m'entende] *at the same time* that I speak. The
signifier, animated by my breath and by the meaning-intention . . . is in
absolute proximity to me. The living act, the life-giving act, the *Leben-
digkeit*, which animates the body of the signifier and transforms it into
a meaningful expression, the soul of language, seems not to separate itself from itself, from its own self-presence. It does not risk death in the
body of a signifier that is given over to the world and the visibility of
space. It can *show* the ideal object or ideal *Bedeutung* connected to it
without venturing outside ideality, outside the interiority of self-
present life. (Derrida 1973: 78–79)

Hence the metaphysical status of voice in Western philosophy. But if
all meanings are made by the marking of difference—if A is A because it is not B—then speaking itself is a form of marking. Even at
the level of sound, [a] is [a] because it is not [b]. Speaking becomes,
in this view, a form of writing: "If 'writing' means inscription and
especially the durable instituting of signs (and this is the only irre-
ducible kernel of the concept of writing), then writing in general covers the entire domain of linguistic signs" (Derrida 1976: 44). *Scribo
ergo sum.*

At least four considerations, however, impede Derrida's revolu-
tionary attempt to decenter voice. First of all, "writing"-through-
speaking is an embodied act: it happens not in the cerebral cortex
alone but in the lungs, the larynx, and the mouth. "No doubt our
voices carry the trace of some 'archiwriting,'" Paul Zumthor concedes; "but the trace may be considered 'inscribed' in another manner in this discourse, one all the less temporal as it has been rooted
in the body and is open to memory alone" (Zumthor 1990: 18). A
second consideration concerns hearing. Hearing orients me in two
dimensions: in time and in space. While it may be true that I hear
myself speak in the very moment, in the very place that I am speaking, I am nonetheless surrounded by speech that comes from various
distances, across various lapses of time. My own speech takes its
place in this larger social system of sound. Speech is not experienced
totally in terms of its *here*ness. Nor is speech the only kind of sound
I can make and hear. This third consideration may be the most important of all. Whoops, clucks, tsks, moans, cries of [o:]—my voice can
produce these and scores of other extralinguistic sounds. Zumthor's
response to Derrida seems just: "what stands out all the more for me
is the broad function of human vocality, for which speech certainly
constitutes the principal manifestation—but not the only one and
perhaps not even the most important" (1990: 18). Finally, anybody
who has ever tried to sing has learned how "to make strange" her

own voice. Or rather, she has *remembered* how "to make strange" her own voice, since every human speaker has gone through a five-year process learning how to control the complicated apparatus of lungs, larynx, and mouth in order to produce intelligible speech (Fry 1977: 101–124).

Voice is not, in all circumstances, self-present. Whatever Aristotle may have said about the direct transformation of ideas into sounds, Renaissance rhetorical training was designed to make the student self-conscious about his own voice. That self-consciousness was, and remains, a professional actor's stock in trade. With due respect (in all senses of the word) to Derrida, the O-factor accepts the logic of deconstruction, but it refuses the supposed transparency—or rather the supposed transitivity—of voice deconstruction needs in order to assert its own *différance*. On the contrary, the O-factor insists on all the things that make the voice strange: lungs, larynx, and tongue, the positions of [a], [i], and [u] in the mouth, the dampening effect of wet earth, the resonating effect of lath and plaster, the differing ways of signifying voice in differing forms of graphic media. Harold Love arrives at a similar position with respect to Derrida. Acknowledging the epistemological primacy of *différance* in the *making* of meanings, Love insists on the phenomenological differences among oral utterance, chirography, typography, and "electronography" in the *experiencing* of meanings: "The notion of 'presence,' whether or not regarded as philosophically sustainable, provides us with a method of discriminating between modes of signification as being more or less distanced from a *presumed* source of self-validating meaning" (1993: 144, emphasis added). Presence, in this understanding of the word, is not the timeless, universal sensation Derrida makes it out to be but a cultural variable. Presence is what a given culture *takes to be* presence.

Listening, as opposed to looking, seems especially apt with respect to early modern England, as a collectivity of cultures that depended so extensively on face-to-face communication. Even the dominant political culture was, in Walter Ong's terms, a "mixed" culture, "literate but with a strong oral residue" (1965: 145–146). "Orality" and "literacy" are not, after all, polar opposites. As Michel de Certeau points out, they are defined in terms of each other, in terms that vary according to cultural conditions. Regardless of what these conditions happen to be, "orality" and "literacy" are, furthermore, never commensurate: one term is always dominant, marking the other as subordinate (de Certeau 1984: 133). The rhetorical basis of Renaissance criticism suggests that the dominant term in early modern culture, even at the highest levels, was in fact "orality," not "literacy." After distinguishing various degrees and kinds of literacy—black-letter versus

Roman-letter, secretary-hand versus italic-hand, English versus Latin, the ability to read versus the ability to write—Keith Thomas reaches this conclusion:

> Early modern England, therefore, was not an oral society. But neither was it a fully literate one. Although many people could read and write, many others could not: and although documents and books played an increasing role in social life they by no means monopolized the means of communication and record. Indeed it is the interaction between contrasting forms of culture, literate and illiterate, oral and written, which gives this period its particular fascination. (1986: 98)

By and large, the artifacts that survive from early modern England ask to be heard, not seen: compared, say, to Renaissance Italy, the number of buildings, paintings, tapestries, pieces of furniture, and utilitarian objects are few. What we have, in great abundance, are *verbal* artifacts. Our knowledge of early modern England is based largely on words, and all evidence suggests that those words had a connection to spoken language that was stronger and more pervasive than we assume about our own culture. Can we be so sure—especially in a culture where "orality" and "literacy" were reciprocally defined in ways quite different from today—that speaking and listening were experienced as uncomplicated acts of self-presence?

○ AS AN ACT OF COMMUNICATION

First ○, then [o:], then "Oh."

Listening precedes speech. Listening prepares speech. Listening predicts speech. A child learns to recognize particular sounds and to attach significance to them long before it develops the neural and muscular capability of executing those sounds itself. From the very beginning, however, a child uses sound as a way of projecting its body, its *self*, into the space around it. [o:] is a primal cry, and we never forget its bodily trace, as Cicely Berry insists in *The Actor and His Text*: "For language, as well as being highly sophisticated, is also primitive in essence. We may have technological jargon of every kind, we have legal language, language of sensibility and emotion allied to literature and art. . . . Yet words evolved out of noises which were first made to communicate basic needs; they were in fact signals. And we still have that sense memory within us—that resonance if you like" (1988: 19–20). Just such a resonance sounds out in Hamlet's pain, in the final moments of life when breath fails and words devolve first into cries and then into silence:

O I dye *Horatio:*
The potent poyson quite ore-crowes my spirit,

> I cannot live to heare the Newes from England,
> But I do prophesie th'election lights
> On *Fortinbras*, he ha's my dying voyce,
> So tell him with the occurrents more and lesse,
> Which have solicited. The rest is silence. O, o, o, o. *Dyes.*
>
> (F1623: 5.2.304–311).*

So sounds Othello's "Oh *Desdemon!* dead *Desdemon:* dead. Oh, oh!" (F1623: 5.2.288). So sounds the "O, o, o, o" that reverberates with "no, no life," with "no breath," with "come no more," perhaps even with "undo" in Lear's final speech, as transcribed in the quarto *Historie of King Lear* (Q1608: 24.300–304).† The semantic emptiness of these O's on the printed page—modern editors of *Hamlet* and *Lear* have generally preferred alternative versions of these two speeches, versions that lack the O's—stands as testimony to their embodied fullness. As Joel Fineman remarks with respect to *Othello*, the insistent sound of [o:] has the effect of undermining the traditionally admired power of literary language to create visionary presence (1991: 150–152). In what Fineman hears as an instance of Lacan's *objet petit a*, and thus an opening into the Real *beyond* language, one can also hear bodily memory of the Imaginary *before* language.

As a burst of energy from within, [o:] is an act of aggression, a projection of one's body into the world. It is, in the most basic sense of the word, an *environmental* gesture. In crying [o:] I extend my person into "the about-me." Both the Italian and the German equivalents of "environment" carry, perhaps, a stronger sense of this unmarked spatiality than the English word now does: *ambiente* waves a hand toward to the air around the speaker, while *Umgebung* invokes the "givens" that surround the speaker. The environment certainly includes plants and animals, but it also includes air, ink, fiber-optic cable—and other people. As a cry from within, [o:] seeks resonance from without. It seeks a listener. It seeks communication. For describing acts of communication, the metaphor that seems most readily to come to mind is a "chain." This linear model might be delineated thus:

$$\text{psyche} \leftrightarrow \text{body} \leftrightarrow \text{[o:]} \leftrightarrow \text{medium} \leftrightarrow \text{society}$$

Here [o:] mediates between the individual speaker on the left and

---

*Quotations from Shakespeare's plays throughout this book are taken from the earliest printed texts, as reproduced in *Shakespeare's Plays in Quarto* (1981), cited as Q + the relevant date, and in *The First Folio of Shakespeare* (1968), cited as F1623. For ease of cross-reference to modern editions, I have provided act, scene, and line numbers from *The Complete Works* (1988).

†For expert opinion on how the phonemes of early modern English sounded I depend throughout the book on Charles Barber's *Early Modern English* (1997: 103–141).

society on the right, in each case via a physical medium: lungs, larynx, and mouth on the left, molecules of air on the right. The trouble
is that psyche, body, society, and media are connected to one another
in all sorts of complicated ways. What we need is a model that recognizes and maps these complications.

Combining observations from systems theory, cybernetics, phenomenology, and biology, Niklas Luhmann has supplied just such a
model. Luhmann invites us to see body, psyche, media, and society
as four autonomous systems (Luhmann 1989: vii–xviii).* Each is self-
contained. Each operates on its own terms. In more ways than one,
that proposition seems counterintuitive. How can the psyche be con-
ceived without the body? What is society if not the collectivity of
subjects who perform it? Luhmann's tack is easiest to see with respect
to the body. As a system of living cells, the body maintains itself and
reproduces itself. In all its transactions with other organisms the body
implicitly seeks one thing and one thing only: to keep going. When
the human body encounters another organism, be it a virus or an
elephant, the body adapts itself to that organism in such a way that
its own integrity is not threatened. "Autopoiesis" ("self-making") is
the term Humberto Maturana and Francisco Varela have given to this
biological imperative. Luhmann's move is to take the physiological
concept of "autopoiesis" and to apply it to the psyches of individuals,
to society, and to systems of communication. Body, psyche, society,
and media become, in effect, four autopoietic systems, and the relationships among them something vastly more complicated than a
unilateral transfer of information from one system to another. Luhmann's proposition will perhaps seem less dubious if we stop to consider the individual parts. Since its original formulation by Freud, the
psyche has been disclosed to us as a self-contained system—an
energy-system, in fact—and one in an inherently troubled relationship with its *Umgebung*. By Marx and his successors, society too is
configured as a relentlessly self-maintaining system for producing

---

*In making such a proposition, I am granting media a separate status it does not
enjoy in Luhmann's original model. For Luhmann, society *is* communication: "Social
systems use communication as their particular mode of autopoietic reproduction.
Their elements are communications that are recursively produced and reproduced
by a network of communications and that cannot exist outside of such a network.
Communications are not 'living' units, they are not 'conscious' units, they are not
'actions.' Their unity requires a synthesis of three selections, namely, information, utterance, and understanding (including misunderstanding). This synthesis is produced
by the network of communication, not by some kind of inherent power of conscious-
ness, nor by the inherent quality of the information. Information, utterance, and understanding are aspects that for the system cannot exist independently of the system; they
are co-created within the process of communication" (1990: 3).

and distributing material goods. Even language becomes detached from its psychic and social referents in Saussure's linguistics. And in the work of Marshall MacLuhan, Walter Ong, and Elizabeth Eisenstein, different modes of media are endowed with a power over psyches and societies that implicitly turns those media into self-generating, self-reproducing systems of meaning.

For the purposes of ○, Luhmann's model of communication complicates any unilateral relationships we might attempt to set up among body, psyche, media, and society. Communication becomes a process, not of *transferring* meaning, but of *negotiating* meaning. The factors in this process of negotiation are body, psyche, society, and media as autonomous systems. Just as two organisms adapt to one another, just as the human body adapts to a virus and the virus to the human body, so body adapts to psyche, so psyche adapts to society, so society adapts to the means of communication at its disposal, so means of communication adapt to body. Communication happens when one system adjusts to another, when two or more systems achieve a *modus operandi* among them, when all the systems are functioning vis-à-vis each other. Meaning is made in the connections among systems, in what Maturana and Varela call "zones of consensuality" (Gumbrecht and Pfeiffer 1994: 170–182). Systems interpenetrate each other: the unity of one system becomes a function within another system. In Luhmann's own coinage, systems "resonate" with one another. When, for example, English adventurers landed on the shores of Virignia, they were rather surprised by the warm reception they received from the native people they encountered. Powhatan Indians supplied the English with food and shelter, as they would have done for any strangers. When the same Indians later turned hostile, the English failed to see that their own aggressive behavior was the cause. To seek hospitality as migrating strangers was one thing; to take up residence was, from the Indians' point of view, something else again. Failing to realize that their own senses of land and ownership were very different from the Indians', the English could only see the natives' original friendliness as treachery (Hulme 1994: 147–152). One system "reads" another only in its own terms.

"Resonance" may be only a figurative term for Luhmann, but the physically interpenetrating qualities of sound make the term literally true in oral communication. The sound-system of the English language resonates in the body. So does the psyche resonate in sound, society in sound, society in psyche. Each of these self-contained systems exists in a state of mutual adaptation to the others. As the body resonates in the psyche, as the psyche resonates in society, so resonates the medium in the message, the message in the medium. The medium is emphatically *not* the message. Rather, one is adapted to

the other. One resonates in the other. Rendered as a diagram, Luhmann's model might look something like this:

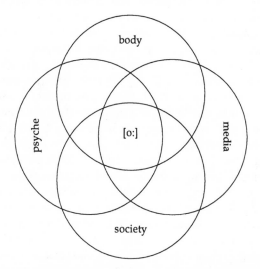

For a given system, whatever is not that system is part of the
environment of that system. For a given society, whatever is not communication is environment. The body functions as environment for
systems of communication. So does the psyche. So does language as
a system of signifiers. It is the set of environmental relationships of
the elements—of the language system, of air as the carrier of sound,
of society as a set of protocols for receiving and issuing information—
that allows us to speak of an "ecology" of voice, media, and community. In its literal sense of *oikos* ("house") + *-logy*, ecology is a study
of the conditions of inhabiting. If language, media, and society are
configured as environments, each one functioning as such for the others, then we can identify three consensual zones, three sites of resonance, that together constitute an ecology of communication: (1) the
zone between language and society, (2) the zone between language
and media, and (3) the zone between media and society. Let us attempt to chart these horizons one by one.

In the twentieth century three models have been put forward for
relating *language and society:* one proposing that social structures
shape lingusitic structures (Durkheim), one proposing that linguistic
structures shape social structures (Sapir, Whorf, Ong), and one proposing that linguistic structures and social structures are analogous
(Saussure, Lévi-Strauss, Derrida). The ecological model set in place
by Maturana, Varela, Luhmann, and Gumbrecht replaces these unilateral models with an interactive model, one recognizing that lin-

guistic structures and social structures interact in complicated, mutually dynamic ways. Between *language and media* most commentators
have, once again, assumed a one-way relationship. Communication
is imagined to be a conduit—a railroad, perhaps—that connects the
messenger-sender with the message-receiver: ideas at one end are
loaded onto some sort of carrier—a piece of paper, electrical wires,
molecules of air—and are unloaded at the other end. The ideas are
imagined to keep their integrity through the whole transaction.

If, however, language and medium are autopoietic systems, we
should expect radical transformations as we pass from one system to
the other. Take, for example, an e-mail message. Common sense tells
me that when I boot up the computer, log onto my server, and call
up a message I am making direct contact with the sender's ideas.
And yet a long and complicated series of mechanical transformations
separate me from the sender's ideas. First she used her brain to think
of the ideas, then she set to work with her fingers. Her finger motions
became electrical impulses, which became a visual image (after potentially a great many translations and transmissions over phone
lines and computer networks), which my eye took in, which my brain
decoded. No fewer than four—and possibly many more—transformations of the message are interposed between the sender's brain
and my own. Face-to-face speaking would seem to be a much less
complicated operation, and yet even here three transformations are
interposed between the speaker's brain and mine: her speech mechanism, sound waves in the air, and my hearing mechanism. In Dennis
Fry's summation,

> Anything that we say takes on a different form as it passes through
> these successive stages. We have seen that it starts out as a string of
> language elements of various kinds in the brain of the speaker. The
> speech mechanism, that is the complicated system of muscles we use
> when talking, converts these into a sequence of movements; the move-
> ments set up minute changes of pressure in the air, the sound-waves of
> speech, and these in turn impose movements on the hearing mecha-
> nism of anyone who is within earshot. The listener's brain uses the in-
> formation transmitted by these movements to get back to the string of
> language elements that make up the message. We normally think of
> speech as being a unitary kind of activity but it is clear that it really
> consists of a series of changes or transformations. (1977: 21)

With its single-minded focus on binary difference-marking, cultural studies as a discipline too often assumes direct brain-to-brain
communication, or at best brain-to-eye-to-brain communication. ○
encompasses brain-to-tongue-to-air-to-ear-to-brain communication,

with a special interest in the middle part of that chain, on tongue-to-air-to-ear. In varying degrees, systems of communication in early modern England maintained a contact with the human body that seems remarkably different from communication systems today. Rather than imagining a rigid distinction between oral culture and literate culture, between aural media and visual media, we should imagine a continuum between speech and vision. At one pole of the continuum are communications that have little or no direct contact with letters: ballads from oral tradition, St. George plays, morris dances. At the other pole are communications that have little or nothing to do with voice: texts in legal French, treatises on mathematics, books on geometry. In between are ranged broadside ballads, scripts for the stage, sermons and orations, familiar letters, manuscript commonplace books, accounts of the New World.

Language and society, language and media, media and society: it is the third of these horizons, the site of resonance between *media and society*, that most challenges common-sense ideas about communication. In Luhmann's model, a society is made up, not of individual subjects, not of the actions those subjects perform, but of systems of communication. An action-based theory of society (a society is what it *does*) becomes a communication-based theory of society (a society is what it *communicates*). Any communications system, Luhman proposes, involves three elements: (1) information, (2) utterance, and (3) understanding (1989: 3). If, as Ong argues, early modern England was a society with a strong oral residue, then we may expect that sounded communication in early modern England would have produced discrete versions of all three links in the chain of communication. Clear enough is the status of sound as a physically distinctive kind of utterance—sound as opposed, say, to handwriting or to print. What may not be so clear is sound as a distinctive kind of information, different in kind from the information presented by movements of the fingers, by electrical impulses, by dots of light on a monitor, by inked impressions on a piece of paper, by marks inscribed by the sender's hand. The result of a distinctive utterance offering a distinctive kind of information can only be a distinctive kind of understanding.

Just what do I hear when someone speaks? What am I being informed about? What do I understand? Let us consider the exchange recorded in figure 1.2.

[ho:ld rait]

Out of the continuous stream of sounds that strikes my ear I isolate the phonemes [h], [o,], [l,], [d], [r], [ai], and [t]. I combine these phonemes into morphemes that form two English words, "hold" and

Figure 1.2. Fruit-pickers, from Thomas Fella, visual commonplace book (Folger MS V.a.311, 1585–1598). Reproduced by permission of the Folger Shakespeare Library, Washington, D.C.

"right." Sorting through all the things these two words might signify, I catch the speaker's semantic meaning. I do what I am told, and I tell what I do:

[so ai du].

But I have probably consciously attended to only half the sounds I have heard. What information have I gained from the other half? What have I understood? That is not easy to say, since what I have understood has as much to do with space and time as it does with semantic meaning. When he speaks, the interlocutor's voice fills the auditory field. My bones reverberate to his [o:]. In my reply I interject myself in that auditory field. His bones reverberate to my voice. We create an acoustical community of two. The understanding I have within this community is an *embodied* understanding: it is grounded in the speaker's body and in mine. Thomas Fella, who sketched this image of two speaking men in his visual commonplace book, seems to have realized as much by transcribing the exchanged speeches just where they belong physically, psychologically, and phenomenologically: in the air that surrounds the speakers.

Observers have given a variety of names to this psychophysical effect. "Proprioception"—"the body's specific awareness of itself"—is the term David Appelbaum sets in opposition to "cognition" (1990: 77). In answering the question "Can Thought Go On Without a Body?" Jean-François Lyotard contrasts the "determinate thought" that computers can do with the "reflective thought" that uses the resources of the human body. The "reflective thought" of the body proceeds analogically, not logically. It is based on configurations, not binaries:

> A field of thought exists in the same way that there's a field of vision (or hearing): the mind orients itself in it just as the eye does in the field of the visible. . . . what makes thought and the body inseparable isn't just that the latter is the indispensible hardware for the former, a material prerequisite of its existence. It's that each of them is analogous to the other in its relationship with its respective (sensible, symbolic) environment: the relationship being analogical in both cases. (1991: 293–294)

A social dimension to this embodied understanding is set in place by Paul Zumthor, who talks about the "sociocorporeal" quality of oral performance. Keeping the body at the center of attention, Zumthor distinguishes the here-and-now "work" that is being performed from the "text," its linguistic component. Texts can be read; works must be heard and seen. It is not only the performer's body that distinguishes "work" from "text" but the *listeners'* bodies. Every act of speaking and listening is an *existential* moment that affirms (1) the selfhood of the speaker, (2) the selfhood of the listener, and (3) the culture that conjoins them (Zumthor 1990: 60–63).

The O-factor focuses on this existential moment. It attends to voice, not "voice." Academic discourse, operating as it does through books, articles, and conference papers (as opposed to the discussion afterward), assumes that words are disembodied signifiers. Like all autopoietic systems, academic discourse is equipped to read other systems only on its own terms. Academic *discourse:* the very word, in Foucault's formulation, points toward something incorporeal, an abstract force to which individual bodies become subject. For all that, we have not lost contact with the sounded word. How could we? We are surrounded by it, immersed in it, every day of our lives. What we lack is not contact with the sounded word, but a sensitivity to sound, a curiosity about how it operates, how it affects us, how it interacts with various forms of media.

But first we must learn to listen. In more ways than one, that is easier *said* than done. The whole enterprise of Western philosophy, especially since Descartes, has granted superior power to speech, to the speaker's ability to construct rational arguments and to impose them on objects in the world around him. This imperative is almost irresistible, as Corradi-Fiumara concedes in *The Other Side of Language: A Philosophy of Listening:*

> When a thinker turns his attention to the humble, though exacting, la-bour of listening it is almost as though he were impeded or held back from this inappropriate operative level by some élitist power (possibly envious of its strength); it deviously insinuates to the thinker who tries to listen authentically that he could well be a member of a high and ra-tional order, so to speak, capable of discerning essences and lingusitic paradigms from the myriad of interactions; an élitist power which has become internalized seems to ask rhetorically why there should be any need to listen, when one has not only achieved a mastery of language but also of the metalanguages whereby one can soar to the level of the relations that exist between discourse and reality, or among different types of discourse.

Against such imperial ambitions Corradi-Fiumara argues for a "coex-istence" of subject and object: not a logical relation of subject to object but an *ecological* relation *between* subject and object. What we have now, she concludes, is "an incomplete rationality (eloquent but deaf)" (1990: 58–60). We can cultivate that ecological understanding by lis-tening for resonances among voice, media, and community.

○ AS A POLITICAL ACT

The kind of listening advocated by Corradi-Fiumara is anything but passive. If common sense persists in thinking so, it is only because of an implicit comparison with speech. Speaking is unmistakably *active:* it entails an exertion of force. Think about your [oː]. Your chest muscles pushed air out of your lungs past your larynx; your vocal cords set up a series of vibrations between air molecules; your throat and mouth shaped those vibrations in certain ways; your pursed lips broadcast the air waves into the environment; the air waves penetrated the ears of anyone who happened to be within earshot. The force put forth in speaking has repercussions in hearing: the physiological [oː] impacts the phenomenological [oː]. In his attempt to argue the bodily basis of abstract thought Mark Johnson pays special attention to the experience of speech, particularly the goal-directed "perlocutionary" aspect of speech isolated in speech-act theory. As something that happens *in* the body and *to* the body, speech is apprehended, Johnson contends, via a gestalt of *force.* Constantly doing things to objects outside ourselves, constantly being acted upon by such objects, we know force as a very basic bodily experience. Speech is a bodily experience of just that kind. Within the gestalt of force, Johnson distinguishes seven common "force structures" that also apply to speaking and hearing: compulsion, blockage, resistance, diversion, removal of restraint, enablement, and attraction. In speaking, I compel others; in hearing their speech, I am compelled. My speech can be blocked by all sorts of obstructions, physical, psychological, and social; I can resist the speech of others by raising those obstructions myself. My speech can be diverted as a result of such blockages or resistance; so can the speech of others. Through enablement I feel the potential for compelling, even if I choose not to speak. Finally, I can be attracted by another's speech, just as I can feel a pull toward a particular object or a particular person (1987: 41–64).

Although grounded in phenomenological philosophy of the twentieth century, Johnson's arguments simply restate the underlying assumptions of Renaissance rhetoric. From Roman antiquity through at least the early eighteenth century, anyone who attempted to rationalize poetry, drama, and prose fiction—or even essays and treatises—wrote as if all the canonical kinds were exercises in persuasion. Before cataloguing the varieties of poesy—pastoral, elegiac, comic, lyric, heroical—Sidney offers this *raison d'être* for them all: "as vertue is the most excelle[n]t resting place for al worldly learning to make his end of, so *Poetry* being the most familiar to teach it, and most

Princely to moue towards it, in the most excellent worke, is the most excellent workeman" (E3–E3^v). The rhetorical persuasion that Sidney has in mind here—the teaching, the moving—has not lost its grounding in the spoken word (Rhodes 1992: 8). By Samuel Johnson's time, if not before, the rhetorical model of criticism had become a cerebral construct; for Sidney and his peers it still had a sounded reality that existed apart from the page and hence, to some degree at least, *on* the page. The transmutation of [o:] into "Oh" is not absolute. The gestalt of force in speech was registered even on the page. Even for the relatively small number of people in early modern England who could write and read, [o:] and "Oh" were both subsumed in ○.

The disproportionate power of this literate minority in representing early modern England to posterity serves as a reminder that ○, in all its guises, is ultimately a political performance. The gestalt of force governs relationships at every level: between two speakers, among all the speakers in a community, between one community of speakers and other communities. Between two individual speakers, first of all, an exchange of sounds always proceeds according to an implicit set of rules. "Prithee," "by your leave," "if it please your honor": tags like these in Shakespeare's scripts make momentarily explicit the protocols of distance, power, and rank that are always at work covertly in determining who gets to say what when (Magnusson 1992: 395–396). If the speakers happen to be social equals, the tags are no less important: to call someone "cousin" is to establish a bond of intimacy that may or may not have anything to do with blood kinship (Wrightson 1982: 46).

Beyond individuals, the gestalt of force is at work among all the speakers in a community. It helps, in fact, to affirm their identity as a community. In an oral culture, as Paul Zumthor points out, acts of communication never lose this social reference. The *embodiedness* of an oral performance—of something like a Robin Hood play—embraces not only the individual performers' bodies but the social body. To the extent that it relies on a physical voice, an orally produced work "resists any perception that might sever it from its social function, from its place within a real community, from an acknowledged tradition, and from the circumstances in which it is heard." Even a text that is written down, if the realization of that text is oral, cannot escape this situatedness within a communal present: "If we concern ourselves with works that are to be recited but are only handed down in written form, then we must assume that both dimensions of effect are simultaneously present in the text" (Zumthor 1994: 221). Shakespeare's scripts occupy just such a status: the written *text* comprises a set of cues to an oral *work*.

Oral performances of stories, poems, and plays—oral performances of art made out of words—is only a ritualized instance of the identity marking and group affirmation that go on in everyday exchanges of speech. Dell Hymes, building on his field work among Native Americans, calls a social group who shares this identity a "speech community." Though it may owe its cohesiveness in the first instance to factors of space and time, a speech community is something that continues to be *performed*:

> Tentatively, a *speech community* is defined as a community sharing rules for the conduct and interpretation of speech, and rules for the interpretation of at least one linguistic variety. . . . Native conceptions of boundaries are but one factor in defining them, essential but sometimes partly misleading. . . . Self-conceptions, values, role structures, contiguity, purposes of interaction, political history, all may be factors. . . . The essential thing is that the object of description be an integral social unit. Probably it will prove most useful to reserve the notion of speech community for the local unit most specifically characterized for a person by a common locality and primary interaction.

Within this definition Hymes includes not just speech exchanges but all "channels" of communication: writing, song, even "speech-derived whistling, drumming, horn calling, and the like" (1972: 53–55).

The fact that our only access to the oral cultures of early modern England comes via written texts points up the political differences that separate one speech community from another. In a word: a small number of writers and readers control the representation of tens of thousands of other people. Does the power of these literate observers over representation correlate with their political power? Received wisdom in new historicism assumes as much. If, however, Raymond Williams's typology of "residual," "dominant," and "emergent" cultures is fleshed out with data on literacy from David Cressy and Keith Thomas, the situation seems rather more complicated. The "residual" culture of agrarian labor may have been, by and large, an oral culture. The "emergent" culture of merchant capitalism may have been, by and large, a literate culture. But the "dominant" culture of aristocratic rule was, by most historians' accounts, mixed (Williams 1973: 3–16; Cressy 1980: 19–41, 118–141; Thomas 1986: 97–131). To say who or what was in control of graphic media varies, then, according to which particular medium we happen to be talking about. Bakhtin's distinction between "epic" and "novel," between closed forms of representation and open forms, establishes two poles between which we can plot the "sociolinguistic politics" of early modern England (1981: 3–40). Different speech communities stand in different relationships to different media, according to how open or closed those media are to the polyphony of dialects and argots, songs and sonnets, whoops and hollers that filled the ambient air of early modern England.

An awareness of differences among various media in how voice is transcribed can help us to reconstruct the distinctive experiences of speech communities that happen to be known to us only through the written records of other communities. The intersection of voice, media, and community offers a way of recovering the subjective experience of people who did not have direct access to print. Early modern women, for example, seem to have had a vital connection with folk ballads. In singing snatches of ballads, Ophelia is adding her voice to those of the singing women workers Thomas Deloney says he overheard about 1595 in one of the earliest factories in England. Sixteenth-century ballads like "Geordie," which survived in the Appalachian Mountains of North America through the first third of the twentieth century, give women a voice, literally as well as figuratively—a voice by and large denied to them by the literary high culture of Renaissance England. The Algonkian-speaking nations of North America's east coast constitute another group of speech communities that remained remote from the collectivizing power of print. With an awareness of what we ought to be looking for, we can use narratives like John Smith's *General History of Virginia* at least to approach the borders of the aural field these people once inhabited.

○ demands political reading, in pursuit of political listening. The cryptic relationship between women and ballads and between Native Americans and narratives of voyage and discovery directs us to listen for two kinds of clues: (1) polyphony and (2) silence. For Bakhtin, *heteroglossia* is a figurative concept: it refers to the way multiple registers of social speech are inscribed in texts. With the graphic media of early modern England we can give this figurative concept a physical and phenomenological application. We can ask not just *whose* voice or voices are inscribed in a text, but *how* those voices are inscribed. What are the graphic means by which the human voice is invoked? What are the politics that govern those vocal graphemes? Our ultimate goal is to counter the tyranny of Cartesian philosophy, with its privileging of visual experience, its ambition to speak with a single authoritative voice. ○ asks us to listen for multiple voices, for competing voices, even for noise. Cartesian philosophy is suspicious of voice because of the possibility that something might remain hidden, unseen, unsaid. For an exacting thinker, face-to-face speaking imposes a certain opacity, a sense that there may be more to what the speaker is saying than meets the eye. With writing, the truth is *there*, right in front, totally susceptible to the reader's surveillance (Ihde 1976: 153–154). Lyotard catches the same distinction from a slightly different perspective when he contrasts "textual" or "discursive" thinking with "figural" thinking (1991: 286–300). Textual thinking depends upon a transparency of signifiers: it operates "horizontally," in a "flat" space marked with binary differences. Figural thinking accepts the opacity of signs: it operates "vertically," in a "dense" space filled with analogues. What the Cartesian writer fears most is the unpredictable, "that-which-cannot-be-said-in-advance." Speech, because it comes from an unknowable "beyond," from inside someone else's body, challenges the Cartesian writer's position of detached assurance. Writing, by contrast, confirms that position:

> The written word "lacks" the sounded significance which already gives a degree of context to the word, and this makes the unsaid less opaque. Writing fails to convey that minimal sense, although the "re- duction" can be compensated for by adding words which also sound- lessly replace, in their own way, what was lost. Writing creates the pos- sibility of a *word without voice*. It opens the way to the forms of unvoiced word which secretly dominate whole areas of the understand- ing of language. (Ihde 1976: 154)

The Cartesian writer confronts silence with a highly ambivalent attitude. On the one hand, he exploits silence: writing secures the truth by voiding the voice, by turning ideas into disembodied ci- phers. On the other hand, he distrusts silence. Out of silence might come resistances, blockages, and diversions. In a word, the Cartesian writer does not know how to *listen*. "Taking soundings" is Roland Barthes's explanation for what goes on in listening. There are, Barthes proposes, three distinct kinds of listening: to indices, to signs, and to signifiers. Listening to indices is a capacity man shares with animals: it requires only an alertness. The specifically human begins with lis- tening to signs, with deciphering, with reading codes. Human lis- tening comes into its own, however, with listening to signifiers, to the unsaid, to the unconscious. It is this kind of listening that creates an intersubjective space between listener and speaker:

> The voice, in relation to silence, is like writing (in the graphic sense) on blank paper. Listening to the voice inaugurates the relation to the Other: the voice by which we recognize others (like writing on an enve- lope) indicates to us their way of being, their joy or their pain, their condition; it bears an image of their body and, beyond, a whole psy- chology (as when we speak of a warm voice, a white voice, etc.). Some- times an interlocutor's voice strikes us more than the content of his discourse, and we catch ourselves listening to the modulations and harmonics of that voice without hearing what it is saying to us. This dissociation is no doubt partly responsible for the feeling of strange- ness (sometimes of antipathy) which each of us feels on hearing his own voice: reaching us after traversing the masses and cavities of our

own anatomy, it affords us a distorted image of ourselves, as if we were to glimpse our profile in a three-way mirror. (Barthes 1985: 254–255)

"Free listening" attends to the otherness of one's own voice and to the otherness of other speakers: it refuses the illusion of a self-possessed voice just as it refuses the social categories that place a listener above or below a speaker: it "disaggregates, by its mobility, the fixed network of the roles of speech" (Barthes 1985: 259). The strange-making effect Barthes describes in sensory experience urges us toward a strange-making effect in conceptual thinking: we have not cornered a subject simply by saying, "I am going to create a site for reading X against Y." The image of visual mastery is seductive, but it suppresses other ways the body has of knowing. In particular, it suppresses listening, with listening's capacity to catch a polyphony of sounds and voices, with listening's openness to the potentialities in silence.

As the ear hears a voice by collecting sound waves coming from a variety of directions, so the O factor stands at the convergence of four critical methodologies: one informed by physiology, one by phenomenology, one by systems theory, and one by sociolinguistics. Each of these methodologies has its own intellectual history, its own set of assumptions, and its own self-imposed limitations. As a study of one set of the material conditions that shape communication, physiology as a critical methodology proceeds along lines laid out by Marx, Benjamin, and Williams. Possessed of certain vocal equipment, situated within the medium of air, "men make their own history, but they do not make it just as they please; they do not make it under circumstances chosen by themselves" (Marx 1972: 437). In this case, the circumstances are set in place by the human body. The phenomenology of listening offers a way of accounting for subjectivity within this materialist matrix. Articulated most thoroughly by Don Ihde (1976), a phenomenology centered specifically on listening is derived from the more general investigations of Husserl, Heidegger, and Merleau-Ponty. Systems theory, in the rigorously materialist directions pursued by Luhmann (1989) and Gumbrecht, adopts Husserl's idea of "horizons" of experience but places that concept within two quite different models: the sociological theories of Parsons and the biological analyses of Maturana and Varela (Luhmann 1989: vii–xviii). Finally, sociolinguistics as developed in the ethnographic studies of Dell Hymes finds textual applications in the work of Bakhtin and political applications in the work of de Certeau. If not individually, then collectively these four methodologies provide a critique of the objectifying imperative that has come to govern textual studies. Secure in their viewing tower, scholars writing under that imperative may see this project as another futile exercise in logocentricism. If it is Presence that this book is after, it is not the Presence of the Word, but of *sound*. Of sound in the larynx, in the mouth, in the bones, tissues, and cavities of the skull. Of sound in the ear and in the gut. ○ is not about ontology, but phenomenology. It is concerned not with "voice," but voice. ○ is not about metaphysics, but materialism—the materialism of the human body, of sound waves, of plaster, lath, and thatch, of quill pens, ink, and paper, of lead type. In the last instance, ○ is about people—people whose voice-based cultures are available to us only if we come at them by indirections.

Although a project in historical reconstruction, attending to the O factor is grounded in some decidedly contemporary concerns. The media anxieties of the late twentieth century—the fear of a paradigm shift occasioned by a media shift from print to digital electronics, and the thrill of such a shift—have their counterparts in the complex interactions among oral traditions, manuscript culture, and the new technology of print in early modern England. Above all, there are parallel questions about the *politics* of media. Who has access to which media? Who gets represented by whom? What difference does that make in the long run? For early modern Europe those questions have long since been answered by Peter Burke, Kevin Sharpe, Ronald Hutton, and other historians of popular culture. In Britain at least, communication systems based on print have all but obliterated the voice-based systems with which they have come into contact. Ballads, morris dancing, and folk plays have enjoyed major revivals in the twentieth century, but the last inheritors of ballads via oral transmission died in the 1920s. The "fall" of merry England, with its traditional folk festivities, had happened, by Ronald Hutton's estimate, two centuries earlier, the result of Protestant dogmatism and Enlightenment reform (Hutton 1994: 69–152). For the postmodern world, the answers are not yet clear. If communication systems exist in an environmental relationship to each other, then an ecology of voice, media, and community—an ecology based on listening—may provide some hope for the future.

Mapping the Field

In William Baldwin's gentle satire *Beware the Cat* (1584) there comes the remarkable moment when the narrator suddenly can hear everything within a hundred miles of London. Geoffrey Streamer manages this feat by cooking up and eating the body parts of a cat, a hare, a fox, a hedgehog, and a kite—reserving their ears as stuffing for two fried pies or "pillows" and their tongues for two jellied lozenges. He consumes the other body parts in the form of broth and a cake, applies the ear-stuffed "pillows" to his own ears, puts the tongue-infused lozenges under his own tongue, puts his left foot on the fox's tale, and *voilà*—or rather *entendrez là*—the "filmy rime" at the bottom of his ear hole was so purged "that the least moouing of the ayre, whether stroke with breth of liuing creatures which we call voyces, or with the moouing of dead, as windes, waters, trees, carts, falling of stones &c which are named noyses, sounded so shril in my head by reuerberacion of my fined filmes, that the sound of them altogether was so disordered and mo[n]strous: that I could discern no one from other." The only thing Geoffrey says his "refined" ears could hear with absolute distinction was the music of the spheres, ringing out so high above the normal range of human hearing that the lowest of the sounds, that of Saturn, far excelled the highest-pitched bird calls or the wind's whistling or organ pipes.

Only the curvature of the earth's surface prevented his horizontal hearing from stretching as far as his vertical hearing: "While I harkned to this broil, laboring to discern bothe voices and noyces a sunder, I heard such a mixture as I think was neuer in *Chaucers* house of fame, for there was nothing with an hundred mile of me doon on any side, (for from so far but no farther the ayre may come because of obliquation) but I herd it as wel as if I had been by it, and could discern all voyces, but by means of noyses vnderstand none." He proceeds to catalog them all:

> Lord what a doo women made in their beds: some scolding, some laughing, some weeping, some singing to their sucking children which made a woful noyse with their co[n]tinuall crying, and one shrewd wife a great way of (I think at S. Albons) called her husband Cuckolde so lowd and shrilly: that I heard that plain, and would faine haue I

heard the rest, but could not by means of barking of dogges, gru[n]t-
ing of hoggs, wauling of cats, rambling of ratts, gagling of geese[,]
humming of bees, rousing of Bucks, gagling of ducks, singing of Swan-
nes, ringing of pannes, crowing of Cocks[,] sowing of socks, kacling of
he[n]s[,] scrabling of pe[n]nes, peeping of mice, trulling of dice, corling
of froges, and todes in the bogges, chirping of crickets, shuting of wick-
ets, skriking of owles, flittring of fowles, rowting of knaues, snorting of
slaues, farting of churls[,] fisling of girles, with many thinges else as
ringi[n]g of belles, cou[n]ting of coines, mounting of groines, whispering of loouers, springli[n]g of ploouers, groning and spuing, baking
and bruing, scratching & rubbing, watching and shrugging, with such
sorte of commixed noyses as would deaf any body to haue heard. . . .
(24–25^v)

It is not just volume, pitch, and plenitude that make what Geoffrey Streamer hears seem so "disordered and monstrous" but the lack
of a physical context for making sense of it all. A friend's house at
the end of St. Martin's Lane, just by the city wall at Aldersgate, is the
site of Geoffrey's experiment. The "broil" that he hears comes from
all over: from indoors, from outdoors, from the city, from the country,
from St. Albans twenty-five miles up the road, from places seventy-
five miles beyond that. Ordinarily, listening is an intensely *situated*
experience. The listener stands at the center of a sphere that extends
in all directions. Just how far he is able to hear depends on all sorts
of contingencies, as does the degree to which his range of hearing
and his range of vision happen to coincide. In all situations, however,
it is the rootedness of the listener's body in space and time, in the
here and now, that enables him to "place" what he is hearing. Isolated
from the sources of sound, Geoffrey has trouble getting his bearings.
The sounds lack locality. They lack temporal sequence. His first impulse is to try to separate "voices" from "noises." The distinction he
gives to these terms—voices as sounds coming from creatures that
have breath, noises as sounds made by inanimate objects—goes back
to Aristotle's *De Anima*. However differently we ourselves might define the terms on acoustic or psychological grounds, Geoffrey points
us toward the importance of attending to the totality of sound—human and nonhuman, intentional and random—in finding perceptual
order amid the sonic chaos that surrounds us. The act of imposing
order is, at bottom, a map-making project.

BEATING THE BOUNDS

Once a year, on the Monday, Tuesday, and Wednesday before Ascension Day, being the sixth Thursday after Easter, people all over
early modern England formally marked their aural boundaries. Ro-

gationtide (from the Latin *rogare*, to seek), "Gang-Days" (from the Anglo-Saxon *gangen*, to go on foot), "beating the bounds"—whatever the local name for the custom, members of the parish would proclaim the church, the fields and woods, the houses and streets as *their* place in the world at the same time that they implored blessings on the newly springing crops or on whatever economic activity happened to sustain the community. The church's bells were rung, ritual procession was made along the boundaries of the parish, boundary-markers were noted, a communal meal was often enjoyed by those who had made the walk. Suppressed under Edward VI, lavishly encouraged under Mary and Philip, the custom persisted through the reigns of Elizabeth, James, and Charles, although stripped of the banners and prayers of pre-Reformation times (Hutton 1994: 34–37, 85, 99, 105–106, 175–176). Even so circumspect a priest as George Herbert could commend "Procession" to a country parson,

> because there are contained therein 4 manifest advantages. First, a blessing of God for the fruits of the field: Secondly, justice in the preservation of bounds: Thirdly, Charity in loving walking, and neighbourly accompanying one another, with reconciling of differences at that time, if there be any; Fourthly, Mercy in releeving the poor by a liberall distribution and largesse, which at that time is, or ought to be used.

Anyone who refuses to join in should be reprehended as "uncharitable" and "unneighbourly" (Herbert 1941: 284). The ringing of bells, the open-air prayers of pre-Reformation times, the chatting of neighbors while they walked: all of these practices served to map the villages, towns, and cities of England in ways that could be heard as well as seen.

The royal realm itself could be mapped in similarly aural terms. Queen Elizabeth's summer "progresses" enacted, on a royal scale, something like the local perambulations that villages, towns, and cities carried out locally at Rogationtide. In 1575, for example, the "Black Book" kept at Warwick records that "it pleased her Majesty to make her Progres into Northamptonshire, Warwickshire, Staffordshire, Worcestershire, and so to returne to Woodstock in Oxfordshire" (Nichols 1823, 1: 417–148). A "progress" it may have been, but it was anything but linear. From Grafton Regis in Northamptonshire to Kenilworth in Warwickshire to Lichfield, Chartley Castle, Stafford Castle, and Chillington in Staffordshire to Hartlebury Castle, the City of Worcester, and Elmley Castle in Worcestershire to Sudeley Castle in Gloucestershire to Woodstock in Oxfordshire, the route traced by Elizabeth and her courtiers between late June and early September came within thirty miles of describing a complete circle (Nichols 1823, 1: 417–600; Dasent 1894, 8: 400–402, 9: 3–20). In other years she made similar progresses into—and *around*—Surrey and Middlesex (1559), the West Midlands (1565, 1572), Somerset (1574), East Anglia (1578), Sussex and Hampshire (1591), and the upper Thames valley (1592) (Bergeron 1971: 11–64). At most of the places she visited there were speeches, pageants, and plays that spoke on behalf of the local inhabitants. If Michael Drayton's *Poly-Oblion* is a project in chorography, in *writing* the land, progress entertainments are an exercise in "choroloquy," in *speaking* the land.

The devices that Robert Dudley, Earl of Leicester, arranged for Elizabeth's amusement at Kenilworth Castle in July 1575 offer an especially eloquent example. At one point in the proceedings, the queen was returning to the castle from the hunt when out of the woods bounded a Savage Man (George Gascoigne in disguise), who sounds the dimensions of the acoustic field in a dialogue with Echo (likely the voice of a boy hidden from sight):

> Ho *Eccho: Eccho,* ho,
>> where art thou *Eccho,* where?
> Why *Eccho* friend, where dwellest thou now,
>> thou woontst to harbour here.
>>>> *Eccho* answered.
> *Eccho.*   Here. . . .

Through a series of such exchanges the Savage Man is led, by degrees, to the true audience for all his speeches, to the true center of the acoustic field in which he finds himself: the person of Queen Elizabeth. One look upon the lady is enough to civilize the Savage Man on the spot. He casts away his club and falls on his knees. The landscape itself, as the Savage Man explains, is endowed with body and voice:

> The windes resound your worth,
>> the rockes record your name:
> These hils, these dales, these woods, these waves,
>> these fields pronounce your fame.
>>> (Gascoigne 1907–1910, 2: 96–101)

And so they do in the fictions out of Arthurian romance and Ovidian mythology that Gascoigne and his collaborators had planted in the garden, meadow, and woods surrounding the castle.

To secure that harmonious effect, however, Gascoigne has tacitly excluded sounds that might be heard as extraneous to the fiction at hand, not least the cacophony that had dominated the countryside during the just-concluded hunt. Another witness to the Kenilworth festivities, Robert Langham, remembers what he heard before the Savage Man's appearance: "the earning of the hoounds in continuans of their crie ... the galloping of horsez, the blasting of hornz, the halloing and hewing of the huntsmen, with the excellent Echoz between whilez from the woods and waters in valleiz resounding" (Langham 1983: 44–45). Maintaining the Arcadian fiction, at least in print, meant ignoring one major source of disruptive sound: the actual inhabitants of Warwickshire, who also contributed to the sports Leicester had commissioned for the queen's entertainment. The published text over which Gascoigne assumed authority, *The Princelye pleasures at the Courte at Kenelwoorth* (1576), makes only passing mention of the bride ale, the morris dancing, the mock tilt, and the choreographed battle between men of Coventry and an army of invading Danes that figure so largely in Langham's account (Wall 1993: 111–167). Perhaps that was because the queen herself at first paid scant notice to these actual country pastimes, translated for her amusement not only in geographical space but out of sequence in ritual time. Parish ales, complete with mock kings and queens, morris dancing, and running at the quintain, were firmly enough associated with early summer to be called "May ales," and the Coventry Hock Tuesday play took its name from the annual date of its performance, nine days after Easter (Hutton 1994: 26–33). At the court at Kenilworth, the queen was otherwise occupied. While the country sports were being staged in the castle's great courtyard, Langham notes, Elizabeth chose to remain indoors: "For her highnes behollding in the chamber delectable dauncing indeed: and heerwith the great throng and unruliness of the peopl, waz cauz that this solemnitee of Brydeale and dauncing had not the full muster waz hoped for: and but a littl of the Coventree plea her highnes allso saw." Two days later, however, the Coventry play was staged all over again, "wherat her Majesty laught well" and rewarded the players with two bucks and five marks in money (Langham 1983: 55). The contrast is eloquent. Within the hall, there was delectable dancing indeed; without, an unruly throng. Within the hall, there was music; without, sounds that were heard, so Langham implies, as noise. Through what it includes and what it excludes, *The Princely Pleasures at the Court at Kenilworth* inscribes a political geography of sound.

On the grounds of Kenilworth Castle in July 1575 representatives of all the "degrees," "ranks," "sorts," or "orders" of people in early modern England came together. William Harrison provides a version of the standard social anatomy in the *Description of England* prefaced to Holinshed's *Chronicles* (1577): "We in Englande deuide our people commonlye into foure sortes, as [1] Gentlemen, [2] Citizens or Burgesses, [3] Yeome[n]: and [4] Artificerers or labourers." The first category turns out to be the most complicated, demanding internal distinctions among titled nobility, knights, esquires, and all others "whome their race and bloode doth make noble and knowne"—to wit, lawyers, university graduates, physicians, professors of the liberal sciences, soldiers with the rank of captain, indeed anyone with the ability to "liue ydlely and without manuell labour." Citizens and burgesses, which Harrison identifies by and large with merchants, possess the freedom of the cities they inhabit and enjoy representation in Parliament. In the countryside, yeomen are distinguished from other inhabitants by maintaining freehold on land to the value of forty shillings a year or by serving as farmers to gentlemen. If clever enough, they can amass enough money to buy the land of improvident gentlemen and secure gentle status for their children by sending them to one of the universities or inns of court. Finally, at the bottom of the social hierarchy come "day labourers, poore husbandme[n], and some retaylers which haue no free lande, copy holders, & al artificers, as Taylours, Shoomakers, Carpenters: Brickemakers, Masons, &c" (Holinshed 1577, 1: 103–105).

Using Harrison as a guide to the folk assembled at Kenilworth, we might distinguish nobility (the queen, Leicester, and other members of the court) from citizens (certain of the men of Coventry and probably Robert Langham) from laborers, husbandmen, and artisans (participants in the bride ale, morris dancing, and mock tilt). The place of yeomen in the festivities is not altogether clear. Did they take part in the country sports or did they stand back and observe as would-be gentlemen? No less ambiguous is Gascoigne's place in the proceedings. Born into a respected Bedfordshire family and educated at Trinity College, Cambridge, he nonetheless spent time in lawyering, in writing for pay, in soldiering, and in debtors' prison before he finally achieved some success in catering to the court's need for verse by the yard. He died two years after the Kenilworth revels (Prouty 1942: passim).

"Plain" is the word Gascoigne (or his printer) chooses to explain why such short shrift has been given to the homely practices staged for the courtly gaze at Kenilworth: "And nowe you have asmuch as I could recover hitherto of the devices executed there: the countrie shewe excepted, and the merry marriage: the which were so plaine as needeth no further explication" (Gascoigne 1910, 2: 106). Just which meaning, or meanings, of "plain" is being invoked here is not altogether unclear: obvious (*OED* II.4)? simple in composition or preparation (*OED* III.10)? unsophisticated, ordinary, common (*OED* V.13–14)? flat, level, even, belonging to a plain (*OED* I.1)? In the late 1570s the social senses of the word were just coming into use. Certainly none of the country sports sorted very well with the heroic designs Gascoigne and his collaborators had on Robert Dudley's demesne. In Bakhtin's terms, fictions out of Arthurian romance and Ovidian myth aspire to a version of "epic": timeless, coherent, univocal. Events that don't fit the grand fictions devolve, in Bakhtin's terms, into a version of "novel": here-and-now, topsy-turvy, polyvocal (1981: 3–40). The *locus amoenus* cannot be a *locus vulgus*. This sense of social exclusivity is reinforced by the first edition of *The Princely Pleasures*, which carries a printer's preface to the "studious and well disposed yong Gentlemen" who have been "desyrous to be partakers of those pleasures by a profitable publication" (Nichols 1823, 1: 486). On the other hand, "plain" may imply no more than that the country show and the merry marriage were self-evident, manifest, patent. In poststructuralist terms, there was nothing there to *read:* nothing to write up, nothing to textualize, nothing to comment upon.

In *A Letter whearin, part of the entertainment untoo the Queenz Maiesty . . . iz signified* Robert Langham proves to be more game. The titular recipient of Langham's letter, Humphrey Martin, "a Citizen, and Merchaunt of London," implies for the reader a subject position a tad below Gascoigne's "studious and well disposed yong Gentlemen." If Gascoigne's account is high mimetic, Langham's—on the subject of country sports at least—is decidedly low mimetic. Where Gascoigne sees epic, Langham sees novel—and a comical novel at that. In the course of *A Letter,* Langham devotes more than a dozen pages to the shows put on by the Warwickshire countrymen—activities Langham variously describes as "gay gamez," "good sport," and "good pastime" (1983: 52, 55).* Langham's evocation of the bridegroom at the mock tilt is typical of his method, in which the most frequent adjectives are "lively" and "good." Decked out in borrowed finery (including his father's jacket and his mother's muffler), the groom goes first in the round of riding with a pike against a plank on a pole: "The Brydegrome for preeminens had the fyrst coors at the Quintyne, brake hiz spear *treshardiment:* but his mare in hiz mannage did a littl so titubate, that mooch adoo had hiz manhod too sit in his sadl, and too skape the foyl of a fall." To such antics Langham adopts a tone of amused condescension: "though heat and coolnes upon sundry occazions made him sumtime too sweat, and sumtime reumatik: yet durst he be bollder too blo hiz noze and wype hiz face with the flapet of hiz fatherz jacket, then with hiz motherz mufflar. Tiz a goodly matter, when yooth iz manerly brought up in fatherly loove and motherly aw" (Langham 1983: 51).

---

*Although David Scott has argued that "Robert Laneham" is a pseudonym and that the actual author of the pamphlet is William Patten, Teller of the Exchequer, Customer of London Outward, and a friend of Lord Burghley, I accept R. J. P. Kuin's identification of "Laneham" with Robert Langham, who had been admitted to the Mercers' Company in 1557. See Langham (1983: 12–15) and David Scott (1977).

Country bumpkins is how the bridegroom and his fellows appeared to Langham. For an account taken from the social level of the Warwickshire participants themselves we have only the financial records of the Coventry corporation. Under the mayorship of Simon Cotton, butcher, from 1574 to 1575 the accounts contain this note: "In this yeare the Queene came to Killingworth castle againe & recreated herselfe there xij or 13 daies. Att which time Coventry men went to make her merry with there play of hockes Twesday & for there paines had a reward & venison also to make yem merry." The play itself figures as so many shillings and so many pence dispensed for props and food:

paide to the players ffor one Reherse	xviij d
paide ffor breakffaste ffor the company at the same Reherse	xviij d
paide ffor hyrynge off one harnys and ffor poyntes	vj d

(Ingram 271)

From the lowliest perspective of all—that of the bride and bridegroom, the morris dancers, the lads who ran at the quintain—the textual record is nonexistent: a blank, a silence. Robert Dudley's closest neighbors live on solely in the words of their social superiors.

SPEECH COMMUNITIES

Nobility, citizens, laborers and artisans: in their day-to-day lives the contributors to the Kenilworth revels functioned not as representatives of abstract "ranks" or "sorts" or "orders" but as members of communities, as people who day by day saw each other in the same places and talked to each other in forms of speech that they recognized as their own and interpreted according to an implicitly understood set of rules. In Dell Hymes's coinage, they constituted distinct "speech communities." In the sixteenth century, as today, the boundaries of such communities can be traced not only in space (according to the geographical location the group happens to occupy) but in time (according to the frequency with which they interact with one another). A common locality and primary interaction are the most visible markers of a speech community (Hymes 1972: 52–66, 1974: 54). Stephano Guazzo in The Civile Conversation (English translation 1581) sees linguistic difference as a fact of social life after the Fall: "it is to be considered, that after the first confusion of tongues, many sortes of languages have by the divine power of God remained in the worlde: whereby not onely one Nation was knowne from another, but also one Countrie, one Citie, one Village, and (which is more) one streete from another" (1967, 1: 140).

The linguistic localizations that Guazzo describes approximate what Benedict Anderson in *Imagined Communities* has call "uni-sonance": an "experience of simultaneity" among speakers of the same language in the same place at the same time. That phenomenon, in Anderson's estimation, is a major contributor to a community's sense of its own identity (1991: 132).

Queen and nobility, citizens, husbandmen and artisans: each of these speech communities stood in a different relationship to sound and to its inscription in graphic media. All of the writing at Kenilworth, let it be noted, was done by men of a middling sort. Gascoigne and Langham were not exactly "burgesses," because they were employed by the court, but they were not quite gentlemen either, because they had to work for a living. Nor were they "artificers," because they did not work with their hands. What they worked with was the written word. The Queen and Leicester were "above" the task of writing, as the artisans of Coventry and the Warwickshire farmers were "below" it. A great deal of the traditional canon of Renaissance literature in English was written by people from this middling social station. Likewise, all three ranks stood in a different relationship to the devices as dramatic events. For Elizabeth and Leicester, *The Princely Pleasures* were designed, performed, and remembered as just that: as sights seen and speeches heard. For Dudley's Warwickshire neighbors the bride ale, morris dancing, and mock tilt were likewise choreographies of sound and body movement. Only for the middling ranks—for Gascoigne, for Langham, for the men of Coventry—did these events require some form of textualization. At the farthest remove from the events themselves stands Gascoigne's "true copy or rather brief rehearsal"; in the closest proximity, the Coventry corporation's payments for breakfast, harnesses, and pikes. In between stands Langham's *Letter*.

Familiar letters of the sort Langham writes were imagined by early modern readers to be transcriptions of the sender's speech. Erasmus gave the idea Europe-wide currency in *De Conscribendis Epistolis* (1522):

> if there is something can be said to be characteristic of this genre, I think that I cannot define it more concisely than by saying that the wording of a letter should resemble a conversation between friends. For a letter, as the comic poet Turpilius skillfully put it, is a mutual conversation between absent friends, which should be neither unpolished, rough, or artificial, nor confined to a single topic, nor tediously long. Thus the epistolary form favours simplicity, frankness, humour, and wit. (1985: 20)

It is this sense of a letter as a transcription of speech, as half of a conversation, that helps to explain why Langham has chosen such idiosyncratic spelling. In *A Letter* the words are made to *look* as they *sound*. Langham's orthography is a mild, rather unsystematic version of the more thoroughgoing phonetic spelling reforms Sir John Cheke, Sir Thomas Smith, and John Hart had been advocating in the 1550s and 1560s (Kuin in Langham 1983: 29–32). In *An Orthographie, conteyning the due order and reason, howe to write or paint th['']image of mannes voice, most like to the life or nature* (1569) Hart summarizes the project succinctly. Just as all material things are made out of the four elements of earth, air, fire, and water, so are words and sentences made out of "voices," out of the minimum units of meaningful sound later linguistics would call phonemes. The *desideratum* is a perfect match between phonemes and graphemes: "The simple voyce is the least part or member of speach, and the letter wée may wel call a maner of painting of that member for which it is written, whose quantitie and qualitie is presented to the vnderstanding thereby, and so the divers members of the speach ought therfore to haue eche his seuerall marke" (1955, 1: 175–176). The trouble is—in early modern English, at least—that voices and marks are *not* perfectly matched. Hart attempts to remedy that situation by limiting which letters can represent which sounds and by inventing, where necessary, new marks or letters. Langham takes the less radical course of using the normal alphabet, but he departs from the spelling conventions that distance the letters from speech sounds. In more ways than one, Langham seems to have taken seriously Erasmus's contention that a letter is the conversation of absent friends.

Not surprisingly, Langham is much more interested than Gascoigne in getting down on paper what the devices he describes happen to have sounded like. His account of the Coventry Hock Tuesday play turns the harnesses and points inventoried by the corporation accounts into gallops, thwacks, and bangs. Mounted on horseback, the Danish landsknights rode first onto the field. Then came the English:

> each with theyr allder poll martially in theyr hand. Eeven at the first entree the meeting waxt sumwhat warm: that by and by kindled with coorage abothsidez, gru from a hot skirmish unto a blazing battayl: first by spear and sheeld, ooutragioous in their racez as ramz at their rut, with furioous encoounterz that toogyther they tumbl too the dust, sumtime hors and man: and after fall too it with sword and target, good bangz a both sidez. . . .

Commemorating an English victory over the Danes in King Ethelred's time, the play involved three onslaughts, the first two won by the

Danes, the last by the English, and concluded with the spectacle of the Danes "led captive for triumph by our English weemen" (Langham 1983: 52–55).

In this detail the citizens of Coventry were enacting a Hocktide custom observed, before the Reformation, all over the country during the second week after Easter. The women on Monday and the men on Tuesday would capture and tie up members of the opposite gender and release them only on payment of a fine, which went toward parish funds. In most places the women raised considerably more money than the men. Suppressed under Edward VI, the custom came sporadically back to life under Elizabeth, although cut off from parish finances (Hutton 1994: 26–27, 100, 119–120; Brand 1870, 1: 197–212). The word "hock" may have been inspired by one or more aspects of the event: seizing and binding (German *hocken*), mocking and scorning (Old English *hocor*), or having a "high time" on a festival occasion (German *Hochzeit*). Langham seems not to have tongue in cheek when he tells Master Martin the citizens of Coventry thought the play might especially please the queen, "for bicauz the matter mencioneth how valiauntly oour english weemen for loove of theyr cuntree behaved themselvez."

According to Langham the play proceeded "in actionz and rymez," but it is the action that dominates his account. When footmen succeeded the mounted soldiers on the field, Langham describes their movements in terms that are positively choreographical: "folloed the footmen, both the hostez ton after toother: fyrst marching in ranks: then warlik turning, then from ranks intoo squadrons, then in too trianglz from that intoo ringz, and wynding oout again" (1983: 54). What mattered was not so much what the men of Coventry *said* but what they *did*. The primary sounds of the Coventry Hock Tuesday play were not words but hooves clattering, alderpoles crashing, harnesses rattling, shields clanging. Performed, according to Langham in "the great coourt" of Kenilworth Castle, these sounds would have been amplified by the buildings' highly reflective stone walls. Small wonder, then, that Elizabeth on the inside should prefer delectable dancing indeed to the throng and unruliness without. Whatever the "rymez" may have been, they existed as cues to body movement. More rhythmic than verbal, they told successive generations of Coventry citizens how to move and what to do. One very good reason the Coventry Hock Tuesday play has not left a more extensive written record is, then, the fact that it was essentially a corporeal event, something that happened with, through, and to the bodies of those who performed and those who watched and listened. In Paul Zumthor's terms, the Coventry Hock Tuesday play, like all oral poetry, was first and foremost a *performance*. As such, it "rejects any analysis that would dissociate it from its social function and from its socially accorded place—more than a written text would. Likewise it rejects dissociation from the tradition that it can explicitly or implicitly claim as its own, from those circumstances in which it makes itself heard" (1990: 28). Gascoigne seems to have realized as much and kept his distance.

As a performance, the combat of the English and the Danes demanded four things: bodies, place, time, and sound. If not entirely absent, all four factors are radically transformed in a written text. The body becomes the hand that writes or holds the book. Place and time become wherever and whenever the reader chooses. Sound fades into the background. By contrast, the Coventry Hock Tuesday play was time- and site-specific, and it happened through bodies, in sound. Normally it happened at a particular point in the ritual year, during the second week after Easter. Furthermore, it was especially identified with Coventry as a place. Langham may be specifying a quite local "here" when he narrates the play's argument,

> how the Danez whylom heer in a tooubloous seazon wear for quietnes born withall and suffeard in peas, that anon by ooutrage and importabl insolency, abuzing both *Ethelred* the king then and all estatez every whear bysyde: at the grevoous complaint and coounsel of Huna the kings cheeftain in warz, on Saint Bricez night, *Ann. Dom.* 1012 (Az the book sayz) that falleth yeerly on the thyrteenth of November wear all dispatcht and the ream rid. (1983: 52)

Jan Vansina has called attention to the mnemonic importance of landscapes in oral history. Battlefields, abandoned towns, and gravesites—whether or not remembered events actually took place there—can become the site of those events through a process Vansina calls "iconatrophy" (1985: 45–46). The land itself offers cues to remembering. Seen in this light, the Coventry Hock Tuesday play is a communal act of historical remembrance, a version of oral history.

Langham, literate Londoner that he was, saw it differently. History was, for him, "as the book says"—although what the book in question might have been is not quite certain. The battle between the citizens of Coventry and the Danes does not figure in Holinshed. For Langham, the play was not so much history as "sport" or "pastime"; for the men of Coventry, on the other hand, history was what the play made happen. What a book might say was a more remote affair. That is not to imply, however, that the performers of the Coventry Hock Tuesday play were illiterate. No better illustration of the complex forms of interplay between oral and literate communication in early modern England can be found than the redoutable "Captain Cox" who musters the Coventry shows in Langham's account: "But aware, keep bak, make room noow, heer they cum. And fyrst captin Cox, an od man I promiz yoo: by profession a Mason, and that right skilfull: very cunning in fens, and hardy az Gawyn, for hiz tonsword hangs at his tablz eend: great oversight hath he in matters of story" (1983: 53). Langham's catalog of Cox's cache of books—virtually a library of deposit for almanacs, broadside ballads, chapbook stories, and folio romances printed in the first half of the sixteenth century—stations the Coventry mason squarely on the bottom rung in the hierarchy of literacies Keith Thomas has described in early modern England. A person might be able to read gothic print even if he could not sign his name. Above "black-letter literacy" ranged ever more rarefied literacies of Roman-letter print, handwriting, and texts in Latin, French, and other foreign languages (Thomas 1986: 97–131; Halliwell-Phillips 1849: 17–35). Cox moves into—and out of—black-letter literacy with ease: "he hath az fayr a library for theez Sciencez, and az many goodly Monumens both in proze and poetry," Langham observes, but "at afternoonz can tallk azmooch *withoout book*, az any Inhollder betwixt Brainford and Bagshot, what degree so ever he be" (1983: 54, emphasis added). Aware! Keep back! Make room! Captain Cox bursts out of two-dimensional print into three-dimensional corporeality. And back again.

Within the broad boundaries of the recognized languages of early modern Britain (English, Welsh, Cornish, Gaelic, Scots), a speech community might be distinguished from other such communities not only by the social station of its speakers but by dialect (Cotswold as opposed to Kent), by variety or register (gentle speech as opposed to servants' speech), by code (legal French, gypsy cant, thieves' argot), or by some combination of these differences. The most extreme example of a speech community in early modern England, perhaps, is provided by thieves, who systematically devised a code outsiders were not supposed to understand—until linguistically mobile writers like Thomas Dekker broke the code. In his pamphlet *O per se O. Or A new Cryer of Lanthorne and Candle-light* (1612) Dekker withholds until the very end just what the title means. As the see-all-tell-all night watchman, Dekker himself cries "Oh," but thieves in the night use the same sound as a badge of communal exclusivity:

> The Fayre is broken vp, and because it is their fashion at the trushing vp of their packes, to trudge away merrily, I will here teach you what *O per se O* is, being nothing else but the burden of a Song, set by the Diuell, and sung by his Quire: of which I will set no more downe but the beginning, because the middle is detestable, the end abhominable, and all of it damnable. Thus it sounds:
>
> Wilt thou a begging goe,
>   *O per se, O. O per se, O.*

> Wilt thou a begging goe?
>> Yes, verily, yea.
> Then thou must God forsake,
>> and to stealing thee betake.
>> *O per se, O. O per se, O.*
>> Yes verily yea, &c.
>
> This is the Musicke they vse in their *Libkens* (their lodgings) where thir-
> tie or fourtie of them being in a swarme, one of the Maister Diuels
> sings, and the rest of his damned crew follow with the burden . . . all
> their talke in their nasty *Libkens*, (where they lye like Swine) being of
> nothing, but *Wapping, Nigling, Prigging, Cloying, Filching, Cursing,* and
> such stuffe. (1612: O4ᵛ–P1)

Beyond dialects, varieties, registers, and codes extends a whole range
of aural possibilities for maintaining communal self-identity. Hymes,
at least, is eager not to limit the means of communication to phone-
mic speech but to include singing, whistling, drumming, horn calling—whatever sound-making keeps the community in aural contact
with one another (Hymes 1972: 53–54).

A professional writer like Dekker provides a good example of
the way one person can move among several speech communities.
Within the "language field" of English, Dekker defined for himself a
"speech field" that included, at a minimum, the London dialect he
grew up with, the rank-specific registers and trade-marked varieties
of *The Shoemaker's Holiday, Old Fortunatus,* and *The Honest Whore,* the
thieves' code of *O per se O* and *The Roaring Girl,* and the aureate register of James I's triumphal entry into London in 1604. Within this
speech field Dekker moved along a "speech network" that connected
the various speech communities with which he interacted. Dekker's
mobility is altogether typical of "the writing sort"—people like
Nashe, Shakespeare, Jonson, and Middleton.

The denizens of court, city, and country who gathered together
at Kenilworth in 1575 may all have occupied the same language
field—early modern English—but they differed greatly from each
other in the breadth of their speech fields. The local Warwickshire
farmers clearly belong to the narrowest speech field, to the least extended speech network, but the field of court speech would have
been nearly as narrow. However many regional *dialects* Elizabeth's
courtiers may have spoken—the Earl of Oxford's spelling, for example, attests an East Anglian accent (Nelson 1995)—they were compelled to interact with one another in the same highly ritualized
register. It was the middling group—the likes of Gascoigne and Langham—who moved comfortably within the largest speech field, along
the most extended speech networks. As with writing, so with speech:

the greatest mobility was enjoyed by people at neither of the social extremes.

SOUNDSCAPES

Speech figures as only one component in the aural identity of geographical places. When Baldwin's Geoffrey Streamer attempts to sort out all the sounds he suddenly can hear, separating "voices" from "noises," he is attempting to map what R. Murray Schafer and Barry Truax call a "soundscape" (Schafer rpt. 1993: 3–67; Truax 1984: 42–56). The field of sound Geoffrey attempts to chart is preternaturally large: it consists of all of southern England. Soundscapes, like speech communities, are usually much more local than that. Outside a scientific laboratory, particular sounds are never heard in isolation but as elements in a larger environment of sound. In a given locality sounds present continuities over time. Certain sounds— the rushing waters of a river, the rumble of automobile traffic, the creaking of a water-powered mill, the clatter of a factory, the hum of fluorescent lighting—emerge as "keynote" sounds, as constants against which other sounds are heard and interpreted. Truax likens the situation to a figure/ground relationship in visual perception (Truax 1984: 21–22). As a maker of sound himself, the listener is never separate from this environment. Rather, he projects sound into the environment—clumping along as he walks, jingling keys, whistling, talking, singing, coughing, chewing, spitting, even exhaling—at the same time that he listens for what the environment communicates to him. And he depends on that interaction. If the environment is so noisy that he can't hear his own footfalls, much less his own speech, he is apt to feel isolated. A soundscape consists, therefore, not just of the environment that the listener attends to but of the listener-*in*-the-environment. It constitutes an ecological system, and like other such systems it can be balanced or unbalanced, viable or dysfunctional (Truax 1984: 19–20, 57–58).

As with an individual listener, so with an entire speech community: human-produced sounds take their place in a larger structure of voices and noises, of sounds produced by animals and sounds made by inanimate objects and forces. Truax imagines the relationship in terms of a continuum, with nonverbal human-produced sounds, music in particular, acting as a kind of mediator:

speech → music → ambient sound

As we move from left to right along this continuum we can observe several changes. Most obvious, perhaps, is the increasing repertoire of possible sounds. Out of the huge range of naturally occuring sounds in any given spot on the globe, indeed out of the huge range of sounds it is possible for humans to make, English speech recognizes only forty or so phonemes. At the same time, a move from left to right traces decreasing temporal density. Phonemes are produced by human speakers at the rate of about five per second; patterns of meaningful sound in the environment—thunder, the waxing and waning of chattering cicadas, church bells striking the hours, birdsong starting in the spring and ending in the autumn—generally take much longer than that. In the move from speech to music to ambient sound there is likewise a decreasing strictness in the human-understood "syntax" that gives meaning to sound. The temporal patterning of sound in a given soundscape is not, after all, an exclusively human construct but a product of all the sounds, nonhuman as well as human, that happen to be present in the physical space. Birds are not indebted to human listeners when they call to one another; neither are leaves as they blow in the wind. There is, then, a decreasing specificity of meaning as we move from speech to music to ambient sound (Truax 1984: 43–47). The points on Truax's continuum are not absolute. Poetry, for example, might belong somewhere between speech and music, New Age music somewhere between music and ambient sound.

Moreover, there may even be a point of contact between the extremes of the continuum, between speech and ambient sound. In crying, screaming, moaning, wailing, ululating the human voice emits sounds that are *non*verbal if not *pre*verbal—sounds, that, according to Aristotle, ally the human voice with the voices of all breathing creatures in the soundscapes of the world. Early modern physiology recognized this continuum between animal sounds and human sounds. "Euen in the tongue of a man," Crooke observes, "sometimes it expresseth onelie those things that fall vnder the Sense, as when wee crie for paine, or for Foode and succour; sometimes those things that fall vnder our vnderstanding as in Discourse" (1616: 629). As La Primaudaye points out, the sounds men share with beasts are, in grammatical terms, "interjections"—literally, "things thrown between" (1618: 379). Whatever statement of semantic meaning it may introduce or qualify, the primal [o:] represents a naked, spontaneous *Gestalt* of force, a projection of the crier's body into the world. As such, it suggests a model for the universe of sound that is not a continuum but a circle:*

---

*For this formulation I am grateful to Catherine Fitzmaurice, a participant in the Folger Institute seminar I convened on this subject in 1994.

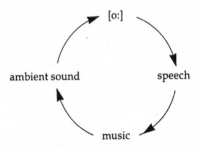

Speech, in this scheme, is not a separate conceptual entity but part of a system of sound that recognizes commonalities among human-produced sounds and sounds in nature. Echo sounds the circumference of the circle.

To understand how early modern speech communities functioned in everyday life, we need to situate them in their distinctive soundscapes. As the inhabitants of a certain geographical space, a *speech* community also constitutes an *acoustic* community. Its identity is maintained not only by what its members *say* in common but what they *hear* in common. Truax has charted the central features in this social soundscape. Certain "sound signals" convey what the inhabitants of the area need to know: bells ring, the tide rises and falls, horns blast, the wind changes direction, hammering starts and stops, birds sing and cease to sing. Sound signals with special importance to the community function as "soundmarks." For early modern England, the bells of the parish church provide a clear example. Each of the sound signals possesses its own "acoustic profile," penetrating a broader or narrower geographic area, standing out to a greater or lesser degree from ambient sounds, taking its place in the mix of sounds as high-pitched or low-pitched, as relatively loud or relatively soft. The community as a whole sits within its own "acoustic horizons," charted not only by the distinctive sounds that emanate from within but by the sounds that come from without. Bells from the next parish serve as a reminder of both differences and commonalities.

As events that happen in time, sounds mark the temporal patterns of a community: the hours as they pass in different activities, day and night as they come and go with the distinctive sounds of setting to work and shutting down, the week as it begins with the altered sounds of Sunday, the seasons as the sounds of different forms of work and recreation succeed one another, the year as sounds heard only on certain occasions return once more. Through sound dominant institutions assert their presence: in early modern England bells signaled mandatory church attendance on Sunday, trumpet blasts heralded a proclamation, guns were fired at regular intervals from castles and fortresses. Features of the natural environment—landforms that catch the wind in certain ways, weather patterns, vegetation, prevailing winds, even humidity—affect the propagation of sound and help give a community its acoustic identity. So, too, does the built environment: the size of buildings, their shape, the materials out of which they are constructed, their placement vis-à-vis each other all determine how reverberant or damped the spaces are, how directly or obliquely sound travels from one place to another within the community. It is not speech alone, then, that gives a speech community its identity, but the whole matrix of sound:

> Our definition of the acoustic community means that acoustic cues and signals constantly keep the community in touch with what is going on from day to day within it. Such a system is "information rich" in terms of sound, and therefore sound plays a signficant role in defining the community [1] *spatially,* [2] *temporally* in terms of daily and seasonal cycles, as well as [3] *socially and culturally* in terms of shared activities, rituals and dominant institutions. The community is linked and defined by its sounds. (Truax 1984: 59–61, emphasis added)

One of the axioms of poststructuralist theory is the situatedness of human subjects in a particular language and, through that language, in a particular culture. *Dico ergo sum:* Lacanian psychoanalytical theory asserts that "I" is a function of language. Truax's idea of acoustic community challenges the primacy of speech as a means individuals use to locate themselves vis-à-vis other people and vis-à-vis the physical world. In Truax's formulation, the voicing of prelinguistic sounds and postlinguistic sounds, and the impinging of nonhuman sounds, all contribute to a given community's sense of self-identity. To some degree, nonhuman sounds are the product of human activities. Think, for example, of the traffic noises in a large city. But to an even greater degree, nonhuman sounds are simply *there,* functioning as a "given" within which the culture in a particular place is constructed. "The wind bloweth where it listeth, and thou hearest the sound of it, but canst not tell from whence it commeth, and whither it goeth" (Geneva Bible, John 3:8).

Describing the distinctive ways in which the Kaluli people in Bosavi, Papua New Guinea hear what goes on in the tropical rainforest that surrounds them, anthropologist Steven Feld argues for the existential force of hearing in the shaping of cultures. The relationship of people to places, he proposes, is reciprocal: "as place is sensed, senses are placed; as places make sense, senses make place." People dwelling in a particular soundscape know the world in fundamentally different ways from people dwelling in another soundscape. For this state of affairs—for "sounding as a condition of and for knowing"—Feld has proposed the term "acoustemology" (1996: 91, 97). As hearing (*akoustikos*) + *logos*, acoustemology recognizes that cultures establish their identities not only through things seen but through things heard and said. By combining Hymes's idea of speech communities with Truax's idea of acoustic communities, and framing them with Feld's idea of acoustemology, we arrive at something like an ecology of speech. At the very least, the acoustemology of early modern England should prompt us to investigate whether people heard things—and remembered what they had heard—in ways different from today. But first let us return to Geoffrey Streamer and help him sort out the chaos of sounds in southern England.

3

The Soundscapes of Early Modern England: City, Country, Court

A sense of future shock reverberates from *New Atlantis* in Francis Bacon's description of the "Sound-Houses" he would install in his utopian college, places *"wher wee practise and demonstrate all* Sounds, *and their* Generation." In addition to musical instruments, the houses contain equipment for altering sounds, for breaking them into their constituent frequencies, amplifying them, and broadcasting them beyond the range of natural hearing:

> *Wee represent* Small Sounds *as* Great *and* Deepe; *Likewise* Great Sounds, Extenuate *and* Sharpe; *Wee make diuerse* Tremblings *and* Warblings *of* Sounds, *which in their* Originall *are* Entire. . . . *Wee haue also diuerse* Strange *and* Artificiall Eccho's, Reflecting *the* Voice *many times, and as it were* Tossing it: *And some that giue back the* Voice Lowder *then it came, some* Shriller, *and some* Deeper; *Yea some rendring the* Voice, Differing *in the* Letters *or* Articulate Sound, *from that they receyue. Wee haue also meanes to conuey* Sounds *in* Trunks *and* Pipes, *in strange* Lines, *and* Distances. (1626: 41)

The realization of Bacon's vision in the twentieth century has altered forever the conditions under which human beings all over the world now hear. Two inventions—electricity and the internal cumbustion engine—make it difficult for us even to imagine what life in early modern England would have sounded like.

To begin with, we have to contend with much louder sounds. A glance at figure 3.1 will reveal how the very loudest sounds that a sixteenth- or seventeenth-century listener might encounter—early modern physiologists specify them to be thunder, cannon-fire, and bells—fall within a range of decibel intensities that would nowadays almost rate as normal events (Paré 1634: 189; Crooke 1616: 585, 588). More subtle changes in hearing have been affected by changes in the matrix of sound, in the "ground" against which individual sounds are heard as "figures." As "keynote" sounds in both urban and rural environments today, internal combustion engines and large-scale electrical apparatus like air conditioners have transformed soundscapes with respect to both the frequency and the intensity of audible sound. Automobile traffic, even at a distance, produces a broad band of frequencies. The result is a masking effect, blotting out other low-

Figure 3.1. COMPARATIVE INTENSITIES OF COMMON SOUNDS

Intensity (in decibels)	Source of sound
180	rocket launching pad
140	gunshot blast, jet plane
120	threshold of pain thunderclap pneumatic hammer at 3 feet amplified rock music performance in large arena
110	interior factory noise
100	chain saw
95	motorcycle at full throttle at 3 feet
90	lawn mover applause in an enclosed auditorium
85	large trucks at 50 feet
80	average city traffic garbage disposal alarm clock at 2 feet
75	human shout at 3 feet moderate surf at 10 to 15 feet telephone ring at 10 feet
70	passenger cars at 50 feet large dog barking at 50 feet vacuum cleaner hair dryer noisy restaurant
60	conversation at 3 feet birds at 10 feet air conditioner at 20 feet automobile at 30 feet
50	light traffic quiet office noise
40	subdued conversation wind in trees at 10 miles per hour refrigerator
30	quiet garden whispered conversation
20	ticking of watch at ear
10	rustle of leaves
0	threshold of audibility

frequency sounds and reducing the distance at which sounds of all kinds can be heard. The "acoustic horizon" shrinks, producing for the listener a relatively constricted sense of space. Electrical hums are more focused, periodic sounds than traffic noises, but they likewise reduce the variety of sounds that can be heard. Against the constant drone of an electricity-driven machine, other sounds have to be more intense to command attention (Truax 1984: 22–23).

Equally disruptive is the capacity of electricity to amplify sound and to convey it across long distances. Cut off from its natural sources of energy, voice and music become disembodied: they are transformed into artefacts (Truax 1984: 46–47). It is just possible today to escape the sounds of automobile engines and electrical equipment—but only in wilderness preserves that are devoid of people as well as machines. Truax's conclusions with respect to the small number of pre-industrial societies still existing in the world would also have been true for all of early modern England:

> few high intensity or continuous sounds exist in the preindustrialized
> world. Therefore, more "smaller" sounds can be heard, more detail can
> be discerned in those that are heard, and sounds coming from a
> greater distance form a significant part of the soundscape. In terms of
> acoustic ecology, one might say that more "populations" of sound ex-
> ist, and fewer "species" are threatened with extinction. (1984: 70–71)

If soundscapes are more than the background to human communication, if soundscapes involve constant interaction between speech communities and their acoustic environment, then we must expect to find in the culture of early modern England fundamental differences from our own culture not only in the range of available sounds but in the degree and quality of the interchanges.

Music provides a particularly telling instance of how such interactions happen. Cultures all over the world, throughout history, seem to have taken delight in listening to sounds in the natural world, devising instruments for replicating those sounds, and imposing on them a human meaning, sometimes by adding words, sometimes by means of instruments alone (Truax 1984: 43). As Don Ihde points out, music may begin as an imitation of natural sounds, but its ultimate reference is to the human listener (1976: 158–159). Music mediates between soundscape and speech: it moves nonhuman sounds in the direction of speech, and speech in the direction of nonhuman sounds. The music of early modern Europe is full of bird songs turned into madrigals (Martin Peerson's "Pretty wantons, sweetly sing . . . Jug, jug, tereu, tereu"), of chickens set to harmonious clucking (Robert Jones's "Cock-a-doodle-doo, thus I begin") (Fellowes 1967: 181, 121). Mediations of natural sounds in music find their counterpart in medi-

Musical quotation 3.1

Derrie ding, ding, ding, Das-son,   I am Iohn Ches-ton, we   weed-don   we

wod-den, we   weed-on, we   wod-den, Bim bom,   bim   bom,   bim   bom,   bim   bom.

(Ravenscroft 1611: no. 17)

ations of speech sounds, especially in the "fancies" by Thomas
Weelkes, Orlando Gibbons, and Richard Dering, who take the street
cries of London and arrange them into a harmonious musical sequence (Brett 1967: 102–147). Thomas Ravenscroft's collection *Melismata, Musicall Phansies Fitting the Court, Citie, and Countrey Humours*
(1611) carries out a similar project, but on three fronts. Among the
"Cittie Rounds" is a canon on cries for shoes, oysters, cockles, straw,
pippins, and cherries, with the night-watchman's warning bringing
all to a close. "Country Rounds" are replete with nonhuman sounds
(see musical quotation 3.1). At quite a different point along the arc
from speech to soundscape are "Court Varieties," a collection of artful
love songs, with some lustier interjections from pages and servants.
Ravenscroft's typology of city, country, and court suggests a plan for
mapping the soundscapes of early modern England that happens to
coincide with the three speech communities assembled at Kenilworth
in 1575. Maps, travelers' accounts, "characters," prints, and passing
allusions in plays, poetry, prose fiction, and pamphlets provide evidence in each case for a repertory of sounds. The challenge is to take
the local specifity of this graphic evidence and to position it within
the far more elusive coordinates of space and time.

CITY

Even without internal combustion engines and large-scale electrical equipment, early modern London was notably full of sound.
When Philip Julius, Duke of Stettin-Pomerania, rode into the city on
September 12, 1602, he and his entourage were astounded by what
they heard:

> On arriving in London we heard a great ringing of bells in almost all
> the churches going on very late in the evening, also on the following
> days until 7 or 8 o'clock in the evening. We were informed that the
> young people do that for the sake of exercise and amusement, and

sometimes they lay considerable sums of money as a wager, who will pull a bell the longest or ring it in the most approved fashion. Parishes spend much money in harmoniously-sounding bells, that one being preferred which has the best bells. The old Queen is said to have been pleased very much by this exercise, considering it as a sign of the health of the people. (Gerschow 1892: 7)

Paul Hentzner, a German jurist who had made the same trip in 1598 and wrote it up in his pan-European *Itinerary*, observes that, in general, English people are "vastly fond of great noises that fill the ear, such as the firing of cannon, drums, and the ringing of bells, so that it is common for a number of them, that have got a glass in their heads, to go up into some belfry, and ring the bells for hours together for the sake of exercise" (1901: 83). No less impressed with the custom was Orazio Busino, chaplain to the Venetian ambassador in London in 1617 and 1618, who notes that the boys made bets "who can make the parish bells be heard at the greatest distance" (1995: 169). Such wagers sound like deliberate attempts to breach the parish's acoustic horizon, to transcend the boundaries marked out in rogation processions.

Church bells functioned as the most obvious "soundmarks" in the acoustically dense soundscape of early modern London. In addition to providing recreation for the youths of the parish, they were actually rung to summon people to services—but with ostentatiously Protestant restraint (Busino 1995: 169). According to Frederic Gerschow, the Duke of Stettin-Pomerania's secretary, bells were never tolled for the dead, even if they were sometimes rung to spread news of a grave illness in the parish (1892: 7). Loudest of all, apparently, was the bell of St. Mary-le-Bow. John Stow, who in the *Survey of London* pronounces the church "more famous then any other Parish Church of the whole Cittie, or suburbs," notes how the bell's ringing signaled rhythms of the workday. Any lateness would prompt apprentices to complain, *"Clarke of the Bow bell with the yellow lockes,/ For thy late ringing thy head shall haue knockes"* (1971, 1: 254, 256). A proverbial association between the bow bell's sound and your true Londoner was already current by 1617, when Fynes Moryson could tell a Europe-wide readership that "Londiners, and all within the sound of Bow-Bell, are in reproch called Cocknies, and eaters of buttered tostes" (Moryson 1617: III2). Was the name "cockney" (from Middle English *coken*, "of cocks") inspired by a cock-shaped weathervane atop the church's belfrey? Or by Londoners' loud loquaciousness? Or by their boastfulness? Whichever the case, the bow-bell rang out over an acoustic community that was also an identifiable speech community, with its own dialect, its own varieties, its own registers.

As a soundscape, early modern London was far too diffuse to be contained by parish boundaries. Within the acoustic horizon of a single parish, for example, might live several different speech communities. Overseas visitors were impressed with the number of foreign-speakers, primarily merchants and Protestant refugees, who made their homes in London. Thomas Platter, visiting from Switzerland in 1599, mentions churches that held services in Dutch, French, and Italian. Platter and his entourage stayed at the Fleur-de-Lys, an inn kept by a Frenchman, in Mark Lane northwest of the Tower, and attended services in a French-speaking chruch. Hentzner gestures toward a center of German-speakers when he points out the hall belonging to the Hanseatic Society (1901: 40). John Chamberlain's letters from the reign of James I twice mention "the Italian ordinary" as a popular eating and gathering establishment (1939, 2: 120, 501). What was true of different languages was also true of different dialects and varieties of English: the custom of servants and apprentices living in the households of their masters brought together under one roof persons from a variety of speech fields.

The distinctive sounds of different trades and workshops were other factors contributing to a soundscape that might or might not fit within the acoustic horizons of parish bells. Dekker's description of *The Seuen deadly Sinnes of London* (1606) conjures up a street scene that sounds as noisy as it looks frenetic:

> hammers are beating in one place, Tubs hooping in another, Pots clinking in a third, water-tankards running at tilt in a fourth: heere are Porters sweating vnder burdens, there Marchants-men bearing bags of money, Chapmen (as if they were at Leape-frog) skippe out of one shop into another: Tradesmen (as if they were dau[n]cing Galliards) are lusty at legges, and neuer stand still: all are as busie as countrie Atturneyes at an Assises: how then can *Idlenes* thinke to inhabit heere? (1963, 2: 50–51)

Stow looks back almost nostalgically to a time when each trade and manufacture had its own distinctive street or ward. However, "Men of trades and sellers of wares in this City haue often times since chaunged their places, as they haue found their best aduantage" (1971, 1: 81). Only Lothbury, still the site of metal-founding in Stow's day, is singled out for its sonic distinctiveness. With their "turning and serating," the makers of candlesticks, chafing dishes, spice mortars, and the like create "a loathsome noice to the by-passers, that haue not been vsed to the like, and therefore by them disdainedly called Lothberie" (1971, 1: 277). La Primaudaye singles out blacksmiths as likely to be "thicke of hearing, beause their eares are continually dulled with the noise of and sound of their hammers & anuiles." People who work with artillery run the same risk (1618: 375). Though their sounds go unremarked by Stow, shoemakers were still tapping away in St. Martin's Lane, and the sounds of horses, oxen, sheep, and swine could be heard in Smithfield, just outside the walls. Despite the dispersal of trades all over the city, early modern London was becoming increasingly segregated with respect to where goods were produced and where they were sold. A. L. Beier's analysis of burial entries and other data suggests that economic activity within the city walls between 1540 and 1600 was devoted 53 percent to the manufacturing of goods, 28 percent to exchanges of goods, and 19 percent to other enterprises. Outside the walls the figures were an overwhelming 70 percent for manufacturing, only 7 percent for exchanges, and 22 percent other. From 1600 to 1700 the contrast between intramural and extramural activity only sharpened (1968: 141–167). In production of decibels these figures are telling: hammering, pounding, tapping, cutting, scraping, and the vocal interjections that go with such activities were more likely to be heard outside the city walls than within.

To the cross-currents of multiple speech communities and dispersed manufacturing should be added the complications of time. Each segment of the day—sunrise, morning, midday, afternoon, sunset, night—would bring its own round of activities, and its own distinctive panoply of sounds, even in the same place. The Exchange in Threadneedle Street provides a good example. Built at his own expense by Sir Thomas Gresham and proclaimed the *Royal* Exchange by a herald and a trumpet blast in 1570, the structure was on every foreign visitor's list of sights. Platter, like the others, was aware that the Exchange was lively at certain hours of the day and not others: before lunch at eleven and afterwards at six the colonnades and courtyard were full of several hundred merchants, "buying, selling, bearing news, and doing business generally" (1937: 157). Hentzner, who notes "the assemblage of different nations" as one of the things to be admired about the Exchange, implies that speech communities from several different quarters could, at these hours, be heard all in one place (1901: 40). Congregation at the playhouses was an event of the afternoon; recreational bell-ringing, an event of the evening.

The factor of time points up the single most important contingency in any attempt to plot the soundscape of the city: people. On a given day, in a given place, at a given hour, people might or might not behave precisely according to habit. Early modern London was not a sociologist's grid but a range of possible paths the inhabitants might take through the day. In de Certeau's terms, the human soundscape of early modern London was created through "tactics" pursued from below, not through "strategies" imposed from above. If a strat-

egy is the attempt of an outside power—a king, a council, a guild—
to exert control over space, a tactic is a subject's counterattempt to
exert control over time, to act contingently, on the spur of the moment
(de Certeau 1984: xix, 35–36). The aural diversity of London was all
on the side of tactics. The soundscape of early modern London was
made up of a number of overlapping, shifting acoustic communities,
centered on different soundmarks: parish bells, the speech of different nationalities, the sounds of trades, open-air markets, the noises of
public gathering places. Moving among these soundmarks—indeed,
making these soundmarks in the process—Londoners in their daily
lives followed their own discursive logic.

For a specific instance let us take the quarter inside Aldersgate
where Baldwin's Geoffrey Streamer is supposed to have had his extraordinary experience of being able to hear everything within a hundred miles. What would Geoffrey, lodged in his friend's house just
within the city wall, have been able to hear during, say, the morning?
A detail from the so-called "Agas" map, probably dating from the
1570s, allows us to place aurally curious Geoffrey with some precision (fig. 3.2). Several major soundmarks present themselves. Closest
to hand, or rather to ear, are the bells of St. Anne's Church, mentioned
as a landmark in Baldwin's narrative and noted in Stow as newly
repaired after a fire in 1548 (1971, 1: 307). Bells from the surrounding
churches would help to define the acoustic horizon of St. Anne's: notably St. Buttolph's just outside Aldersgate, St. Michael's at the western end of Cheapside, and St. Michael's Wood Street. Just visible in
the lower right is the belfry of St. Mary-le-Bow, producing a ring with
an acoustic profile excelling all the others. The workaday significance

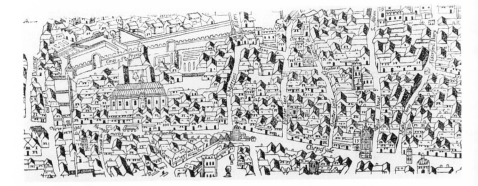

Figure 3.2. Aldersgate and Cheapside. From the "Agas" map of London (probably 1570s). Reproduced by permission of Guildhall Library, Corporation of
London.

of the bow-bell can be appreciated from the church's proximity to Cheapside, one of the major thoroughfares of the city and the site of one of the city's major markets. Traffic noises, the hawking of wares, and the sheer density of conversation would distinguish Cheapside as a soundmark in its own right. As routes in and out of the city, St. Martin's Lane and Wood Street would likewise be full of sound, especially in comparison with the narrower streets and lanes, particularly those running east and west. In St. Martin's Lane shoemakers' tapping on leather and cutlerers' clanging their wares would contribute distinctive low- and high-frequency sounds to the broad-band frequencies of cart wheels turning, horses clopping, and human feet shuffling (Stow 1971, 1: 81; Beier 1968: 141–167). In Baldwin's fiction Geoffrey has taken up residence in St. Martin's Lane so as to be near the printing house that is producing one of his books. The clinks and thuds of the press would, in the story at least, add to the ambient sounds of the morning.

The local soundmarks of St. Martin's Lane, the recurring sounds that marked the area as different from elsewhere, would take their place in the larger soundscape as it was created anew everyday by the city as a whole. Let us attempt to chart that larger soundscape, paying particular attention to keynote sounds, to the natural environment, to the built environment, and finally to a small group of soundmarks recognized by inhabitants and visitors alike as giving London its identity as a single acoustic community. In the absence of motorized traffic and heavy mechanical equipment, what would the keynote sounds have been? What would have provided the ground against which more prominent sounds would figure? At the lowest level of intensity would be the constant sound of running water. The Great Conduit in Cheapside, described by Hentzner as "a gilt tower, with a fountain that plays," is one of a dozen or so running conduits inventoried by Stow (Hentzner 1901: 44; Stow 1971, 1: 17–19). The Great Conduit in Cheapside appears on the "Agas" map just to the left of the belfry of St. Mary-le-Bow. In several places the running waters of built-over rivers and streams could be heard through iron grates (Stow 1971, 1: 175). But the most intense water sounds reverberated up from the Thames. The tight spaces between the narrow arches that carried London Bridge across the river produced so swift a current during the ebbing of the tide—and such a roar of water—that passage for boats was extremely hazardous (Platter 1937: 156).

Street traffic, another source of dispersed keynote sounds, would present a rather different acoustic profile from the broad-band noise of postindustrial cities. In *The Seuen deadly Sinnes of London* Dekker presents the sounds of traffic as positively deafening. Sloth has trouble finding a place, "for in euery street, carts and / Coaches make such a thundring as if the world ranne vpon wheeles" (1963, 2: 50). Be that as it may, other witnesses—and Dekker himself in a different context—suggest a less totalizing sound than thunder. Busino's detailed description of carts and coaches mentions the jingling bells on harnesses that helped to clear a way for horses (1995: 155). Voiced warnings would supplement these signals. Dekker's evocation of the busy streets of Westminster in term time sets in place a range of discrete, individual sounds: "Yea, in the open streetes is such walking, such talking, such running, such riding, such clapping too of windowes, such rapping at Chamber doores, such crying out for drink, such buying vp of meate, and such calling vppon Shottes, that at euery such time, I verily beleeue I dwell in a Towne of Warre" (1963, 4: 25). What Dekker's catalog suggests is not a drone of continuous broad-band sound but a scatter of jingles, bangs, crunches, clops— and voiced *words*. The sound of people talking—not just hawking their wares or clearing a passage for someone important, but *talking*—would, to us, be the strangest feature of the urban soundscape of early modern London. In the absence of ambient sounds of more than 70 dB (barking dogs excepted), the sound of outdoor conversations would become a major factor in the sonic environment. Something of the effect can be experienced today in the historic centers of certain cities in Europe (Bologna comes to mind) that have been emptied of automobile traffic during specified hours of the day. What one notices in the highly reflective corridors of these streets is the audibility of conversations going on a hundred feet and more away. Whispering, or at least dropping one's voice, is actually necessary for privacy. The absence of machines returns the urban soundscape to the human scale of the seventeenth century—the aural equivalent of replacing fifty- and seventy-five-story buildings with two- and three-story buildings.

The reflectivity of London's streets would not be precisely the same, of course, as we might find anywhere else, due to historically specific qualities of both the natural and the built environments. Prevailing winds, for example, would give sound a dominant direction: west to east. The acoustic profile of any given sound, that is to say, would not form a perfect sphere but would be skewed toward the east. Climate, too, would have its effect. Damp, during the wetter months of winter, would have the effect of lowering reflectivity, damping sounds that would be heard as crisper, louder, closer at hand during drier seasons of the year. Helkiah Crooke attends to both factors in his *Microcosmographia: A Description of the Body of Man* (1616). "As the ayre is mooued," he observes, "so also is the Sound carried as wee may perceiue by a ring of Belles farre off from vs: for when the winde bloweth towards vs we shall heare them very lowd:

again when the ayre is whiffed another way, the sound also of the bels wil be taken from vs." A man's voice in winter is "baser" than in Summer, he explains, because the air in winter is "thicker" than in summer and thus is moved more slowly (1616: 607, 610).

Amid these climatic variations, the buildings of early modern London would form, over all, a relatively reverberant environment, but one that presented sharp contrasts between louder spaces and softer. In his *Itinerary* Fynes Moryson describes London in terms of interplay between surface and depth:

> Now at *London* the houses of the Citizens (especially in the chiefe streetes) are very narrow in the front towards the streete, but are built fiue or six roofes high, commonly of timber and clay with plaster, and are very neate and [c]ommodious within: And the building of Citizens houses in other Cities, is not much vnlike this. But withall vnderstand ... that the Aldermens and chiefe Citizens houses howsoeuer they are stately for building, yet being built all inward, that the whole roome towards the streets may be reserued for shoppes of Tradesmen, make no shew outwardly, so as in truth all the magnificence of *London* building is hidden from the view of strangers at the first sight, till they haue more particular view thereof by long abode there. (1967: KKK4)

The unseen depths of London's houses were also, to outsiders at least, un*heard* depths. Crooke, paraphrasing Aristotle's *Problems* 11.37, puzzles over the reason someone inside a house can more readily hear sounds coming from outside than someone outside can hear sounds coming from within. From inside out, he reasons, sounds get lost in the larger space, while from outside in "the sound entering into the house is contracted, gathered, or vnited, and therefore it must needes mooue the Sense more fully" (1616: 700).

With respect to street sounds, wooden timbers and plaster-over-lath are both moderately reflective surfaces; even more so is the window glass that formed an expansive feature of more prosperous houses (Busino 1995: 114; Egan 1988: 52–53). Paving material, on the other hand, might serve to damp street sounds. Although most modern writers have imagined London's streets to be solidly (and reverberantly) cobbled, Busino's description of a mob scene in Cheapside during the Lord Mayor's Show in November 1617 suggests that there was plenty of sound-absorbing mud at hand. Any coachman who tried to make his way through the crowd by using his whip, Busino notes, was pelted with mud. Then someone noticed a Spaniard: "his garments were foully smeared with a sort of soft and very stinking mud, which abounds here at all seasons, so that the place deserves to be called Lorda (filth) rather than Londra (London)" (1995: 114–116). Bearing out Busino's testimony, W. T. Jackman's survey of various paving statutes from the fifteenth and sixteenth centuries would indicate that the paved surface of London's streets was most likely rough-cut stone or cobbles, set in mud (1962: 36–41). In *Henry VI, Part Two*, Gloucester refers to "the Flintie Streets" of London, and his barefoot wife, doing penance, curses "the ruthlesse Flint" that cuts her feet (F1623: 2.4.9, 35). If Cheapside, one of the city's major thoroughfares, was so loosely paved, what would have been the condition of lesser streets and back lanes? In these narrow corridors the reflectivity of close-built houses would be partially offset by the damping of wet earth. When the volume of human traffic is figured in, Moryson's contrast between surface and depth would extend to major thorough-fares versus secondary streets. Lined with timbered buildings, paved with stones, full of people at all hours of the day, spaces like Cheap-side would present the aural "surface" of the city; back lanes, some of them miry with mud, would present its "depth."

On the city's aural surface some places were louder than others. Their acoustic profile was such that it extended deep into side streets and back lanes—and deep into the domestic privacy of London's nar-row houses. Jonson's Morose, whose abhorrence of noise drives him to live in the quiet depths of a street too narrow for coaches, carts, and common noises, provides a handy guide to the most prominent soundmarks on the urban surface. What wouldn't he do to escape a wife who has turned out to be anything but a silent woman? "So it would rid me of her! and, that I did superoratorie penance, in a bellfry, at *Westminster*-hall, i' the cock-pit, at the fall of a stagge; the tower-wharfe (what place is there else?) *London*-bridge, *Paris*-garden, *Belins*-gate, when the noises are at their height and lowdest. Nay, I would sit out a play, that were nothing but fights at sea, drum, trum-pet, and target!" The belfries of churches, huge interior spaces like Westminster Hall, the precincts of the Tower, shopping streets and market places, bear-baiting arenas, public theaters: these, in Morose's experience, were the noisiest places in London. In effect, they estab-lished the acoustic horizon for all Londoners. Again, church bells are the most obvious. Time in a belfry is hard penance for a man who tries to escape church bells by leaving town entirely on Saturdays, Sundays, and holidays (Jonson 1925–63, 5: 169–170, 230). Westminster Hall, partitioned off into court chambers and shops, presented a ca-cophony rivaled only by the aisles of St. Paul's at certain times of the day (Wheatley and Cunningham 1891, 3: 484).

In an age without newspapers, Westminster Hall, St. Paul's, the Royal Exchange, and the court were all centers of hearing the latest—and spreading it by voicing it. The favored hour at Paul's was just before the midday meal. *"Quid novi?"* ("What's the news?") was the greeting among men who had gathered to stroll—and to engage in the aural equivalent of sumptuary self-display. The Duke of Stettin-Pomerania's secretary distinguishes the Paul's crowd as "Gentil-homini," as opposed to the merchants gathered at the Exchange, but native satirists suggest a more heterogeneous throng. The offer of exclusive news might be enough to secure would-be gentlemen a free dinner (Cogswell 1989: 20–35; Jonson 1925–63, 6: 287; Brathwait 1631: 34; Gerschow 1892: 61). The acoustic result of all the walking and talking is delineated in John Earle's character of "Paules Walke":

> It is a heape of stones and men, with a vast confusion of Languages, and were the Steeple not sanctifyed nothing liker Babel. The noyse in it is like that of Bees, a strange humming or buzze, mixt of walking, tongues, and feet: It is a kind of still roare or loud whisper. . . . It is the other expence of the day, after Playes, Tauerne, and a Baudy-house, and men haue still some Oathes left to sweare here. It is the eares Brothell, and satisfies their lust, and ytch. (1629: K1^v–K2^v)

Earle may here be echoing the words Dekker puts into the mouth of "Paules Steeple" in *The Dead Tearme. Or Westminsters Complaint for long Vacations and short Termes* (1608):

> when I heare such trampling vp and downe, such spetting, such halk-ing, and such humming (euery mans lippes making a noise, yet not a word to be vnderstoode,) I verily beleeue that I am the Tower of *Babell* newly to be builded vp, but presently despaire of euer beeing finished, because there is in me such a confusion of languages.

It is varieties and registers of English the steeple is hearing, not the tongues of different nations:

> For at one time, in one and the same ranke, yea, foote by foote, and el-bow by elbow, shall you see walking, the Knight, the Gull, the Gallant, the vpstart, the Gentleman, the Clowne, the Captaine, the Appel-squire, the Lawyer, the Vsurer, the Cittizen, the Bankerout, the Scholler, the Begger, the Doctor, the Ideot, the Ruffian, the Cheater, the Puritan, the Cut-throat, the Hye-men, the Low-men, the True-man and the Thiefe: of all trades & professions some, / of all Countreyes some. . . . (1963, 4: 51)

"Countries" in this case refers not to foreign nations, but to different regions of England. Dekker's Tower of Babel is built entirely of local stone.

The Exchange was, by contrast, a genuinely polyglot soundmark: Hentzner pointedly mentions "the assemblage of different nations" (1901: 40). Georg von Schwartzstät, visiting in 1609, describes the throng of merchants, nobles, and other people as being so great that he could scarcely make his way through (1950: 81). Platter fixes the number at "several hundred." Here, too, the favored hour for gather-

ing—for "buying, selling, bearing news, and doing business generally"—was eleven o'clock, just before the midday meal. A crowd assembled again at six o'clock (Platter 1937: 157). Built in a square, with
a colonnade enclosing an open court, the Exchange was, according to
Nashe, "vaulted and hollow, and hath such an Eccho, as multiplies
euery worde that is spoken" (1904–10, 1: 82). The taverns to which
the denizens of Paul's and the Exchange might repair for dinner were
likewise prominent soundmarks. The partitions between tables Platter remarks may have discouraged overhearing other diners' conversations, but taverns were notably *noisy* places, in more ways than one.
"Men come here to make merry," says Earle in his character of "A
Tavern," "but indeed make a noise, and this musicke aboue is answered with the clinking below" (1628: D1). A "noise" was a group
of musicians who went about from tavern to tavern (Jonson 1925–63,
5: 206; OED, "noise" 5.b). Closer to a diner's ear were the burps and
belches emitted by his fellows. Busino is not the only foreigner to be
aghast. With respect to the high prices charged in London for good
wines he quips, "Thus they hold their very hiccoughs in account, nor
is it considered impolite to discharge them in your neighbour's face,
provided they be redolent of wine or of choice tobacco" (Busino 1995:
133–134; Perin 1809: 512).

The interior gathering-places of Westminster Hall and St. Paul's
had their external counterparts in shopping streets (Morose mentions
London Bridge) and in public markets (Billingsgate is Morose's choice
for noise). Something in between inside and outside was presented
by the Royal Exchange—and by animal-baiting arenas and public
theaters. Like St. Paul's and the Exchange, these places of entertainment defined the London soundscape not only geographically but
temporally: they were places where people came together only at certain times of the day. During those times they became dominant
soundmarks. Outside those times they fell into relative quiet—
except, perhaps, for the animal-baiting arenas. The adjacent ken-
nels constituted one of the sites to be seen, and sounds to be heard,
by foreign visitors (Platter 1937: 169–170). The Duke of Stettin-
Pomerania's secretary fixes the number of animals at 200; Lupold von
Wedel estimates a hundred (Gerschow 1892: 17; von Wedel 1895:
230). The bellowing, barking, and roaring produced by these animals
when pitted against one another are described by Dekker in *The Dead
Tearme. Or Westminsters Complaint for long Vacations and short Termes*
(1608) in nothing short of apocalyptic terms: "No sooner was I entred
but the very noyse of the place put me in mind of *Hel:* the beare
(dragd to the stake) shewed like a black rugged soule, that was
Damned, and newly committed to the infernall *Churle,* the *Dogges*

like so many *Diuels* inflicting torments vpon it" (Dekker 1963, 4: 97–98). The public playhouses nearby—especially with the "fights at sea, drum, trumpet, and target" that Morose expects—offered aural events almost as intense as those to be heard in animal-baiting pits. One of Barnabe Rich's apothegms proclaims, "A Drummer is the pride of noyse, *for* he puts downe all but thunder" (1619: 57). The "excessive applause" Platter situates in the playhouses would rate in decibels somewhere between a modern symphony orchestra playing *fortissimo* in an enclosed hall and outdoor truck traffic at fifty feet.

In sheer intensity of sound, however, nothing could approach the precincts of Tower Wharf, at least on certain signal occasions. Morose's mention of that locality has nothing to do with the river but with the ordnance ranged along the Tower's western wall just above the wharf and high above most of the city of London. One of the sights foreign visitors were expected to note during their tour of the Tower, the guns were discharged on ceremonial occasions. In "An Ode . . . in celebration of her Majesties birth-day" (in Jonson's *Underwood*, no. 67) Jonson invokes the way the Tower cannon "poure / Their noises forth in Thunder" (Chalfont 1978: 184; Jonson 1925–63, 8: 239). The effect would be *imposing,* in every sense of the word. At the moment of their discharge, the cannon would preempt all other sounds, overwhelming parish church bells, surpassing even the Bow bell, asserting aural dominance over the soundscape within which Londoners talked and listened their way through daily life.

With the main contours of the London soundscape set in place, let us trace some of those routes of talking and listening. The challenge is finding a sense of direction. What we hear ranges widely along the arc that stretches from primal cry to speech to music to sounds in nature. How do we make sense of so many disparate sounds? Is it possible to hear the city whole? What presents itself first certainly seems like cacophony. Foreign travelers never failed to be struck by the crowdedness of the streets in what was then Europe's most populous city. "This city of London is not only brimful of curiosities," Platter observes, "but so populous also that one simply cannot walk along the streets for the crowd" (1937: 174). Above the keynote sounds of running water, shuffling feet, clopping hooves, creaking wheels, and buzzing conversation could be heard cries: "What do ye lack?" Shopkeepers made their presence known partly through displays of their goods, partly through painted signs and carved or wrought emblems, but mostly by accosting passersby, just as they still do in parts of the eastern and southern Mediterranean. In John Earle's quip, a shopkeeper, like a book, *utters* his contents. "No man speakes more and no more, for his wordes are like his

Wares, twenty of one sort, and hee goes ouer them alike to all com-
mers" (1628: F2^v–F3). "What do you lack?": uttered over and over, it
would be the aural equivalent of a flashing neon sign.

If fruit happened to be the shopkeeper's stock in trade, loud
chomps might be added to the ambient noise: English people do not
put fruit on the table, Busino notes, but "between meals one sees
men, women and children always munching through the streets, like
so many goats" (1995: 179). Here a street vendor would hawk his
wares. There a ballad-monger would set up in front of a bay window,
and his accomplice would crow out "*A proper new Ballad, to the tune of
Bragadeery round* . . . with varietie of ayres (having as you may suppose, an instrumentall *Polyphon* in the cranie of his nose" (Brathwait
1631: B4–B5). At landing places along the river, watermen would
crowd around potential customers, raising a clamor to be chosen
(Platter 1937: 154). Beggars would add their cries to the din, sometimes amplifying their presence with "clack dishes" in which passersby could place alms. In the vicinity of prisons could be heard the
pleas of the inmates (Guilpin 1974: 56). The secretary to the Duke
of Stettin-Pomerania provides a particularly visceral description of
passing the Clink in Southwark: "When we entered the suburb, the
prisoners with petulant cries [*mit einem muthwilligen Geschrey*] be-
seeched us for alms" (Gerschow 1892: 59).

Moving in and out of these localized sounds would be the cries
of itinerant peddlers. The ritualized cadences of their calls seem positively to demand description as music. Early modern witnesses imply
a gamut that runs from the "treble voic'd *Bel-man*" to "the hollow
charnell voyce" of the metal-man (Dekker 1612: C3^v; Brathwait 1631:
E7^v). For Edward Guilpin, the extremes were set by an Irishman crying "pippe, fine pippe, with a shrill accent" to the "grunt" of the man
selling blacking (1974: 56). In between ranged cries for all manner
of goods and services. Attempts by professional musicians—Richard
Dering in 1599, Thomas Weelkes in 1599, Thomas Ravenscroft in 1611,
Orlando Gibbons in 1614, and a now untraced composer in c. 1614
(all collected in Brett 1967)—to transform London street cries into
consort music for voices and viols provide some evidence of what the
cries actually sounded like. In appropriating the voices of Harrison's
lowest sort of people for the musical pleasure of Harrison's higher
sorts, none of the composers pretends to be playing the ethnomusi-
cologist—Gibbons, for one, sounds indebted to Dering—but in their
points of coincidence all five composers do seem to be transcribing
something like what they heard in the streets. What each composer
does, in effect, is *domesticate* the cries, bringing them in from the
streets one by one and setting them to a musical continuo provided
by a family of viols. Four of the five composers give remarkably simi-

lar renditions of the oyster-seller's cry. Gibbons's version appears in musical quotation 3.2. Where oysters went down, nuts went up (musical quotation 3.3). Inks and pens had their own acoustic profile, as witness Dering's transcription in musical quotation 3.4. Just as idiosyncratic, to judge from three of the musical versions, was the chimney-sweeper's cry (musical quotation 3.5). All five composers give the town crier his moment of prominence. Lost horses (Gibbons measure 57; Dering 202; Weelkes 85) and lost wenches (Gibbons 180; anon. 8) provide the substance of the crier's appeal for information (musical quotation 3.6). Gibbons even renders prisoners' pleas for alms (Gibbons measures 132–141, 228–237 in Brett 1967: 119–120, 124). Like Gibbons (measures 160–165), the anonymous composer provides

Musical quotation 3.2

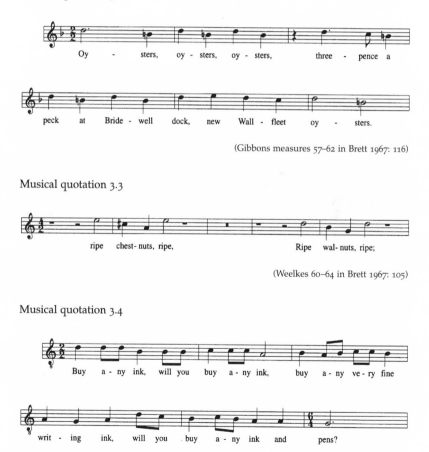

(Gibbons measures 57–62 in Brett 1967: 116)

Musical quotation 3.3

(Weelkes 60–64 in Brett 1967: 105)

Musical quotation 3.4

(Dering 279–284 in Brett 1967: 143)

Musical quotation 3.5

Soop, chim - ney soop, soop, chim - ney soop,

soop, chim - ney soop, mis- tress, soop, with a hoop der- ry der- ry der- ry soop,

from the bot- tom to the top, soop chim- ney soop; there shall no soot fall

in your por - ridge pot, with a hoop der - ry der - ry der - ry soop!

(Dering 170–177 in Brett 1967: 139–140)

a transcription of Thames boatmen clamoring for custom (musical quotation 3.7). For the postindustrial city Steve Reich has produced a parallel piece of artful transcription in "City Life," performed by musicians who follow a score and at the conductor's cues play recorded samples of pile-drivers, taxicabs, and overheard wisecracks (Walters 1996: 10).

As Ravenscroft, Dering, Gibbons, and Weelkes staged the cries for the gentle ear, so a pair of early seventeenth-century prints stage the criers themselves for the gentle gaze. Although each of the hucksters is distinguished from the others by posture, clothing, and the articles he or she carries, all are fitted into the frame of a uniform arcade. In effect, the street vendors are fixed in space and time: they are turned into collectibles (Shesgreen 1990: 12–22). In the soundscape of everyday life peddlers claimed a performative space that was much more expansive. A song inserted into later performances of Heywood's *The Rape of Lucrece* (originally acted in 1608) maps out this dynamic space. Though the song is styled "The Cries of ROME," as required by the decorum of Heywood's play, the cries themselves proclaim their groundedness in the streets outside the Red Bull Theater:

> Thus goe the cries in Romes faire towne,
> First they go vp street and then they go down.
> Round and sound all of a colour,
> Buy a very fine marking stone, marking stone,
> Round and sound all of a colour,
> Buy a very fine marking stone a very very fine.

Musical quotation 3.6

If a-ny man or wo-man can tell a-ny tid-ings of a lit-tle maid-en child, a-bout the age of six, or sev'n and for-ty: This child was lost be-tween the stan-dard and the piss-ing con-duit; if a-ny man can bring a-ny news of her, let him come to the Cri-er, and he shall have four-pence for his hi-re, and that's more than she's worth!

(Anon. 21–49 in Brett 1967: 127–128)

> Thus go the cries in Romes faire towne,
> First they go vp street and then they go downe.
> (1614: K2–K2ᵛ)

Then come bread and meat, prisoners' cries out of Newgate, salt, traps for vermin, kitchen stuff, radishes, lettuce, onions, rock samphire for pickling, mats and hassocks, whiting, hot oatcakes, milk, lanterns and candles. Last of all comes a cry that figures also in Gibbons's setting: the pleas of women prisoners. Up street and down, in and out, hither and thither, from surface to depth: the criers carry out a rogation of the city soundscape. *"Round and sound"*: the song in Heywood's play performs the cry of the writing-stone seller before it identifies the crier or his wares. As sound in the air, separated visually from its source, the line describes the phenomenal effect of cries heard in the streets. Along the arc from primal cry to speech to

Musical quotation 3.7

(Anon. 77–78 in Brett 1967: 129)

music to sounds in nature, street cries occupy a position somewhere between speech and environmental sound. Like speech, they possess semantic meaning; like environmental sound, they are dispersed in space and time.

By composing the cries, by scoring them for voices and viols, Ravenscroft and the others were attempting to hear the city whole. That inspiration shows up particularly in Gibbons's setting, where two or more cries are sung at the same time—fugued, in effect. It shows up, too, in the ways the composers variously manage to bring the cries to closure. Only Ravenscroft avoids closure altogether: he "catches" the cries in the form of a canon ("Row, row your boat" is a famous example), opening up the possibility that the cries could go on forever in all their variety. At the opposite extreme is Weelkes, organist at Chichester Cathedral, who follows a penultimate cry for radishes and lettuce with the command "Now let us sing; and so we make an end: with alleluia" (Brett 1967: 113). The anonymous composer also registers a desire for community, if not unity, by ending with a plea from singers outside on the street to be invited inside, where, after all, the performing singers are most likely situated themselves (see musical example 3.8). A movement from the surface of

Musical quotation 3.8

(Anon. 128–143 in Brett 1967: 132)

street sounds to the depths of domestic quiet likewise concludes Gibbons's and Dering's pieces. In a rehearsal of actual custom the night watchman makes his midnight rounds and bids all householders hang out a lantern, look to their lock, their fire, and their light, "and so goodnight" (Brett 1967: 120, 125, 147). In a sense, the city could be heard whole once a day in the quiet of midnight, in the bell-man's rounds.

On ceremonial occasions there were attempts on a larger scale to hear the city whole. The installation of a new Lord Mayor, for example, gave foreign visitors a chance not only to see the city's visual randomness brought into processional order, but to hear its ordinary chaos of sounds brought into consonance. Music and the shooting of cannon provided the means. Lupold von Wedel, who witnessed such an event on October 29, 1584, records the sounds as well as the pageantry that accompanied the new mayor on his passage by river to Westminster and back again by land to the City. As the mayor stepped into the barge, "a salute of more than a hundred shots was fired, trumpets and musical instruments were heard from all the barges, and there was great rejoicing on the river as far as Westminster." Returning along the Strand, the mayor walked in a procession of several hundred men that was heralded by trumpeters and accompanied by sixteen additional trumpeters and four pipers (von Wedel 1895: 253–254). Busino, watching and hearing another Lord Mayor's installation thirty-three years later, notes the fired salutes and the flourishes from "trumpets, fifes, and other instruments" that marked the mayor's boarding his barge. "The oarsmen rowed rapidly with the flood tide, while the discharges of the salutes were incessant. . . . When the gay squadron had reached a certain point it received a salute from the sakers, which made a great echo. The compliment was repeated even more loudly when my Lord Mayor landed at the water stairs near the court of Parliament, on his way to take the oath before the appointed judges" (1995: 114). In the shrill of brass music and the echoing of ordnance Busino was experiencing a totalizing of the field of sound, an experience of sounds that possessed an acoustic profile broad enough and high enough to stretch to the very horizon of hearing.

Within that horizon there was also an attempt to give the city a unified voice. Between 1432 and 1604 the City of London witnessed no fewer than nine royal entries: those of Henry VI (1432), Katharine of Aragon (1501), Charles V (1522), Anne Boleyn (1533), Edward VI (1547), Mary (1553), Philip and Mary (1554), Elizabeth (1559), and James (1604). On each of these occasions the city became, in visual terms, a series of pageants at precisely the same places along precisely the same processional route. Some of the entries began at London Bridge, others at the Tower, but they all proceeded through

Gracechurch Street to Cornhill to Cheapside to St. Paul's. Some ended there; others continued through Fleet Street to Temple Bar at the city's western boundary. Pageants at fixed points along the route—usually at the conduit in Cornhill, at the Great Conduit, the cross, the Little Conduit in Cheapside, and at Paul's Cross—carried out one or another allegorical program, but always through a combination of tableaux and speeches. At the conduits the pageants often took thematic advantage of running water—or, as the case might be, of running wine. In these emblems-brought-to-life, the city was made to speak.

The most direct of these addresses occurred in Cheapside, usually at the standard or the cross, where the mayor and council were assembled and the city recorder made a speech. Elizabeth's entry in 1559 is altogether typical. At the eastern end of Cheapside there was a triumphal arch showing the eight beatitudes, flanked on each side by "a noyse of instruments." Further along, at the standard, there was another "noyse of trumpettes, with banners and other furniture." Near the cross, above the porch to St. Peter's Church, "stode the waites of the Citie, which did geve a pleasant noyse with their instrumentes as the Quenes Majestie did passe by." Finally at the western end of Cheapside, in front of the Little Conduit, the city recorder spoke for the city as a whole. Giving the queen a crimson satin purse filled with a thousand gold marks, the city recorder "did declare brieflie unto the Quenes Majestie; whose woordes tended to this ende, that the Lorde Maior, his brethren, and Comminaltie of the Citie, to declare their gladnes and good wille towardes the Quenes Majestie, dyd present her Grace with that golde, desyering her Grace to continue theyr good and gracious Quene, and not to esteme the value of the gift, but the mynd of the gevers." The dispersed sounds that usually rang out in Cheapside—gurgling water, groaning carts, jingling horses, chattering strollers, barking dogs, market vendors crying their wares—were composed into the harmonious sounds of music and oratory. On these rare occasions it was possible, in more ways than one, to hear the city whole. When the recorder had finished his speech, the queen made a brief reply, which moved in the crowd "a mervaylous showte" (Nichols 1823, 1: 46–49). From [o:] to speech to music to the total acoustic environment to [o:] the city comes full circle.

COUNTRY

To get our sonic bearings in the country, let us return to Kenilworth. The earliest surviving map of the Kenilworth estates, drawn by Thomas Harding in 1628 (now Public Record Office document M.R. 311), allows us to situate Robert Dudley and his neighbors in a remarkably detailed soundscape. One of the prospects Wenceslas Hollar engraved for Sir William Dugdale's *The Antiquities of Warwickshire* (1656) raises Harding's plat into three dimensions. Hollar's viewpoint can be located on Harding's plat approximately at the point marked H (figs. 3.3 and 3.4). Taken together, the map and the view

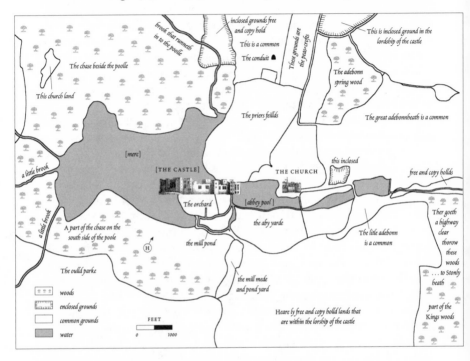

Figure 3.3. (*above*) Map of Kenilworth Castle estates, based on a plat by Thomas Harding (Public Record Office document M.R. 311, 1628).

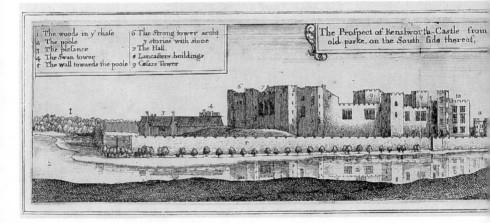

compose a landscape/soundscape that is specific to Kenilworth, representative of the Midlands as a region, and in its general features typical of early modern England as a whole. Most remarkable about the whole ensemble, perhaps, is the close proximity of castle and village. The spot marked 16 in Hollar's view shows "The houses in Kenilworth," ranged along the road to Warwick; the parish church, a former abbey, is indicated a short distance to the right, at the spot marked 17. On Harding's plat, which shows streams and field-boundaries but not roads, the village stands to the upper left of the church, at approximately the point where the field boundary abruptly curves to form a funnel-like opening.

Noble-born Dudley and his common neighbors may have lived their lives as members of distinct speech communities, but they inhabited a single soundscape. Indeed, Dudley had heightened the sense of connection between castle and village by building a new entrance to the castle on the northeast—labeled 11, "The great gatehouse," in Hollar's view—that gave direct access to the village and the road to Warwick (Salzman 1951: 137). On the other side of the castle, however, Dudley had consolidated his privacy by negotiating a land-swap with the villagers, wherein they gave up common land in the chase southwest of the castle in exchange for fields northeast of the castle, one of the parcels being "The priers feillds" in Harding's plat (W. Dugdale 1656: 165). In trying to distance himself from his country neighbors, on one side at least, Dudley was observing an architectural decorum that was new to the sixteenth century. At Woolaton Hall, Nottinghamshire, for example, Sir Francis Willoughby abandoned the old hall in the village and hired Robert

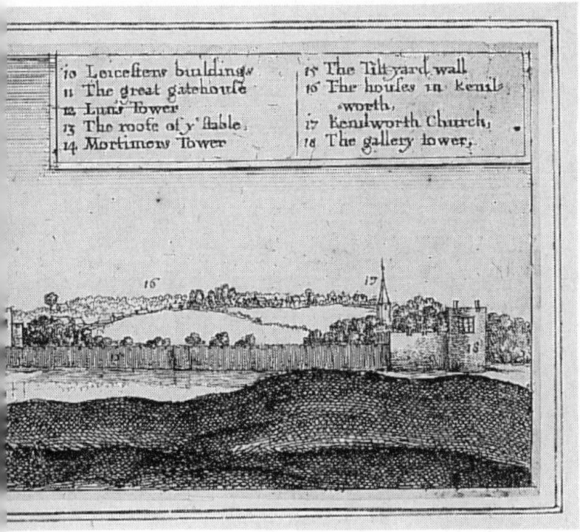

Figure 3.4. Wenceslas Hollar, view of Kenilworth Castle. From Sir William Dugdale, *The Antiquities of Warwickshire* (1656). Reproduced by permission of the Folger Shakespeare Library, Washington, D.C.

Smythson to design an ostentatious new house on a hill some distance away (Girouard 1978: 248). Whatever Dudley may have done to assure visual segregation, he and his neighbors formed a single acoustic community. They heard the same keynote sounds; they recognized the same soundmarks.

On Harding's plat and in Hollar's prospect, the landscape around Kenilworth Castle presents three aspects, each of them possessing its own acoustic properties. To the left in Hollar's view, just beyond the village housetops, is a forest. In the foreground is a meadow, part of "the old park" of the Kenilworth estate. Beyond the castle walls to the left and right of the roofs of the village stretch cultivated fields. In its division into three kinds of land—woods, pasture, and "champian fields"—the Kenilworth landscape is typical of early modern England (St. Clare Byrne 1954: 102–122; Orwin 1967: 53–62; Thirsk 1984: 67). Warwickshire itself was divided into two regions by the River Avon. Northwest of the Avon, in the Forest of Arden, the land was mostly wooded; southeast of the Avon, in the area including Kenilworth, it was mostly under cultivation. All over the Midlands cattle-fattening was an increasingly profitable enterprise, prompting landowners to clear timber and to turn fields into pastures (Butlin 1979: 65–82; Roberts 1973: 188–231).

At Kenilworth, that landowner was Robert Dudley. He had been granted both of the local manors by gift of the queen in 1563 and 1564. Each had its own village and its own lands. The village, fields, and woods visible to the right in Hollar's view belonged to Castle Manor; southeast of the castle, behind the viewpoint in Hollar's prospect, was Abbey Manor, the village of which forms today the main part of the town of Kenilworth (Salzman 1951: 137). In a manuscript of the 1580s, Castle Manor is listed as worth £52 16s 4d per annum and Abbey Manor as worth £70 5s 5d per annum (HMC 1925: 302). Dudley collected this income in the form of fees and rents from his tenants, whose title to the land varied in security from "free-holders," who held virtual title over multiple generations, to lease-holders and copy-holders, who paid an annual rent, to small-holders, who mainly worked for others but had a plot for their own maintenance. In all cases, the tenants lived in the village and went out to work the land (St. Clare Byrne 1954: 102–222). Within the single acoustic community of Kenilworth there were, then, three population centers and three centers of human-produced sound: the castle, the village of Castle Manor, and the village of Abbey Manor.

The acoustic horizon within which those sounds were made and heard was even wider and deeper than in the city. Again, the absence of masking noises from internal combustion engines and electrical equipment would give intensity and presence to the keynote sounds:

wind in the trees, birds, domestic animals, and running water in the several streams Harding delineates and from the dam he shows above the mill pond. Water also spawns frogs, and frogs spawn croaks. "The noise of a Frogge is not great Iwis," Crooke observes, "yet what time they breede they may bee heard many miles out of the Isle of *Elie*" (1616: 694). The background of bird sounds would come not just from the melodious songsters beloved of pastoral poets but from raucous crows. About the former Dekker waxes conventionally eloquent in the encomium of country life he uses to offset the urban horrors of *The Belman of London*: "The melodye which the Birds made, and the varieties of all sorts of fruites which the Trees promised, with the pretty and harmelesse murmuring of a shallow streame, running in windings through the middest of it (whose noise went like a chime of Bels, charming the eyes to sleepe) put me in minde of that Garden whereof our great Grandsire was the keeper" (1616: B^v). For real inhabitants of the countryside, who had ploughing, sowing, and harvesting to do, birds—particularly cawing crows—were more than inducements to sleep. Of a ploughman Wye Salstonstall observes, "A whole flight of Crowes follow him for their food, and when they fly away they give him ill language" (1946: no. 7). Crows are much on Tusser's mind at harvest-time. "Kepe the corne from the crowes," he advises in September. And in November the pests are still cawing: "Except thou take good hede, when first they apere: / the crowes will be halfe, grow they neuer so nere" (1557: A4^v, B1^v). When it came to the country, crows were on the mind even of professional musicians in the city. Ravenscroft's *Melismata* includes under "Country Pastimes" the earliest recorded version of Child ballad number 26 (musical quotation 3.9). In subsequent stanzas the ravens spot a slain knight and contemplate eating him for breakfast, before a doe comes to the corpse's rescue and buries it. Crows, upstart or otherwise, are as assertive in sound as in behavior.

Day and night, summer and winter, domestic animals made their contribution to the soundscape. The Country-man in Nicholas Breton's dialogue *The Court and the Country* (1618) invokes "the Cowe lowing, the Eue bleating, & the Foale neighing" as providing more profitable and *pleasurable* music than the Courtier's "idle note and a worse ditty" (1868: 182). Barking dogs punctuated space and time. Frederic Gerschow, the Duke of Stettin-Pomerania's secretary, was astonished at the number of dogs he and his party saw, and presumably heard, in England in 1602. Here, he says, even "peasants" are able to hunt: "they keep fine big dogs, at little expense, for with a little money they can procure the heads, entrails, and feet of lambs and calves, which in England are always thrown away, with the exception of the tongue" (1892: 47). At least one of Tusser's *Hundreth Good*

Musical quotation 3.9

There were three Rauens sat on a tree, Downe a downe, hay down, hay downe. There were three Rauens sat on a tree, with a downe, There were three Rauens sat on a tree, they were as blacke as they might be, with a downe der- rie, der- rie, der - rie, downe, downe.

(Ravenscroft 1611: no. 21)

Pointes of Husbandrie concerns dogs. If you let your hogs go foraging in October, "giue eie to thy neighbour, and eare to his dogges" (1557: B1). Barking dogs will be ever at their canine rogations.

Against the ground formed by these keynote sounds, several soundmarks declare themselves in Harding's plat and Hollar's prospect: the bells of the two parish churches; the creaks and rattles of the mill; and the distinctive sounds of the different kinds of human activity going on in forest, meadow, and fields. About all of these sounds two things are clear, even at this distance: their audibility and their *legibility*. In an acoustic environment that, apart from barking dogs and the occasional gunshot, lacked any sounds above 60 decibels, all sounds would be present with an intensity quite beyond anything imaginable on the same site today. And in a close-knit social environment those sounds would never be anonymous. Whose dog just barked? Where did that whistle come from? Whom do I hear talking over there? Who might that be that I hear in the woods today? On audible evidence alone a resident of Kenilworth in 1575 would be able to give precise answers to such questions.

Against the background of wind, birds, water, and domestic animals, the sounds of human activity in the Kenilworth soundscape can be plotted in two dimensions: in space and in time. As acoustic spaces forest, meadow, and fields present three different physical conditions for the production and propagation of sound. Large tree trunks without much undergrowth would form a relatively resonant space, potentially full of echoes. Meadowland, lacking any reflective surfaces, would form a relatively damped space. Fields, especially in their varied sixteenth-century configurations, would form a relatively damped space physically but a highly resonant space socially. Let us take the acoustic and social measure of each kind of land. Most textual witnesses see the forest as a place for hunting—just the use to which Dudley was putting the land when "the earning of the hoounds in continuans of their crie . . . the galloping of horsez, the blasting of hornz, the halloing and hewing of the huntsmen" gave way to the Savage Man's dialogue with Echo in 1575. The site of those events is to the left in Hollar's prospect. Saltonstall's character of a gamekeeper amplifies Langham's description of the Kenilworth hunt. To the ear of a gamekeeper, Saltonstall says, "The horne that affrightes other men, is his best musicke: he knowes the changes of the chase, and when a noted Deere is hunted, he windes his fall, and weepes at it. . . . He understands no chamber whispring, but drownes the winds with hallowing, and is answered backe in the same Language" (1631: F3–F3^v).

Hunting is chief among the pleasures that recommend the country in a dialogue between *The English Courtier, and the Cu[n]trey-gentleman* (1586). Vincent, speaking for the country, distinguishes two modes of hunting, one that appeals to the eye and another that appeals to the ear: "bee it your will to hunt with your eye or eare, wee are ready for you as if you please to see with the eye. Wee course the Stagge, the Bucke, the Roa, the Doa, the Hare, the Foxe, and the Badger: Or if you would rather haue some Musicke to content your eare, out goes our dogges, our houndes (I should haue saide;) with them wee make a heauenly noise or cry, that would make a dead man reuiue, and run on foote to heare it" (Breton 1868: 54–55). Vincent, let it be noted, is a simple, downright countryman, the very antithesis of the courtier Valentine. According to Gerschow, the pleasures Vincent describes were available even to "peasants," thanks to the cheapness of keeping dogs fed in England (1892: 47). "O sweet woods, the delight of solitariness": descriptions like Vincent's make us wonder whether Sir Philip Sidney's famous poem testifies to a literary *topos* rather than to social topography.

By comparison with forests, pastures were relatively quiet places. Amid the lowing cows, bleating ewes, and neighing foals evoked by Breton there was only minimal need for a human presence. Saltonstall's character of a shepherd sets in place an isolated man of few words: "if any of his sheepe chance to transgresse the bounds of their sheepe-walke, he whistles out his dogge to fetch them in againe. . . .

To strangers hee's a living *Mercury,* & if he be layd, poynts them out their way with his foote, instead of his hand, and his knowledge sel-dome extends farther than the reach of his eye" (1631: D7^v–D8^v).

"Champian fields" present a far more complicated feature of the Kenilworth soundscape. Harding's plat makes a careful distinction between "common" fields and "inclosed" fields. A huge common ap-pears just to the right of the woods on the middle right of the plat; in the lower left of this parcel an area is marked off from the rest, with the note "this inclosed." Harding likewise marks the difference between "free and copy holld" parcels, held by tenants under long-term agreements, and parcels not so marked, presumably worked by small-holders. "These are inclosed grounds free and copy hold" says the label on a thin parcel just to the left of the upper center. The huge tract of free and copy-held land in the lower right of the plat was presumably a common. Although it postdates Robert Dudley's lord-ship of the manor by forty years, Harding's plat nonetheless indicates the juxtaposition of different methods of farming—and hence of dif-ferent patterns of sound—that characterized Kenilworth in the late sixteenth and early seventeenth centuries. "Open" fields and en-closed fields existed side by side, as did land held "in common" and land held "in severalty." The former is a physical distinction, the latter a legal distinction, and the two did not necessarily coincide (Butlin 1979: 65–82). Under the open-field system each tenant did not work his own separate plot but joined his neighbors in working fields that contained strips variously allotted to each of the workers. An individ-ual worker's holdings were usually spread over several fields and sometimes changed from year to year, so that all the workers held a mixture of desirable and less desirable strips. Under the enclosed-field system each worker received his own parcel, which he could farm as he chose. If the workers held individual title to these enclosed fields, the land was said to be held "in severalty," but it was possible for workers to choose to divide up land held in common and to work it in individual parcels rather than in strips. Not all enclosures were carried out by landlords greedy for the larger profits of pasturage (Butlin 1979: 65–82; Orwin 1967: 30–52). In Warwickshire the region northwest of the Avon was much enclosed by 1600 and probably had never been held in common; southeast of the Avon, in the region including Kenilworth, perhaps a quarter of the arable land had been enclosed by 1607 (Butlin 1979: 65–82). Harding's plat seems to bear out these estimates.

Differences in farming methods made for differences in acoustic community. For a start, the open-field system put workers into close physical proximity to each other. Ploughing, sowing, weeding, har-vesting, grazing on the stubble—all of these activities had to go on at the same time for everybody. Ploughing, for example, required team work. Using a single plough, workers would do a day's plough-ing in one part of the field—often representing one worker's strip—and then move on the next day to the next part of the field (Orwin 1967: 59). The length and width of a worker's allotted strips varied according to contours of the land and soil types, but in the Midlands individual strips seem to have been about 200 yards long by 26 feet wide (Roberts 1973: 188–231). Between the strips were grassy walks called "balks," just wide enough to give workers access to their indi-vidual strips (St. Clare Byrne 1954: 102–122). It was characteristic of the Midlands for one worker's strips to be divided up among two or three fields (Roberts 1973: 188–231).

The social consequence of open-field farming was—or could be—a strong sense of community. John Norden in *The Surueiors Dia-logue* makes a pointed distinction between forests and open fields with respect to civility—or the lack thereof: "people bred amongst woods, are naturally more stubborne, and vnciuill, then in the Cham-pion Countries" (1618: 221). As a prime example of the sense of com-munity fostered by "Champion Countries" Norden cites the Vale of Taunton Deane in Somerset. There, he claims, all of the farming tasks are carried out by groups of people working in unanimity of spirit: "Their hearts, hands, eyes, and all other powers concurre in one, to force the earth to yeeld her vtmost fruit, and the earth againe in re-compence of their loue to her, vouchsafeth them in incredible in-crease" (1618: 228–229). To hearts, hands, and eyes may be added lungs and throats. The sonic consequences of open-field farming were conversation, shouts, and song. Elsewhere Norden speaks of the view of harvest fields from Harrow-on-the-Hill as a sight at which the husbandman "cannot but clap his hands, for ioy, to see this vale, so to laugh and sing" (1593: 11). That Norden was not just being metaphorical is indicated by Paul Hentzner's encounter near Windsor with harvesters making their last trip in from the fields:

> As we were returning to our inn, we happened to meet some country people *celebrating their harvest home [spicilegia sua celebrantes]*; their last load of corn they crown with flowers, having besides an image richly dressed, by which, perhaps, they would signify Ceres; this they keep moving about, while men and women, men and maid servants, riding through the streets in the cart, shout as loud as they can till they arrive at the barn. (1901: 74)

Whether the dispersal of labor in the enclosed-field system encour-aged such a sustained sense of acoustic community is, so to speak, an open question.

Sounds of human activity in the Kenilworth soundscape chart

not only geographical space but seasonal time. Textual witnesses most often cite music and dancing as markers of time in the country. "Every distinct time must bee accommodated to a severall tune," Brathwait says in his character of a piper: "Hee ha's a straine to in-chant the sheepheard in his shearing; another for the husbandman in his reaping; in all which hee ha's a peculiar priviledge for gleaning" (1631: L10ᵛ). Of "A poore Fiddler" Earle observes, "A countrey wed-ding, & Whitson ale are the two maine places he dominiers in, where he goes for a Musician, and ouerlooks the Bag-pipe" (1629: E10). Bret-on's apologist for country life also puts music at the center of seasonal festivals, complete with "daauncing on the greene, in the market house, or about the May-poole." The dancers are happy to offer "a little reward of the Piper" (Breton 1868: 183). His reply is loud: "His *Stentors* voice stretcheth it self to the expression of a *largesse* upon receit of the least benevolence" (Brathwait 1631: L11). When spirits get out of hand, music turns into noise. Brathwait's Ruffian comes into his own on festival occasions, "at May-games, Wakes, Sum-merings, and Rush-bearings. . . . Hee will now and then for want of a better Subject to practise on, squabble with the *Minstrell*, and most heroically break his *Drone*, because the *Drone* cannot rore out his tune" (1631: G4ᵛ).

Music measures days, weeks, seasons, and years. At the end of the day the country alehouse might supplant the parish church as a soundmark: "They drinke here till their mirth and drinke fly out both together, the one in the Chimney and the other in drunken catches, till the streete ring againe, and every pot rises them a note higher" (Saltonstall 1631: E10ᵛ). On Sundays, depending on the tolerance of the local authorities, the churchyard itself might ring out with tabor and pipe and with the shouts of dancers. If challenged, the villagers might reply "tis an old custome, and therefore lawful" (Saltonstall 1946: no. 31). The character of "A Franklin" added to later editions of Overbury's *His Wife* takes the villagers' side in the politics of mirth: "Hee allowes of honest pastime, and thinks not the bones of the dead any thing bruised, or the worse for it, though the Countrey Lasses dance in the Church-yard after Euen-song. Rocke-Monday, and the Wake in Summer, shrouings, the wakefull ketches on Christmas Eue, the Hoky, or Seed-cake, these he yerely keeps, yet hold the[m] no reliques of Poperie" (1616: M4). Periodic fairs would draw peddlers from afar—and the villagers into a knot of noise. "T'would do you good," Saltonstall says of the local gentlewomen, "to heare them bar-gaine in their owne dialect." Hither too might come a ballad-seller, who "may be sooner heard heere than seene, for instead of the violl hee sings to the croud. If his Ballet bee of love, the countrey wenches

buy it, to get by heart at home, and after sing it over the milkepayles"
(1631: E9). Music and dancing would also be inspired by once-a-year
festivals: May games (Dekker 1963, 4: 22; Breton 1868: 183; Brathwait
1631: G4ᵛ), Whitsuntide ales (Earle 1629: E11; Brathwait 1631: L10ᵛ),
sheep-shearing (Brathwait 1631: L12ᵛ), harvest-home (Tusser 1557:
D1; Dekker 1963, 4: 22), and Christmas Eve (Overbury 1616: M4). By
all accounts, harvest-home was the communal high point of the country year. It was also the occasion for bringing the outside sounds of
spring and summer indoors for winter in the form of song. Tusser
advises his husbandman,

> Then welcome by haruest folke, seruantes and all:
> with mirth and good chere, let them furnish thine hall.
> The haruest lorde nightly, must geue the a song:
> fill him then the blacke boll, or els he hath wrong.

(1557: D1)

What Dering, Weelkes, Ravenscroft, Gibbons, and the anonymous composer do for the cries of the city Dering also does for the
cries of the country. "Cries" is perhaps not quite the right word. To
be sure, an itinerant pig- and sheep-gelder puts in a sonic appearance
(musical quotation 3.10). And the country equivalent of the town crier
sounds out his "O yes!" before summoning one and all to the town
hall, where scholars of the free school will put on a play "Where shall
be both a devil and a fool" (measures 144–169). What mostly catches
Dering's ear, however, are the day-in, day-out sounds of country
work, beginning with the neighbors waking one another up—"Jack,
Jack, sleep'st or wak'st? Vast asleep, vather, 'cham vast a sleep" (measures 24–29)—and ending with harvest-home. In between come, first,
barnyard sounds (musical quotation 3.11). Then come two long sequences of hunting sounds, punctuated by a farmer carrying beans
to the court. The first hunt, in pursuit of a hare, includes hounds that

Musical quotation 3.10

(Dering 170–178 in Brett 1967: 154–155)

Musical quotation 3.11

Tig, tig, tig, tig, tig, tig, tig;   Coop, coop, coop, coop, coop, coop, coop, Bid-dy, bid-dy, bid-dy, bid-dy,

bid-dy, bid-dy, bid-dy, bid-dy,

Ho mal, ho mal, ho mal,   ho mal, ho mal, ho mal, ho!

(Dering 38–47 in Brett 1967: 149–150)

go "Yebble, yabble, yebble, yabble, yebble, yabble," horns that go "Ta ra re ro, ta ra re ro," and people that go "Now Wat, Wat, Wat, Wat, look well unto thy scut" (measures 48–90). The second hunt, in pursuit of a partridge, includes not only dogs but whirring hawks (measures 117–140). Just before harvest-home comes a communal effort to induce Mother Crab's bees to hive. "Buzz, buzz, buzz, buzz" drones the bass, while the treble and tenor voices twitter, "Ring out your kettle of purest metal to settle, to settle the swarm of bees" (measures 179–194). The communal effort is an ensemble effort: all five voices sing at once and continue singing together right through the harvest-home to the end. Among the city cries only Gibbons—not even Dering—has different voices singing different texts all at the same time.

Along the arc from [o:] to speech to music to environmental sound to [o:], the sounds of the country occupy a different place from the sounds of the city. At least in the ears of professional musicians like Ravenscroft and Dering and of character-book writers like Saltonstall and Brathwait, the country soundscape is as full of music and environmental sound as speech. At one point in his "country cries" Dering gives up speech altogether for phatic cries and whistles (musical quotation 3.12). In [hei] [ho:] [dzi:] [ri:] [ho:] the singer swoops all the way round the circle: boundaries between speech, music, environmental sound, and the primal [o:] become impossible to mark. Dogs, crows, pigs, sheep, bees, and men bark, caw, grunt, bleat, buzz, and speak within the same acoustic horizon. As Brathwait observes of "A plaine Country Fellow," "He expostulates with his Oxen very vnderstandingly, and speakes Gee and Ree better then English" (1631: F3ᵛ).

Musical quotation 3.12

Hey - ho!   (whistle ··································) Gee, gee!   (whistle ···············

·················) Ree,   hut, hut, hut,   ho!   (whistle ·································)

(Dering 91–100 in Brett 1967: 151–152)

COURT

Along with the Tower of London, Westminster Abbey, and perhaps Oxford and Cambridge, the person of the monarch was something foreign visitors of noble or gentle birth were expected to see while they were in England. Princes and ambassadors might receive a private audience, but the plan for gentlemen was to show up on a Sunday wherever the court happened to be. By gaining admittance to the presence chamber, a visitor could at least see the monarch on her way to chapel or to dinner even if he could not speak with her directly. Lupold von Wedel managed that feat at Hampton Court, on Sunday, October 18, 1584. Two months later at Greenwich he was lucky enough to be present on one of the festival occasions when Elizabeth dined publicly. On this occasion he not only *saw* the protocols of court life; he *heard* them. Queen Elizabeth was surrounded by her courtiers, whom von Wedel delights in identifying and gossiping about. "It is the queen's habit," he says, "to call one of them to her and to converse with him. When she does so, he has to kneel until she orders him to rise. When they leave the queen, they have to bow down deeply, and when they have reached the middle of the room they must bow down a second time." After dinner, the queen "took the son of a count by his mantle and stepped with him to a bow window, where he knelt before her and held a long conversation with her. When he had left her, she took a gentleman, who also knelt down on his knees and spoke with her, after him she called a countess, who knelt down to her in the same manner as the gentleman." During the dancing that followed, Elizabeth continued to call certain courtiers to come and talk with her.

> The queen, as long as the dance lasted, had ordered old and young persons to come and converse with her, who, as I have mentioned, were all obliged to kneel on their knees before her. She talked to them in a

very friendly manner, making jokes, and to a captain named *Ral* she pointed with her finger in his face, saying he had some uncleanness there, which she even intended to wipe off with her handkerchief. He, however, prevented her and took it away himself. It was said that she loved this gentleman now in preference to all others; and that may be well believed, for two years ago he was scarcely able to keep a single servant, and now she has bestowed so much upon him, that he is able to keep five hundred servants. (1895: 263–265)

How von Wedel managed to hear what was said in this particular exchange is not altogether certain, since his report makes it clear that Elizabeth tried to take absolute control of the acoustic environment: she managed to hold everyone's visual attention at the same that she eluded their auditory curiosity. Bay windows, as Lena Orlin has pointed out, were, along with gardens, one of the few places in the built environment of early modern England where two people could expect to carry on a private conversation without being overheard (Orlin 1994). Music would handily have masked the queen's conversations while dancing was in progress.

The eagerness of the entire room to hear what the queen might be saying is indicated by the Duke of Saxe-Weimar's secretary, who records how the duke was received by James at Theobalds in September 1613. When the duke began to address James in Latin, "the Earls and Lords crowded round to hear what his Highness said" (von Rammsla 1865: 150). According to the Venetian ambassador's chaplain, who witnessed a similar meeting between his own master and King James in December 1617, public interviews were *supposed* to be overheard. The contrast, for the Venetians, came sometime later in a private audience with Queen Anne: "it was remarked that all the bystanders drew aside, without listening to every word, as at the public interviews with the king and prince" (Busino 1995: 128). At Greenwich in December 1584 Queen Elizabeth was playing a political game with well established rules. On other occasions she could certainly speak loud enough for all to hear. When she received the Duke of Würtemberg-Mömpelgard at Reading in 1592, the duke's secretary records, the queen "conversed with him on various subjects, and that openly and aloud, so that any in the apartment might understand" (Rathgeb 1865: 11–13). At Greenwich she had other ends in view. Ultimately, however, the acoustics of the room favored neither side in the game. The courtiers may not have been able to hear what the queen was saying to the privileged few, but neither could the queen hear what the rest of the courtiers might be saying about *her*. Hence von Wedel's success in finding out all the gossip about Captain "Ral" but not quite hearing Sir Walter Raleigh's name aright.

One way or another, the monarch remained the focus of court discourse. Dekker, speaking in the person of Westminster, imagines the court as a pyramid with the monarch at the top: "The first and *Capitall Columne* (on which leanes all my strength) is a *Pyramides*, whose point reaches vppe to the Starres: whilest that stands in mine eye, I behold a Maiesty, equall to *Ioves*" (1963, 4: 23). Within the presence chamber the monarch's centrality took shape as a chair. When the monarch occupied the chair of state, every ear hung on what she said. When she was away from the chair, the space still commanded obeisance. Several of the travelers, waiting for the monarch to appear, remark the way servants bowed to the chair of state even in the monarch's absence (Platter 1937: 194–195; Hentzner 1901: 49–50; Valdś-stejna 1981: 81). Once the monarch appeared, her centrality was marked by the kneeling or (in the case of princes) the bowing of everybody else (Platter 1937: 193; Hentzner 1901: 48; Gerschow 1892: 51–53; von Wedel 1895: 250, 264–265). In the very act of speaking to her a suitor lowered the sound of his own voice, physically if not physiologically, and elevated hers. "I am told," Platter says, "that they even play cards with queen in kneeling posture" (Platter 1937: 193). Another mark of Queen Elizabeth's vocal superiority was her command of any language a visitor was likely to speak. As Hentzner describes her appearance in the presence chamber at Greenwich, her visual splendor was amplified by her linguistic bravura:

> As she went along in all this state and magnificence, she spoke very graciously, first to one, then to another, whether foreign Ministers, or those who attended for different reasons, in English, French, and Italian; for, besides being well skilled in Greek, Latin, and the languages I have mentioned, she is mistress of Spanish, Scotch, and Dutch. Whoever speaks to her, it is kneeling; now and then she raises some with her hand. . . . Wherever she turned her face, as she was going along, everybody fell down on their knees.

From the masses of people waiting with petitions outside the hall Elizabeth received the acclamation "*God save the quene Elisabeth*"; her reply, in Hentzner's phonetic English, was "*I thanoke you myn good peupel*" (Hentzner 1901: 48–49; Hentzner 1612: 136). Within the soundscape of the early modern court, the monarch stood as the chief soundmark.

The protocols of speech in the early modern court were made visible as protocols of space. Busino, describing the Venetian ambassador's audiences with King James, Queen Anne, and Prince Charles at Whitehall in December 1617, notes precisely how far into the room the ambassador ventured on each occasion and how far the royal personage advanced to meet him. In all three instances the ambassador and his party had to pass through a succession of outer rooms. When they reached James in his presence chamber, the ambassador made his way through the crowd with some difficulty. "However[,] he obtained room to make the due obeisances. When he reached the centre of the chamber, his Majesty rose from his seat and came to the edge of the royal platform." The ambassador mounted two steps, presented his credentials, and carefully limited his talk, "having previously acquainted himself with the king's humour, who does not relish long speeches, as he is ever intent upon his hunting and enjoying the society of those dearest to him." The contrast with Elizabeth's strategic ways with speech is striking. Similar proprieties were observed during the ambassador's audience with the queen: "After his Excellency had made the proper number of bows, at the right distances, her Majesty rose from her seat and came to meet the mystic lion as far as the extremity of the dais." The prince, in his reception of the ambassador, "advanced as far as the last step beyond the canopy" (Busino 1995: 124–129).

The logic of space in royal palaces was dictated by just such occasions as these. From the medieval plan of a great hall with a single, more private chamber beyond had evolved the sequence of rooms that was to be found at Whitehall, Greenwich, Nonsuch, Hampton Court, and other royal residences. An outer or guard chamber gave access to the presence chamber, beyond which lay first the privy chamber, then a withdrawing room, and finally the royal bedchamber. The move from outer chamber at one end of the sequence to bedchamber at the other was a move, by degrees, toward more and more exclusive access to the monarch's person (Baillie 1967: 169–199; Girouard 1978: 110–116). The door between presence chamber and privy chamber constituted the most important threshold. Into the presence chamber might come all manner of people: courtiers, ambassadors, even foreign travelers like Platter and Hentzner. Beyond the presence chamber only the select few—the ever more select few as the bedchamber grew closer—might pass. The Duke of Saxe-Weimar, for example, made it as far as the first room beyond the presence chamber when he was received by James at Theobalds in September 1613 (Rathgeb 1865: 150). When the Venetian ambassador was granted a private audience with Queen Anne at Whitehall, Busino and other lesser-ranking members of the party were left waiting, "stopped by the tacit intimation *non plus ultra*," while the ambassador proceeded onward "by stairs and unknown passages, which I fancy are not even visited by the sun" (Busino 1995: 130). Waiting was just what the whole scheme was designed to accommodate. Life at court, from the monarch's standpoint, was a round of appearing, speaking, and withdrawing; from the courtier's standpoint, of waiting, listening, and rumoring.

"Discourse & conversation," declares Thomas Gainsford, are "the principall end of a courtiers life" (Gainsford 1616: 20ᵛ). What else is there, after all, for a lady-in-waiting or a gentleman-of-the-chamber to *do* in between withdrawals and appearances of the monarch? And what else should they talk about but the monarch and each other? The nature of courtly talk is suggested by the "Necessary Notes for a Courtier" Nicholas Breton appends to his dialogue *The Court and the Country* (1618):

> Q. What are most dangerous in a Courtier?
> A. To bee inquisitiue of Occurrents, to reueale Secrets, to scorne Counsaile, and to murmur at Superiority.
> Q. What things are most profitable to a Courtier?
> A. A sharpe wit and a quicke apprehension, a smoth speech, and a sound memory. (1868: 209)

Breton indicates here two modes of speaking at court: private whispers and public performance. Knowing just when to deploy "smoth speech" and when "to reueale Secrets, to scorne Counsaile, and to murmur at Superiority" was a thing most requisite in courtiers.

The built environment of the court facilitated these alternations between loud speech and soft speech, between declamation and rumor. With the exception of Nonsuch, it was not the grandiosity of the royal residences that impressed foreign visitors to England but their intricacy. Visitors were escorted through a long succession of rooms full of curiosities. About the treasures of Whitehall the Duke of Stettin-Pomerania's secretary is appropriately awed. About the edifice itself he is less impressed: "The lodgements in this palatio are almost all low, and constructed with many recesses after the monkish way of building" (Gerschow 1892: 23). Busino, visiting the same palace fifteen years later, is even more emphatic: "The palace is not remarkable in itself except for its size, as in case of need it could accommodate more than 600 persons" (1995: 123). About Greenwich he makes a similar comment: "The palace is very large, big enough indeed to accommodate the whole court, but it is not very well arranged, having originally been a monastery" (1995: 161). What Busino implies about both Greenwich and Whitehall is a series of small spaces clustered around a few large spaces. A ground-plan of Whitehall made in 1670 confirms Busino's description (fig. 3.5). Having as its nucleus Cardinal Wolsey's confiscated palace York House, Whitehall was not through-designed but built up over the years by the piecemeal addition of new buildings. Hollar's prospect of c. 1647 gives the impres-

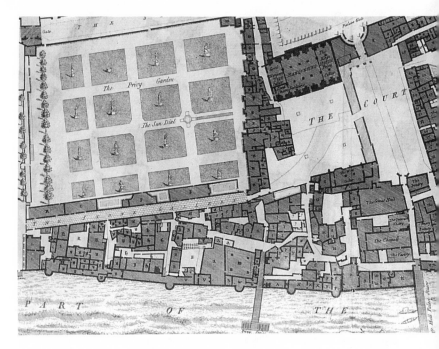

Figure 3.5. Detail of groundplan of Whitehall Palace (1670). Reproduced by permission of the Folger Shakespeare Library, Washington, D.C.

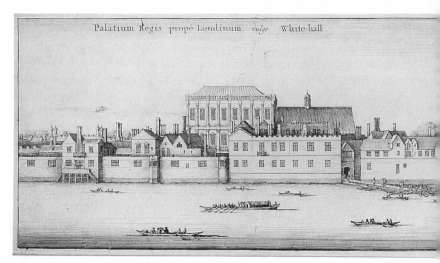

Figure 3.6. Wenceslas Hollar, view of Whitehall Palace (c. 1647). Reproduced by permission of Guildhall Library, Corporation of London.

sion of scores of small houses that happen to be joined together (fig. 3.6). Most ranges are just two storys tall and present a façade of narrow gables between chimneys (G. Dugdale 1950: 11–53; LCC 1930: 10–139).

Inside Whitehall four sorts of spaces presented themselves to a visitor: courts, long galleries, large ceremonial chambers, and smaller rooms. Each of these spaces had its own social functions—and its own acoustic properties. Like most grand houses in Tudor England, Whitehall was built around a series of courts. Two of these open spaces figure prominently in the detail from the 1670 plan: the Privy Garden to the left and the space labeled "The Court" just to the right of center. The first space, described by von Wedel as "a grass plot surrounded by broad walks below and above, enabling many persons to promenade there," afforded a place for whisperings. The second space, known in the late sixteenth and early seventeenth centuries as the "Sermon Court" or the "Preaching Court," was, by contrast, a place for declamation. Von Wedel describes an outdoor pulpit with a sounding board above it and notes that when the queen commanded outdoor preaching the court as well as the windows round about would be full of auditors (1895: 236).

On two sides of the Privy Garden long galleries, a major feature of Tudor and early Stuart country houses, afforded pleasurable indoor walks while at the same time giving access to other ranges of the palace. "The Stone Gallery" is indicated on the ground plan; above it was the "Long Gallery" or "Matted Gallery." At right angles to these two galleries, running along the first floor was the "Privy Gallery." The "Shield Gallery," one of the sights seen by all foreign visitors, ran along the river front above the Privy Stairs. Especially in the recesses provided by bay windows these extraordinarily long and narrow rooms afforded some measure of privacy—and an acoustic space for rumor.

The large ceremonial chambers of Whitehall were situated between the Preaching Court and the river in the block containing spaces labeled "The Great Hall" and "The Chapell." Since most of the ceremonial chambers were on the first floor, not the ground floor, they do not figure in the 1670 plan. Nonetheless, verbal descriptions and works accounts situate the sequence of Outer Chamber, Presence Chamber, Privy Chamber, Withdrawing Room, and Bedchamber in an L-shaped area that begins on the river side of the Preaching Court, above the room labeled "The Kings Gallery," proceeds counterclockwise above the rooms labeled G and H, and ends in the range along the river front (LCC 1930: 10–139). It was through this sequence of rooms, as far as the Presence Chamber—as far as the *second* step of the dais in the Presence Chamber—that the Venetian ambassador made his way in 1617. Although outside the series, and less important than it once had been, the Great Hall continued to be used for court functions like plays and banquets. Banqueting houses provided another site for major gatherings and major sound events. A succession of more or less temporary buildings gave way in 1619–22 to Inigo Jones's permanent structure, shown on the opposite side of the Preaching Court from the state rooms. Finally, the Chapel Royal near the Great Hall accommodated aural events of two distinctive kinds: singing and preaching. Taken altogether, the large ceremonial rooms of Whitehall Palace functioned as centers of both social activity and sound—the court equivalent to Cheapside, the Exchange, St. Paul's, and the amphitheaters of the South Bank. In such spaces interplay between declamation and rumor was at its most complicated.

Rumor held greater sway in the fourth kind of space, the warren of smaller rooms used for courtiers' lodgings, offices, and service activities. The 1670 ground plan shows these rooms as they were configured during the reign of Charles II: small, dispersed, enclosed. All three qualities are caught by the early modern term for such spaces when they were used as lodgings: "closets." Not all followers of the court were fortunate enough to have such a room. Hentzner at Greenwich and Rathgeb at Reading mention tents being used as temporary lodgings (Hentzner 1901: 46, Rathgeb 1865: 14). In Gainsford's image, the life of "a meaner courtier" resembles that of "Humble-bees"—note the pun—"which flie abroad the pleasant fields all day, and then retire to a cowshard at night: so they frequent the pallace, and sometimes are in presence of the King; but how they lie and rest in their lodging, it is pittiful to relate, and barnes & stables are good resting places" (1616: 19^v–20).

The accumulated evidence about the late Tudor and early Stuart court—travelers' accounts, character writings, the 1670 ground plot of Whitehall Palace—defines a built environment sharply differentiated into grand open spaces and small enclosed spaces. Those architectural spaces find their social correlatives in declamation and rumor, their acoustic correlatives in loud and soft. Courts and ceremonial chambers constitute the first sort of space; lodgings and service rooms the second. Galleries seem to have functioned as a combination of the two: the long room itself presented a large open space, but recesses along the window side offered small, relatively enclosed spaces. The contrast between loud and soft would have been heightened by the furnishings travelers noticed in their rounds of England's royal palaces. Platter, Hentzner, Gerschow, and Busino were all struck by the peculiarly English custom of strewing the floor with loose straw, Busino specifying that it was "very deep." Only where the queen was to walk, Platter observes, were carpets laid down. Gershow noticed not loose straw but plaited straw mats (Platter 1937: 192; Hentzner 1901: 46; Gerschow 1892: 23; Busino 1995: 124). Hence, probably, the "Matted Gallery" at Whitehall. These floor coverings—especially in combination with the tapestries Platter and Hentzner noted on the walls—would serve to damp the ambient sound.

Such a muffling effect helps us to understand why Platter and Busino would notice not just the presence of elaborate clocks among the royal treasures but the kinds of sounds those clocks made in striking the hours (Platter 1937: 203–204; Busino 1995: 127). The damping of ambient sound also suggests that the virginals, positive organs, and mechanical music boxes the travelers saw were intended to be heard as well as seen. Von Wedel claims that at Whitehall, "Almost in every room there was a musical instrument with silver-gilt ornaments and lined with velvet" (1895: 237). Velvet, let it be noted, is also a damping material. The uses to which these instruments might be put is suggested by the amorous songs that make up the "Court Varieties" section in Ravenscroft's *Melismata*. Rathgeb, in the privileged company of the Duke of Saxe-Weimar, actually got to hear the queen perform on the virginals: during what seems to have been a private audience the French ambassador "so far prevailed upon her that she played very sweetly and skilfully on her instrument, the strings of which were of gold and silver." Platter saw the same or a similar instrument at Hampton Court (Rathgeb 1865: 12; Platter 1937: 203–204). The relative quiet implied on these occasions may have extended to the outdoors. At Greenwich, Busino remarks an aviary "situated at the end of some flower beds, not very far from the palace, and purposely, so that the song of these numerous warblers may be heard there" (Busino 161). Splashing fountains were famous features of the gardens at Greenwich and Nonsuch (Platter 1937: 191; Hentzner 1901: 78; Busino 1995: 161); an indoor fountain graced the "Privy Gallery" in Whitehall during James's early reign (G. Dugdale 1950: 51). Within an environment muted by rushes, carpets, mats, tapestries, and velvet covers, certain keynote sounds emerge in the soundscape of the court: running water, birds, striking clocks.

Quiet at court was very much, however, a sometime thing. Platter, Hentzner, and the rest were able to tour the palaces and hear the clocks precisely because the court happened not to be in residence. When the court actually inhabited a building, the soundscape would have been dramatically altered. Six hundred people talking—even in small, dispersed, enclosed rooms—would have added murmuring voices to the keynote sounds. When the court gathered together in the presence chamber, in the preaching court, or in the great hall, diffuse keynote sounds would have yielded place to concentrated, high-volume sounds. Beyond the natural effect of so many people talking at once was the aesthetic effect of the loudest possible music. Platter, Hentzner, and von Wedel all describe the playing of music—*brass* music—before and after dinner. Eight trumpeters clad in red called the assembly to dinner while von Wedel looked on at Hampton Court in 1584. Afterward two drummers and a piper "made music according to the English fashion" (1865: 251). At Greenwich in 1598 Hentzner remembers that while dinner was being brought in "twelve trumpets and two kettledrums made the hall ring for half an hour together" (1901: 51). Platter specifies trumpets and shawms as after-dinner entertainment at Nonsuch in 1599 (1937: 195). The totalizing effect of brass music outdoors has already been remarked with respect to the Lord Mayor's procession; indoors, even when damped with tapestries, rushes, and layers of clothing, brass music would have ranked second only to cannon fire as the loudest sound imaginable.

In such moments of totalized sound the acoustic horizons of the court—usually multiple and eccentric—contracted to the confines of a single room: rumor gave place to declamation. On certain other occasions there were more sophisticated attempts to speak for the court in a unified voice. The court equivalent to city pageants and country harvest-home festivals was masques. Plays might be hired in from professionals like the King's Men; masques, like their city and country counterparts, were performed, in large part, by the inhabitants of the soundscape themselves. As in the city and the country, the primary medium for achieving acoustic unity was music, and for precisely the same reason: music modulates between speech and environmental sound.

A particularly eloquent instance is Ben Jonson and Inigo Jones's "Pleasure Reconciled to Virtue," which Busino witnessed in company with the Venetian ambassador on Twelfth Night, January 6, 1618. The venue was one of the most resonant spaces in all of Whitehall Palace. Constructed in 1606 to replace a decaying relic of Elizabeth's reign, the first Jacobean banqueting house was a wooden theater fitted out within a brick box. Occupying the same site on the 1670 plan as the banqueting house erected by Jones in 1619–22, the structure burned a year after the performance of "Pleasure Reconciled to Virtue." Busino, who had to wait for two hours for the king to arrive, had plenty of time to survey the ranks of boxes, the superimposed colonnades of Doric and Ionic columns, nine to a side, set about six feet in from the wall, and the deeply coffered ceiling, complete with carved pendants and thirty *putti*. "The whole is of wood," Busino notes, "including even the shafts, which are carved and gilt with much skill" (1995: 137; Charlton 1964: 15–16). The result of so much vibrating wood and so many reflective surfaces in an area about forty feet by ninety feet would have been an intensely "live" ambience—all the more so when filled with, by Busino's estimate, more than six hundred people all talking at once. What silenced them was the king's entry, announced by fifteen or twenty cornets and trumpets playing "a sort of recitative" (1995: 139). The aural assault continued in Jonson and Jones's masque, which is scripted to open with an apparition of Comus "to a wild Musique of *Cimbals Flutes, & Tabers.*" A song sung by Hercules' bowl-bearer—"*Roome, roome, make roome for y* bouncing belly, / first father of Sauce, & deuiser of gelly*"—fills the room not so much with music as with noises of bodily excess. "I know it is now such a time as the *saturnalls* for all the world," the Bowl-Bearer exclaims, "that every man stands vnder the eaves of his owne hat; & sings what please him, that's the ryte, & yᵉ libertie of it." He himself has seized the occasion by singing something appropriate to the belly, something that can be measured in yards of gut: "a Ballad, and yᵉ Belly worthie of it I must needs say, and 'twere forty yards of ballad, more: as much ballad as tripe." The primal [o:] in this case issues from an orifice other than the mouth: "Beware of dealing wᵗʰ yᵉ belly, the belly will not be talkd to, especially when he is full: there is no venturing vpon *Venter* then; he will blow yoᵘ all vp: he will thunder, indeed la: Some in derision call him the father of farts: But I say he was yᵉ first inventoʳ of great ordynance: and taught vs to discharge 'em on feastivall daies" (Jonson 1925–63, 7: 479–481). The antimasque that is scripted to follow, a dance of a tun and bottles, was accompanied, according to Busino, by the same blaring cornets and trumpets that had announced the king's entry (1995: 140).

Jonson does not explicitly say so, but blown instruments were proverbial in Platonic lore for their gutsiness, their mindlessness, their distance from *logos*. By contrast, stringed instruments admitted the possibility of the singing human voice. It was the varying lengths of the strings, proportionate to the mathematical principles of the universe, that allowed arche-poets like Apollo and Orpheus to charm the material world (Smith 1979: 81–108). Hence, perhaps, the twenty-five to thirty viols Busino specifies as accompanying the dancing of Prince Charles, the Marquis of Buckingham, and ten other nobles when they appear to dispel the figures of the antimasque. Hence, certainly, the instrument carried by the singing figure of Daedalus, who joins Mercury in presiding over the masque proper. The detail does not appear in Jonson's script, but Busino describes Daedalus (without catching his precise identity) as "a guitar player [*un Chittarone*] in a gown, who sang some trills, accompanying himself with his instrument" (1995: 141; Jonson 1925–63, 10: 583). What Daedalus invites the noble performers to do is to dance an emblem of cosmic accord. That is something that happens in space, in vision; it also happens in time, in sound. Gesturing toward the ladies waiting to be asked to dance, Daedalus articulates the arresting idea that music and dance might supply the place of "those silent arts ... Designe & Picture":

> Begin, begin; for looke, y* faire
> do longing listen, to what aire
> you forme your second touch,
> that y*i may vent y*ir murmuring hymnes
> iust to the tune you moue your limbes,
> and wish y*ir owne were such.
> (Jonson 1925–63, 7: 489)

In effect, Daedalus asks the dancers and their audience not just to *see* cosmic accord but to *hear* it. (What Inigo Jones thought of this moment is not recorded.) The primal farts of Comus and his crew, the speeches of Mercury, the music of strings and voices, the harmony of the spheres are subsumed into ○. Busino's description of the nobles' dancing catches this quality of constancy in mutability. The dancers, he notes, made their entrance in the formation of a pyramid: "After they had made an obeisance to his Majesty, they began to dance in very good time, preserving for a while the same pyramidical figure, and with a variety of steps. Afterwards they changed places with each other in various ways, but ever ending the jump together" (1995: 141–142). Whatever dilations they may have performed, they came full circle: they ended as they had begun.

Amid the profusion of sounds, the king remained not only the chief audience but the chief soundmark. He made his aural presence felt. Once the masquers had claimed their partners among the ladies, Busino reports, the nobles proceeded to perform dances of every country imaginable: "Last of all they danced the Spanish dance, one at a time, each with his lady, and being well nigh tired they began to lag, whereupon the king, who is naturally choleric, got impatient and shouted aloud Why don't they dance? What did they make me come here for? Devil take you all, dance." The Marquis of Buckingham rose brilliantly to the occasion: "Upon this, the Marquis of Buckingham, his Majesty's favourite, immediately sprang forward, cutting a score of lofty and very minute capers, with so much grace and agility that he not only appeased the ire of his angry lord, but rendered himself the admiration and delight of everybody." Afterward, says Busino, the king rewarded Buckingham "with marks of extraordinary affection, patting his face" (1995: 142–143). Jonson's script ends the whole affair by commanding the masquers to return whence they came, to a mountain crowned with the figures of Virtue and Pleasure. First Mer-

cury and then "2. *trebles*, 2. *tenors, a base,* and ye whole Chorus" re-
mind the masquers that they were sent forth to "walk" with Pleasure,
not to "dwell" (Jonson 1925–64, 7: 490–491). Outside the fiction, the
evening ended rather differently. According to Busino, the king, the
ambassadors, and the performers left the banqueting house, passed
through a number of chambers, and entered a hall where waited a
table laden with seasoned pasties and sugar confections, all served
on glass platters. On these blandishments for the belly the whole
party descended like so many harpies. The table collapsed. The last
sound Busino records is the crash of glass, reminding him "precisely
of a severe hailstorm at mid-summer smashing the window glass"
(1995: 143).

4   Re: Membering

Language most shewes a man: speake that I may see thee. It springs out of the most retired, and inmost parts of us, and is the Image of the Parent of it, the mind. No glasse renders a mans forme, or likenesse, so true as his speech. Nay, it is likened to a man; and as we consider feature, and composition in a man; so words in Language: in the greatnesse, aptnesse, sound, structure, and harmony of it.

To Ben Jonson's literate view, language is a body. It possesses, first of all, *stature.*

Some men are tall, and bigge, so some Language is high and great. Then the words are chosen, their sound ample, the composition full, and absolution plenteous, and powr'd out, all grave, sinnewye and strong. Some are little, and Dwarfes: so of speech it is humble, and low, the words poore and flat; the members and *Periods*, thinne and weake, without knitting, or number. The middle are of a just stature. There the Language is plaine, and pleasing: even without stopping, round without swelling; all well-torn'd, compos'd, elegant, and accurate.

Different varieties of speech each present a different *physiognomy.*

The next thing to the stature, is the figure and feature in Language: that is, whether it be round, and streight, which consists of short and succinct *Periods*, numerous, and polish'd; or square and firme, which is to have equall and strong parts, every where answerable, and weighed.

Then comes the *skin*, something that can be not only seen but felt.

The third is the skinne, and coat, which rests in the well-joyning, cementing, and coagmentation of words; when as it is smooth, gentle, and sweet; like a Table, upon which you may runne your finger without rubs, and your nayle cannot find a joynt; not horrid, rough, wrinckled, gaping, or chapt.

Last to present itself to the listener's view is an *articulated structure*, something that has a certain weight, tactility, and locomotion:

After these the flesh, blood, and bones come in question. Wee say it is a fleshy style, when there is much *Periphrasis*, and circuit of words; and when with more then enough, it growes fat and corpulent; *Arvina ora-*

tionis, full of suet and tallow. It hath blood, and juyce, when the words
are proper and apt, their sound sweet, and the *Phrase* neat and pick'd.
. . . Juyce in Language is somewhat lesse then blood; for if the words
be but becomming, and signifying, and the sense gentle, there is Juyce:
but where that wanteth, the Language is thinne, flagging, poore,
starv'd, scarce covering the bone: and shewes like stones in a sack.
(1925–63, 8: 625–627)

Although the controlling images in Jonson's conceit are visual
("shewes, "see," "glasse"), the phenomenon he is attempting to describe is fundamentally aural: *"speake that I may see thee."* And the
link between these two sets of phenomena, visual and aural, is the
human body. The body produces sound; the body can be seen and
read. As signs of speech, Jonson's three factors—stature, skin, and
structure—are anything but arbitrary: they are actually part of the
mechanism that produces speech. Their relationship to speech is a
matter, not of metonymy, but of synecdoche. Many of the images Jonson uses can be heard as well as seen: "greatnesse" can be loudness
as well as bulk, "harmony" occurs among sounds as well as visual
elements, qualities of "ample" and "full" can be apprehended by the
ear as well as the eye, "absolution" can be "powr'd out" of a speaker's
mouth as well as from a chalice, "swelling" is both an increase in
musical volume and a distention of the flesh. The most arresting of
these visual/aural puns are images of body parts, of *members*. Thus,
speeches that consist of short, succinct periods are said to be "round,
and streight." In Vives' *De Ratione Dicendi* (1555), the text in which
Jonson made this particular "discovery" in *Timber*, "round, and
streight" is *"rotunda, & teres,"* "circular" or "spherical" and "rounded"
or "well turned" (Jonson 1925–63, 11: 271)—terms that are more usual
in rhetorical writers because they accord so well with the idea of a
rhetorical "period" as a circular figure, a gesture of speech that
ranges through a variety of dependent elements before it returns to
the subject at hand and reaches closure. "Streight" is Jonson's affectation for an effect that most rhetorical writers see, and hear, as "circular." Roundedness and well-turnedness are qualities of both figures
of speech and figures of flesh and bone, and Jonson himself envisions
periods as "members" of the discourse. Deriving from the Latin *mem-
brum*, a "member" is, materially, literally, a limb or other body part
and only metaphorically a part of a whole or one of the individuals
who compose a community. What Vives and Jonson attempt is an
inscribing of speech on a surface; the volume they choose is the human body. In the acoustemology of early modern England the body
is the site of three activities having to do with sound: voicing, listening, and recalling.

VOICING

Torso, neck, head: speaking engages three of the body's major members. (In most cultures it also involves the arms and hands, if not the feet and legs.) Seeing, by contrast, engages only the neck and head. In that respect, seeing is more readily locatable in the body than voicing is, just as the objects it attends to are more locatable in space than sounds are. Vision involves the projection of an inverted image on the retina inside my eye, but I experience that image as "out there"—and right side up. Voicing, on the other hand, is both "out there" and "in here": when I speak the sound waves reverberate through my body as well as in the air around me. Not all sounds are located in the body in quite the same place. Try again to pronounce the sequence of sounds [u] as in "hook," [a] as in "ah ha," [i] as in "hit" and notice the way the sounds move forward in your mouth. Now compare these sounds with [o:]. That round, open sound owes its intensity to reverberations lower down in the body, in the chest. By contrast, sounds like [m] and [n] feel higher in the body, since the reverberations in this case are centered in the nasal passages.

What early modern writers have to say about the voice confirms these bodily experiences of sound, but when the writers attempt to explain the phenomena they do so in physical, physiological, and psychological terms quite specific to their own time and place. In the course of the sixteenth and seventeenth centuries propositions about voice originally argued by Plato, Aristotle, and Galen were often repeated, but increasingly they had to be revised as early modern anatomists investigated the human body in finer and finer detail. Take, for example, the three-part vocal apparatus of lungs, larynx, and mouth. Helkiah Crooke in *Microcosmographia. A Description of the Body of Man* (1616) consolidates both ancient and modern authorities to provide a minutely detailed account of lungs, windpipe, vocal cords (then only recently anatomized), uvula, tongue, and palate that would not differ siginificantly from a modern account of those members and how they function in the production of speech (1616: 390, 623–645). As up to date as his anatomizing may be, Crooke nonetheless maintains a philosophical idea of voice that goes back to Aristotle's *De Anima*: "Voice is a kind of sound characteristic of what has soul in it [*enpsychos*, literally "in-soul-being"]; nothing that is without soul utters voice." Voicing may have its *material* causes, Aristotle implies, but its *efficient* cause is the soul. The sound itself is produced by the impact of air against the "windpipe," but "what *produces* the impact must have soul in it and must be accompanied by an act of imagination, for voice is a sound with a meaning, and is not merely the result of any impact of the breath as in coughing" (*Works* 2.8.420b, emphasis added).

For us, "soul" is a theological or perhaps psychological phenomenon; for Aristotle and the early modern writers who looked to him as an authority, it was a physiological fact. The Latin word *anima* catches the double sense of "soul" as "spirit": both a physical substance (particles of air) and a psychological concept (something a subject feels within himself). Aristotle assumes that "soul" is dispersed throughout the body: what causes air to be forced against the windpipe is "the soul resident in these parts of the body" (*Works* 2.8.420b). Galenic medicine envisioned soul coursing through the body in the form of "vital spirits," defined by Crooke as "a subtle or thinne body always mooveable, engendered of blood and vapour, and the vehicle or carriage of the Faculties of the soule" (1616: 174). Crooke himself would distinguish "vital spirits" from "animal spirits," which he relegates to the conveying of sense experience to the brain and the heart (1616: 514–516).

Suddenly assaulted by all the sounds within a hundred miles of London, Baldwin's Geoffrey Streamer clearly has Aristotle's definition of voice in mind when he tries to distinguish air movements caused by "windes, waters, trees, carts, falling of stones &c which are named noyses" from the "stroke with breth of liuing creatures which we call voyces" (1584: 24). Most early modern commentators follow Aristotle in attributing voice to all sensible creatures, but they find its apogee in articulate speech. "Al voice is not speach," Pierre de La Primaudaye acknowledges in *The French Academie* (1577, English translation 1618): "For the name of voice generally taken comprehendeth all sounds & things which bring any noyse to the eares. Neuertheles it is more properly and specially attributed to those sounds, which al sortes of liuing creatures are able to make with their throat to signifie any thing therby." Only mankind, however, can make *articulate* sounds.

> But in men, voices framed into wordes are signes and significations of the whole soule and minde, both generally and specially, namely of the fantasie and imagination of reason and iudgement, of vnderstanding & memory, of wil and affections. Wherefore it is an easie matter to iudge by his speach how all these parts are affected, namely, whether they be sound, or haue any defect in them. . . . We see then how the voice and speach of man lay open his whole heart, minde, and spirite. But the voices of beasts haue no significations, but onely affections, I meane such as are in men, and which the Grammarians call Interiections, because they are not framed into speach, nor well distinguished as others are. (1618: 378–379)

Crooke attributes mankind's powers of articulation not just to reason but to the shape and positioning of the tongue, which, compared to the tongues of animals, is "the best proportioned and most at libertie" and hence enjoys "the greatest perfection in the deliuery and variety of the voice." Nonetheless, the human tongue is capable of nonlinguistic sounds: "sometimes it expresseth onelie those things that fall vnder the Sense, as when wee crie for paine, or for Foode and succour; sometimes those things that fall vnder our vnderstanding as in Discourse" (1618: 639).

A distinction of a different sort is set in place by Francis Bacon in one of the experiments with sound collected in *Sylva Sylvarum* (1626). After discussing pitch, volume, and harmony, Bacon turns to an aspect of sound less subject to physical explanation: "exterior" sound versus "interior" sound.

> In *Speech* of *Man*, the *Whispering*, (which they call *Susurrus* in *Latine*,) whether it be louder or softer, is an *Interiour Sound*; But the *Speaking out*, is an *Exteriour Sound*; And therfore you can neuer make a *Tone*, nor sing in *Whispering*; But in *Speech* you may: So *Breathing*, or *Blowing* by the *Mouth*, *Bellowes*, or *Wind*, (though loud) is an *Interiour Sound*; But the *Blowing* thorow a *Pipe*, or *Concaue*, (though soft) is an *Exteriour*. (1626: no. 288)

With respect to speech, modern linguistics would describe the distinction here as one between "voiced" phonemes ("exterior") and "unvoiced" phonemes ("interior"). Bacon himself explains the difference in physical terms: "the *Interiour* is rather an *Impulsion* or *Contusion* of the *Aire*, than an *Elision* or *Section* of the same. So as the *Percussion* of the one, towards the other, differeth, as a *Blow* differeth from a *Cut*" (1626: no. 288).

If voice is caused by the soul, acting in the form of an aerated fluid, it is shaped by the vocal apparatus in ways determined by the body's four humors. Early modern physiology understood volume of voice to be an effect not just of air pressure but of heat. Levinus Lemnius in *The Touchstone of Complexions* (English translation 1576) summarizes medical opinion that goes back to Galen:

> They therefore that haue hoate bodyes, are also of nature variable, and chau[n]geable, ready, pro[m]pt, liuely, lusty and applyable: of tongue, trowling, perfect, & perswasiue: delyuering their words distinctly, plainlye and pleasauntlye, with a voyce thereto not squekinge and slender, but streynable, comely and audible. The thing that maketh the voyce bigge, is partlye the wydenes of the breast and vocall Artery, and partly the inwarde or internall heate, from whence proceedeth the earnest affections, vehemente motions, and feruent desyers of the mynde. (1576: 45^v)

The possessors of such voices are, needless to say, men, whose bodies were reckoned by Galenic medicine to be hotter than women's bodies.

One result of women's relative coldness was a "smaller" voice. *"Why haue women such weake small voices?"* goes one of the questions in *The Problemes of Aristotle, with other Philosophers and Phisitions.* The answer is physiologically precise: "Bicause their instruments and organs of speaking, by reason they are cold, are small and narrow: and therefore receiuing but little aire, causeth the voice to be smal and effeminate" (Aristotle 1597: I8ᵛ).

To early modern ways of considering the matter, the word "voice" meant, first and foremost, a concatenation of bodily members: muscles, gristly tissues, fluids, and "soul." Voice as a political concept—that is to say, "voice" as it signifies in poststructuralist theory—certainly existed in early modern English. Appeals for the votes of the Roman plebs make "voice" one of the most frequently sounded words in *Coriolanus.* But even in that play—perhaps *especially* in that play—"voice" never loses its physiological grounding.

LISTENING

Classical and medieval thinkers sometimes give visual form to the functioning of the human mind by imagining it as a walled city— a city that is circular in shape. The topos is elaborated on an epic scale in Bartolomeo del Bene's philosophical poem *Civitas Veri sive Morum* (1609) (figs. 4.1 and 4.2). Five gates in the city walls correspond to the five senses. The gate of hearing—creaking on copper hinges, adorned with musical instruments, presided over by Apollo and the Muses—stands to the right in del Bene's image, toward the north or *tramontana*, against the echo-producing mountains that define one horizon of the Italian imagination. It is, in fact, through the gate of hearing that del Bene, his patron Queen Margaret of Savoie, and two of her ladies in waiting, along with the poem's reader, enter the city. Their guide is Aristotle, whose *Nicomachaean Ethics* provides the philosophical program for del Bene's poem. The fact that the party enters through the gate of hearing, and not through the gate of seeing, points up the primacy of hearing over seeing, at least where learning is concerned (Vigne 1975: 79–84). At the center of the city stands a citadel crowned with five temples, dedicated to Science, Art, Prudence, Reason, and Wisdom. Five axial roads, lined with temples of the virtues, connect this central mound with the five "external senses" at each of the city's gates. Access to the mound itself comes via three ladders up the sides of the citadel, representing the three "internal senses" of common sense, imagination, and memory—faculties that receive, order, and retain sense impressions coming from outside. In the wild valleys between the roads stand houses of the vices, enveloped in fogs of error.

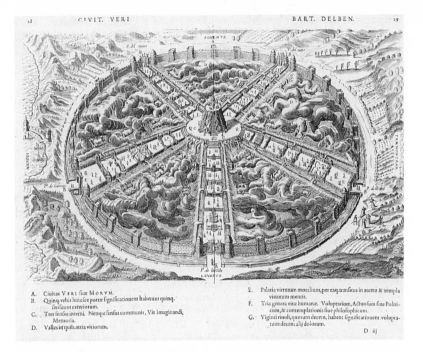

A.   Ciuitas VERI siue MORVM.
B.   Quinq. vrbis huiusce portæ significationem habentes quinq. sensuum exteriorum.
C.   Tres sensus interni. Nempe sensus communis, Vis imaginandi, Memoria.
D.   Valles in quib.attia vitiorum.

E.   Palatia virtutum moralium,per eaq.transitus in arcem & templa virtutum mentis.
F.   Tria genera vitæ humanæ. Voluptarium, Actiosum siue Politicum,& contemplationis siue philosophicum.
G.   Viginti riuuli,quorum decem, habent significationem voluptatumdecem alij dolorum.

D iij

Figure 4.1. The soul as a walled city. From Bartolomeo del Bene, *Civitas Veri sive Morum* (1609). Reproduced by permission of the Folger Shakespeare Library, Washington, D.C.

A.   Apollo habitu pastoritio,cithara canens in hyperthyro portæ ahenea circumq. eum
    Musæ cum suo quæq. instrumento, Qua effigie declaratur esse hæcporta audiendi.

Figure 4.2. The Gate of Hearing. From Bartolomeo del Bene, *Civitas Veri sive Morum* (1609). Reproduced by permission of the Folger Shakespeare Library, Washington, D.C.

Early modern opinion about hearing vis-à-vis the other senses, especially sight, was divided. On the one hand, Plato had delcared in no uncertain terms that sight provides the most direct access between external objects and the soul. In the *Timaeus*, for example, the dialogue's namesake declares that eyes were the first sense organ to be placed on human heads by the creating gods: "So much of fire as would not burn, but gave a gentle light, they formed into a substance akin to the light of every-day life; and the pure fire which is within us and related thereto they made to flow through the eyes in a stream smooth and dense, compressing the whole eye, and especially the center part, so that it kept out everything of a coarser nature, and allowed to pass only this pure element" (*Dialogues* 2:26). Here is the origin of the idea, elaborated in John Donne's "The Exstasie," that seeing results from beams of light sent out from the seer's eye, not from beams radiated to the eye by the object seen: "Our eye-beames twisted, and did thred / Our eyes, upon one double string" (1965: 59). In Plato's view, sight is less *material* than hearing and hence offers more direct communication between actual objects and the Real images of those objects that pre-exist in the mind:

> When the light of day surrounds the stream of vision, then like falls upon like, and they coalesce, and one body is formed by natural affinity in the line of vision, wherever the light that falls from within meets with an external object. And the whole stream of vision, being similarly affected in virtue of similarity, diffuses the motions of what it touches or what touches it over the whole body, until they reach the soul, causing that perception which we call sight. (*Dialogues* 2:26)

Bacon, considering the claims of sight versus hearing in *Sylva Sylvarum*, would give primacy to hearing, not sight, but for precisely the same reason: sounds have more direct access to the spirits within. Bacon's reasons for supposing that to be the case are not ontological, as Plato's are, but physiological. Sight, Bacon will concede, is "the most spiritual" of the senses (1626: no. 873), but music and other sounds have a power to excite men's passions far beyond things seen. Why should that be so? "The *Cause* is, for that the *Sense of Hearing* striketh the *Spirits* more immediately, than the other *Senses;* And more incorporeally than the *Smelling:* For the *Sight, Taste,* and *Feeling,* haue their Organs, not of so present and immediate Accesse to the *Spirits,* as the *Hearing* hath" (1626: no. 114).

Just how the waves of air striking the ears communicated with the soul was open to several interpretations. The older explanation, going back to Aristotle and confirmed by Galen, situated the nexus between external sounds and internal perception in the "implanted air" of the inner ear. Fixed ideas about the soul dictated that this internal air be different in kind from the external air of the listener's environment. As Crooke explains it,

> This ayre of which *Plato* among the Phylosophers made first mention, is seated in the eares from the first originall of our generation in the wombe of our mothers. I meane as soone as there were emptie cauities hollowed in the bones, all which are filled with this ayre. And therefore the ancient Phylosophers and Physitians, yea *Aristotle* himselfe in the eight chapter of his second booke *de Anima,* and the 83 text called it . . . *Inbred.* Others call it *Congenit, implanted, complanted* and *inaedificated.* (1616: 606)

Thin and pure so that it might register the slightest external sounds, lacking any sound of its own, immovable against the external agitations that strike it, plentiful enough to allow for full reception, separated by a tympanum from the impurities and temperature variations of outer air, this "inbred" air is the medium through which external sounds become internal motions of spirit.

Early modern anatomy complicated the picture by demonstrating how the tiny bones of the inner ear function in transmitting sound waves and, even more importantly, how nerve fibers connect the inner ear with the brain. Physiologists like Crooke were clever enough, however, to reconcile these discoveries with the received idea of an aerated fluid as the main conductor of sensation. Crooke's way is to identify the "implanted air" with "the Animal spirite which also is aiery" and to position it as a necessary medium between the outer air and the recently anatomized nerves (1616: 606–610). Ambroise Paré, one of the first early modern medical authorities to describe the process of hearing with precision, likewise affirms that the air inside the body is infused with "animal spirits." The air within the eardrum, says Paré following Aristotle, "truly is not pure and sole aire, but tempered and mixed with the auditory spirit" (1634: 191). That said, Paré proceeds to offer an account of the functioning of hammer, anvil, and stirrup against the tympanum that modern physiology would modify in only minor ways.

However they may be set in motion, it is animal spirits that convey sensations through the body, from the sensing members (in this case, the ears) via the brain to the heart. In this way, spirits function like the axial roads in Del Bene's circular city. Stimuli judged by the brain to be pleasurable dilate the heart, while painful stimuli make it contract, producing changes in the body's temperature and hence in its balance of humors. "Animal spirits" were imagined to operate in something like the way modern medicine has demonstrated that nerves do, but with a much more holistic effect. Bacon, no less than the physiologists, must find room for spirits in the transmission of

sound. Even Descartes managed to preserve the idea of animal spirits, albeit explained in a totally mechanistic way (Gouk 1991: 104–
105).

For all his attention to spirits, Bacon is keenly interested in the
materiality of hearing. The powers Bacon attributes to sound have to
do with solidly material causes. He is careful, first of all, to distinguish sound itself (vibrations in the material source) from its propagation (waves in a material medium like air) (1626: no. 124). Considering, in a later experiment, why unpleasant sounds are more grating
than unpleasant sights, Bacon concludes that sounds are more physically assaultive than visually perceived objects:

> The *Cause* is, for that the *Obiects* of the *Eare*, doe affect the *Spirits* (im-
> mediately) most with *Pleasure* and Offence. We see, there is no *Colour*
> that affecteth the *Eye* much with *Displeasure:* There be *Sights*, that are
> *Horrible*, because they excite the *Memory* of *Things* that are *Odious*, or
> *Fearefull*; But the same *Things Painted* doe little affect, by a *Participation*,
> or *Impulsion* of the *Body*, of the *Obiect*. So is *Sound* alone, that doth im-
> mediately, and incorporeally, affect most. (1626: no. 700)

It is sound's greater dependency on the material medium of air that
accounts for many of the ways in which it differs from sight: sounds
fade while colors remain constant (1626: no. 211), visible objects remain discrete while sounds commingle (nos. 224 and 270), sight
moves in direct lines while sounds move in "arcurate" (arc-like) lines
(no. 270), sight requires distance while sound works best in physical
proximity (no. 272). Above all, the power of sound resides in the fact
that it is a matter of motion, of waves of air physically striking the
members of hearing. Sound is communicated through motion, ac-
tively impressing itself on the senses of the listener, whereas sight
seems to be incorporeal, to lack motion, and thus to be communicated directly:

> The *Species* of *Visibles* seeme to be *Emissions* of *Beames* from the *Obiect*
> *seene;* Almost like Odours; saue that they are more Incorporeall: But
> the *Species* of *Audibles* seeme to Participate more with *Locall Motion*, like
> *Percussions* or *Impressions* made vpon the *Aire*. So that whereas all Bod-
> ies doe seeme to worke in two manners; Either by the *Communication*
> of their *Natures;* Or by the *Impressions* and *Signatures* of their *Motions;*
> The *Diffusion* of *Species Visible* seemeth to participate more of the for-
> mer *Operation;* and the *Species Audible* of the latter. (Bacon 1626: no.
> 268)

If a saw being sharpened or one stone being ground against another
or a shrieking noise puts listeners' teeth on edge, while unpleasant

sights have no such effect, "the *Cause* (chiefly) is, for that there be no *Actiue Obiects* to offend the *Eye*" (1626: no. 873).

On balance, Bacon's explanation of listening recognizes both spiritual and material elements. Indeed, Bacon went so far as to attribute "spirit" to *all* kinds of matter, which he divided into two kinds: "tangible" and "pneumatic." All bodies of tangible matter—even inanimate objects—are infused to varying degrees with a pneumatic substance or "spirit" possessing an airy, fiery nature. Thus, in explaining why a pipe serves to amplify sound, Bacon observes that "When the *Sound* is created betweene the *Blast* of the *Mouth,* and the *Aire* of the *Pipe,* it hath neuerthelesse some *Communication* with the Matter of the Sides of the *Pipe,* and the *Spirits* in them contained" (1626: no. 167; see also nos. 136 and 150). Communication, in Bacon's scheme, becomes the transmission of "spirit" from one body to another.

Toward the end of *Sylva Sylvarum* he distinguishes eight varieties of transmission, along a continuum from material ("the *Thinner,* and more *Airy Parts* of *Bodies,*" as in odors and infections) to immaterial ("*Emission* of *Immateriate Vertues*" not just from one individual to another but among members of an entire species) (nos. 904–911). In terms of this continuum, vision and hearing belong at the material end, less material than odors and infections but more material than works of imagination, because their transmission involves a material medium, usually air. Of the two, hearing is the more powerful, and its effects are especially notable in music (nos. 114, 700). The kind of communication that happens between a reader and a printed text— that is to say, the kind of communication studied by poststructuralist theory—belongs to a much less material place along Bacon's continuum. Fifth out of eight is "the *Emissions* of *Spirits* . . . Namely, the *Operation* of the *Spirits* of the *Minde* of *Man,* vpon other *Spirits.*" Communication of this sort happens in two ways: through "*Operations* of the *Affections,* if they be Vehement," and through "the *Operation* of the *Imagination,* if it be Strong" (no. 908). To understand voicing and listening in early modern culture we have to keep our sight much more focused than we are accustomed to on the material realities of metal, wood, air, and the members of the human body. Almost as an afterthought in his final account of the senses in *Sylva Sylvarum* Bacon suggests the intimate, physical connection between voicing and listening. As a demonstration of the visceral effects of grating, unpleasant sounds, he invites the reader to try the following experiment: "As for the *Setting* of the *Teeth* on *Edge,* we see plainly, what an Intercourse there is, betweene the *Teeth,* and the *Organ* of the *Hearing,* by the Taking of the End of a Bow, betweene the *Teeth,* and *Striking* vpon the *String*" (no. 700). In voicing and in listening the members of the human body function as a single entity. So do they too in *re*-membering.

RECALLING

Ben Jonson's need to "write" speech on the body stems from the fugitive character of the sounds that issue from the mouth. La Primaudaye in *The French Academy* grasps at this disembodied evanescence. "External speech," he observes, begins as "internal speech," in the speaker's mind. What was formerly "hid and couered" becomes manifest as voice:

> Now when this voice and speach is propounded with the mouth, as it is inuisible to the eyes, so it hath no body wherby the hands may take holde of it, but is insensible to all the senses, except the hearing; which neuerthelesse cannot lay hold of it or keepe it fast, as it were with griping hands, but entring in of it selfe, it is so long detained there whilest the sound reboundeth in the eares, and then vanisheth away suddenly.

No sooner is the speaker's voice heard than it dies away. Traces remain, however, in the minds of both the speaker and the listener:

> But albeit the sound & the voice passeth so suddenly, as if presently it flew away, hauing respect to the outward speach, neuerthelesse the internall speach remaineth, not only in the spirit, hart & thought that ingendred it, being not in any sort diuided, cut off, or separated, but also it filleth all the hearers, by reason of the agrement that is betweene the spirits & mindes of men, & the speach that is bred there, and because it differeth not much from the minde, & from the thought where it first began and was bred. And thus the thoughts and counsailes of the minde & spirit are discouered & manifested by speach. (1618: 378–379)

Between the mind of the speaker and the mind of the listener, separated only by the thin, highly transitive medium of air, La Primaudaye imagines communication to be a direct, intimate process—uncomplicated by the materiality of less transitive media. Nonetheless, the mental traces left behind by speech were, according to early modern physiology, densely material.

Working from Galen's proposition that the brain is cold and moist, physiologists imagined memory in graphically physical terms. What was needed for a good memory, they reasoned, was a tempering of the brain's congenital moistness with a degree of dryness and hardness. Too moist a brain readily receives impressions but quickly loses them; too dry and hard a brain retains old impressions but is impervious to new ones. In Ambrose Paré's formulation,

> you shall not easily imprint any thing in dry bodyes, but they are most constant reteiners of those things they have once learned; also the motions of their bodyes are quicke and nimble. Those who have a moist braine doe easily learne, but have an ill memory, for with like facility

as they admit the species of things and imprint them in their minds, doe they suffer them to slide and slip out of it againe. So Clay doth easily admit what Character or impression soever you will, but the parts of this Clay which easily gave way to this impression, going together againe, mixes, obliterates and confounds the same. (1634: 166)

If too moist a brain is like clay, too dry a brain is "euen like vnto a piece of Leade, Yron or Steele, which will not easelye suffer the poynte of anye engrauinge Toole to enter and pearce into it" (Lemnius 1576: 120ᵛ). Early modern physiology invited people to think of their memory as something physical and graphic: a trace in the brain tissues that could practically be seen and touched.

According to one school of thought, those traces could be located in a specific part of the brain. Greek medicine and Arabian medicine were in agreement that the brain has three faculties—imagination, reason, and memory—but they differed over whether those functions were dispersed through the whole of the brain, as Galen argued, or lodged in specific ventricles, as Avicenna and Averroës proposed. The Arabian position had a certain logic to recommend it: the functional sequence of sense experience → cogitation → recollection was imagined to coincide with a physical sequence of front → middle → back. Quite literally, memory was always looking backward. One of the "problems" in natural philosophy attributed since antiquity to Aristotle shows the consequences of such thinking:

> Question. *Why doeth a man when he museth, or thinke on things past, looke downe towards the earth?*
>
> *Answer.* Bicause the cell or creeke which is behinde, is the creeke or chamber of memorie, and therefore that looketh towards heauen, when the head is bowed downe, and so that cell is opened, to the end that the spirits which perfit the memorie should enter in. (1597: A7ᵛ)

Dissection of the human brain seemed to confirm a sequence of discrete functions in its revelation of a sequence of tissues that appeared soft → temperate → hard (Crooke 1616: 504–506).*

---

*Crooke himself sides with Galen: "*Galen* the Prince of this Sect conceiueth that the principall faculties are all established in one place, occupied about the same Images or Notions, do vse the same instrument, to wit, the brain; but their manner of working saith he is diuers. In the eight booke *de usu partium* . . . hee comprehendeth the three principall faculties and teacheth that they reside in the whole braine; and in the same booke he sayeth that the principall faculties haue their resience or residence not onely in the ventricles but also in the whole body of the braine" (1616: 505). Texts by Galen that seem to make the contrary argument Crooke takes to be suppositious works or else testimony to the fact that certain functions may be stronger in certain parts of the brain than in others.

In all three sequences—functional, physical, textural—the tissues of memory, situated just where the brain joins the spinal column, were imagined to communicate directly with the rest of the body. Hence the relative hardness of those tissues. There are two reasons, La Primaudaye observes, why the lower brain is more solid and firm than the rest of the brain:

> First, because it is the fountaine of the marrow in the back bone, of which those sinewes are deriued that giue the strongest motions to all the members of the body. Therefore also it was requisite, that they should be of a more firme and solide matter then the rest that are taken from the substance of other parts of the braine: which are not to sustaine so great stresse. Secondly, forasmuch as the memory is as it were the Register & *Chancery Court* of all the other senses, the images of all things brought and committed vnto it by them, are to be imprinted therein, as the image and signe of a ring or seale is imprinted and set in the waxe that is sealed. (1618: 417)

That is to say, memory mediates between the senses and bodily actions, between bodily actions and the senses. With respect to the sense of hearing in particular, memory transforms air waves into embodied action. It remembers sound in various parts of the human body: in the other ventricles of the brain, in the ears, in the hands, in the eyes, in the body as a kinaesthetic whole.

In most situations the brain's three faculties were imagined to function in this order: first fantasy or imagination gathers sense impressions; then reason processes those impressions; finally memory stores the result. Even writers skeptical of the possibility of precisely locating the faculties—writers like Juan Huarte de San Juan in *The Examination of Mens Wits* and Helkiah Crooke in *Microcosmographia*—nonetheless agree on what the faculties are and how they work (Huarte 1594: 52–55; Crooke 1616: 505–506). In certain situations, fantasy might not only supply sense impressions for cogitations but take cogitations and turn them into sense impressions. That is the basis for the recommendations made by Cicero, Quintilian, and the anonymous author of *Rhetorica ad Herennium* that orators should turn key ideas into "images" (*imagines*) and lodge them in "places" (*loci*) that might take shape as a house, a colonnade, a recess, or a vault. Frances A. Yates's *The Art of Memory* (1966) traces the extraordinarily long and interesting after-life of this ancient idea. Let it be noted how apt the scheme appears if memory itself is imagined to be lodged in a "cell," "creek," or "chamber." The decidedly visual cast of the whole exercise is maintained in Cicero's *De Oratore*, where sight is affirmed as "the keenest of all our senses" (1967: 2.87.357).

Quintilian, ever the practical skeptic in his *Institutio Oratoria*, endows the images with sound. The images, he says, are *voces*, "voices" or "utterances," even as he quotes Cicero's statement that the images function like letters inscribed on wax (11.2.21 in 1921–22, 4: 230–231). The surest way to make the scheme work is to say the speech *sotto voce* as one tries to parcel it out into images. In Quintilian's view, the "house of memory" is an acoustic as well as a visual space:

> The question has been raised as to whether we should learn by heart
> in silence [*ediscere tacite*]; it would be best to do so, save for the fact
> that under such circumstances the mind is apt to become indolent,
> with the result that other thoughts break in. For this reason the mind
> should be kept alert by the sound of the voice, so that the memory
> may derive assistance from the double effect of speaking and listening.
> But our voice should be subdued, rising scarce above a murmur.
> (11.2.331 in 1921–22, 4: 231)

Thomas Wilson makes the same point in *The Arte of Rhetorique*. "Among all the senses," he observes, "the eye sight is most quicke, & co[n]teineth the impressio[n] of things most assuredly, the[n] any of the other senses do. And the rather when a manne bothe heareth and seeth a thinge (as by artificiall memorye he dothe almost se thinges liuelye) he dothe remember it muche the better" (1557: 116–116ᵛ).

In effect, Yates's book marks out two divergent paths leading out of the house of memory. One path leads into the theater. Palladio's Teatro Olimpico in Vicenza was perhaps inspired by the model "memory theater" that Giulio Camillo transformed from paper and ink into wood (Yates 1966: 129–172). The other path leads into silence. With Giordano Bruno and Robert Fludd, houses of memory became refuges for occult knowledge—knowledge that was unprintable if not unspeakable (1966: 199–307, 320–341).

If the brain served as the general "Register & *Chancery Court* of all the other senses," specifically *aural* experience might be located elsewhere in the body, most obviously perhaps in the ears. "And David spake the words of this song unto the Lord, what time the Lord had delivered him out of the handes of all his enemies, and out of the hande of Saul" (2 Samuel 22:1): Thomas Trevelyon's illustration of these and other biblical verses about music shows two "Syng[i]ng men" sharing the bass part (marked "B") of a four-part song. The man on the left also grasps the counter-tenor part (marked "CT"). In the air around them float books containing the other two parts: medius (marked "M") and tenor (marked "T") (fig. 4.3). Music, in Trevelyon's view, belongs in two places at once: on the page and in the air. Trevelyon's graphic shorthand is borne out by music printed

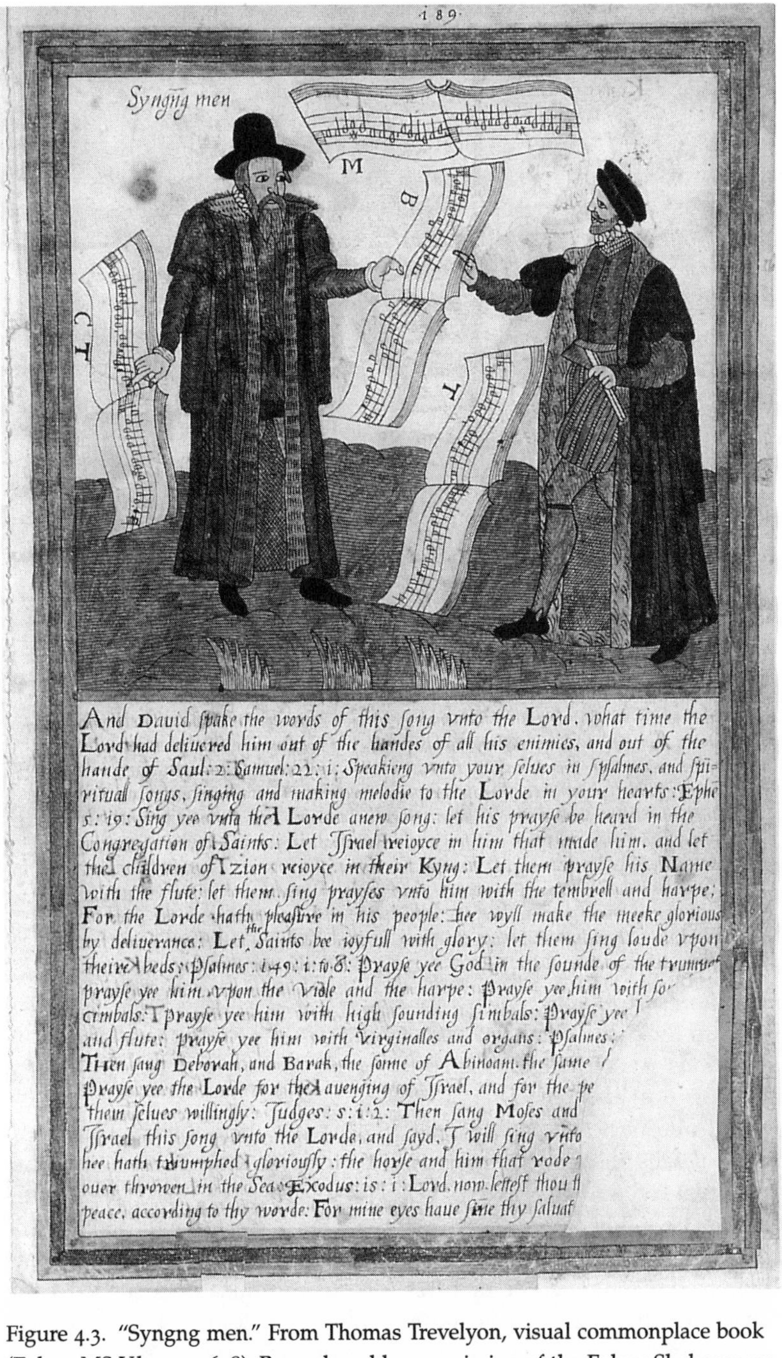

Figure 4.3. "Syngng men." From Thomas Trevelyon, visual commonplace book (Folger MS V.b.232, 1608). Reproduced by permission of the Folger Shakespeare Library, Washington, D.C.

for actual use: in four-part vocal scores of the sort that he illustrates the four parts were often printed in separate booklets, one for each part, or, if all together, in separate parts printed at ninety-degree angles to each other on a page, so that the four performers could face each other as they sang (Boorman 1986: 227–228). With their (to us) quirky orientation in space, these four-part scores never let us forget that music is not what is printed on the page, but what is heard in performance. However elaborate the score, music belongs, first and last, just where Trevelyon places it: in the air. The written score serves no other purpose than to provide cues for what the performers, guided by their ears, should do with their bodies to produce the required sounds. In the last analysis, all musical performances result from "playing by ear." Music, in the sense of a written score, serves as metonymy for music as an acoustic event, as the cause for a particular range of effects. All schemes of musical notation must negotiate this ontological difference.

Music in early modern England had a complicated relationship to graphic transcription. Despite treatises like Thomas Morley's *A Plain and Easy Introduction to Practical Music* (1597), the art of music was more typically transmitted person to person, singer to singer, lutenist to lutenist. Hence the sufficiency of the tag that accompanies most printed ballads, "To be sung to the tune of. . . ." Only a handful of sixteenth- and seventeenth-century broadside ballads do supply a printed tune. More typical is a ballad version of "The Spanish Tragedy" (Roxburghe Collection, British Library, 1.364–365; [Anonymous] c. 1620), which specifies that both sets of stanzas are to be sung "To the tune of *Queene Dido*" (fig. 4.4). For a literate historian, such a phrase all too often points toward silence. For musicians in early modern England, it pointed toward the known and the familiar, toward traces of sound that were already in the reader's brain, lungs, larynx, and mouth. Standing outside that cultural soundscape, we have access to such tunes only when a well-placed composer like William Byrd happened to take up the tune as the theme for a set of variations which he subsequently wrote down in a score. "Queen Dido" is not, alas, one of those tunes.

"To the tune of *Queene Dido*" is a *somatic* notation: in shorthand form, it tells the reader just what to do with his or her body. Lute entabulature is notation of just this sort: in a series of pictures of the lute's neck it tells the performer just where, and when, to place his or her fingers in certain positions on the frets. For a musically literate lutenist the sequence of pictures also communicates a sense of the whole. In this regard it is like all musical notation. That is to say, putting music into the form of a table provides two perspectives on the music: as a moment-to-moment physical activity and as a totality,

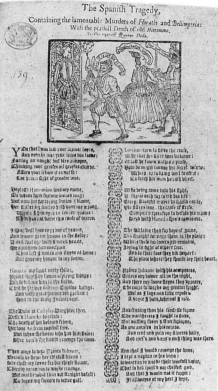

Figure 4.4. Broadside (recto) of the ballad of "The Spanish Tragedy" (c. 1620), BL shelf mark C.20.f.7. Reproduced by permission of The British Library, London.

as an aesthetic whole.* Musical scoring is something more than the sum of the discrete points in time. In both dimensions, however, the scoring necessarily remains incomplete. The graphemes on the page can record only experiences that the performer knows already. Even if a piece of music is new, the musician's ability to "read" that music depends on what he or she knows from past experience. To the graphic cues the musician brings a bodily "feel" for the relationship between one pitch and another, for the rhythmic patterns that distribute the pitches in time, for the turns of voice that might embellish a single pitch, for the subtle changes in speed that group different notes within a single measure of time, for the changes in volume that shape a phrase, for the changes in volume and speed that signal closure.

All of these things necessary for musical performance are simply not there on the page: they have to be supplied by the performer out

*I owe this observation to Dennis Helmrich, a professional musician and vocal coach.

of his or her own bodily experience. Even if the score specifies a ritard, the performer has to decide just how much. The choices in such cases are informed not only by cultural variables but by the physiological constants of human perception and the physical qualities of particular sounds. With respect to gaps on the page, "reading" a musical score is like reading a verbal text: the reader has to fill in the gaps out of his own experience (Iser 1978: 164–170). In the case of music, the gaps are filled not so much by the reader's *cognitive* experience as by his *somatic* experience.

The same is true of all systems of graphemes. Because words have semantic meanings, we forget that they also encode bodily experience—at the very least the expulsion of air, the adjustments of muscle, the shaping of tongue that it takes to pronounce those words. Poetry in particular takes shape under the aegis of the ear. In the cycle of sounds that begins with the primal [o:] and continues through speech, music, and ambient sound, poetry occupies a position somewhere between speech and music. As such, it operates within a different time frame than speech, less quick in the production of phonemes per second. At the same time, it is less trammeled than speech by the dictates of ordinary syntax. As well as semantic meaning, poetry admits somatic meaning, most readily noticeable as rhythm. To Samuel Daniel's ear, it is rhythm more than rhyme that lodges poetry in the body and makes it possible for both poet and listener to remember sounds: "All verse is but a frame of wordes, confinde within certaine measure; differing from the ordinarie speach, and introduced, the better to expresse mens conceipts, both for delight and memorie." Distinctive patterns of rhythm, he observes via Aristotle, are to be found in all languages. For the speakers of each language, it is the ear that puts the sounds of poetry in place: "they fall as naturally already in our language as euer Art can make them; being such as the Eare of it selfe doeth marshall in their proper roomes, and they of themselues will not willingly be put out of their ranke; and that in such a verse as best comports with the Nature of our language." Rhyme, prominent in English but absent from ancient Greek and Latin poetry, is "an excellencie added to this worke of pleasure" (1965: 131–132). The way poetry is written down, framed into into lines and stanzas, works like musical notation: its ultimate appeal is to the ear.

Hands can also function as sites of recalling. In marks made by his or her own hand a speaker has the power to member speech in an immediate, physical form. The sounding body actually touches the surface on which fugitive words can be *re*membered. Phenomenologically speaking, handwriting asks to be experienced in two quite opposite ways. The point of reference in both cases is the writer's

Figure 4.5. Basic strokes of handwriting. From Guilantonio Hercolani, *Lo Scrittor' Utile, et brieve Segretario* (1574). Photo courtesy of John M. Wing Foundation, The Newberry Library, Chicago.

body. Handwriting can be apprehended, first, via its *connection* with the writer's body: the hand that I write is *mine*. But it can also be apprehended via its *separateness* from the writer's body: what I have written out is not mine but *other*. That ambiguity is captured in the terms Guilantonio Hercolani applies to the three basic strokes of handwriting in his manual *Lo Scrittor' Utile, et brieve Segretario* (1574) (fig. 4.5). *Corpo* ("body"), *traverso* ("cross"), and *taglio* ("cut") are properties of both the body that writes and the body that is written. In writing according to Hercolani's dictates, I create a body apart from my body: "[em]body," "cross," and "cut" are actions I do with my hand, but they are also aspects of the object I create on paper. The letter possesses a "body," a "cross," a "cut." Once I have removed my hand from the paper, the body I have created takes on a life of its own: it registers my voice, but it remains mute. It stands precisely at the point where my body enters the horizon of my culture's signifying systems.

What handwriting registers is not only semantic meaning but sound. The problem, as Richard Mulcaster points out in *The First Part of the Elementarie Which Treateth Chefelie of the right writing of our English tung* (1582), is that graphemes are not phonemes. Speech remembered on the page is not the same thing as the members of speech: "letters resemble the joynts in *sound,* but ar not the same with the things resembled." To remedy these discrepancies mankind brought to the factor of sound two other factors: "*Reason* to observe where the sternnesse of *sound* were to be followed, and where to be qualified," and "*Custom* to confirm that by experience in the pen, which *reason* doth observe, and note in the *sound.*" The "rule of right writing" to which Mulcaster adheres is, then, a combination of acoustics, logic, and convention (1582: 104–105). Structuralist linguistics attends to the second of those factors, to the logic of writing systems. Poststructuralist linguistics attends to the third, to the arbitrariness of such systems. To read early modern handwriting we need all three.

Mulcaster's tripartite model catches both of the phenomenological dimensions of hand: its property of being mine and its property of being other. To the extent that handwriting registers sound, it belongs to the speaker/writer: sound undisciplined by reason and custom belonged, in Mulcaster's phrase, to a lawless state "wherein everie mans brain was everie mans book, and everie privat conceit a particular print" (Mulcaster 1582: 104). To the extent that handwriting adheres to the demands of reason and custom, it mutes the subject's voice. In Mulcaster's terms, it turns "a *particular* print" into a *public* text. Goldberg's *Writing Matter from the Hands of the English Renaissance* (1990) attends to the supposed brutalities of these two regimes; the O factor asks us to listen for the voice that early modern readers and writers say they heard through it all. At issue here is not the ontological wrongness or rightness of their convictions but the phenomenology of those convictions. It ought to be possible to construct a "cultural poetics" of handwriting no less than a cultural poetics of gender, sexuality, and nationhood.

Sound registered by the writing hand is not just a selection from the repertory of sounds that make English English but a selection from the repertory of sounds that distinguish the writer from all other speakers of English. The writing hand leaves its mark on the page; the mark is said to be the writer's "hand." For the physical act of writing, hand serves as synecdoche, just as tongue does for the physical act of speaking. (Hand and tongue, that is to say, are both conspicuous parts of whole actions that involve other sets of muscles and nerves.) For the thing written, "hand" serves as metonymy, just as "tongue" does for the thing said. ("Hand" and "tongue" are the products of certain processes, the effects of certain causes.) The casualness with which early modern writers can connect hand and tongue seems, with or without Derrida, remarkably uncomplicated, if not naive. Erasmus in his treatise *De recte pronuntatione* (1528) is only the most famous authority to observe that handwriting, under the right circumstances, could serve as an encoding of the writer's own voice:

> recognition of the handwriting can add a note of conviction, or at any rate an element of pleasure, to a letter. One should remember that the

apostle Paul sent his Letter to the Gallatians all in his own handwriting. It would impress anyone to receive a handwritten letter from a
king. Why, even when we get letters in their own hand from friends
and fellow-scholars, how we welcome them and seem to be listening to
their very voices and to be looking at them face to face. There is no
need to labour the point.

(Tell that to Derrida.) In sum: "Just as individual voices differ, so does
every handwriting have something unique about it" (1978: 390–391).
With its professed goal of cultivating the individual subject, Human-
ism prizes "something unique" (*quidam singulorum*) as a criterion of
good handwriting as much as neatness or correctness.

Erasmus's Latin sets "voice" and "hand" into eloquent alignment:
"habet enim singulorum vt vox, ita manus quoque quiddam suum ac
peculiare" (1969: 1.4.34). Gervase Markham echoes Erasmus's senti-
ments in one of the model letters in *Hobsons Horse-load of Letters: Or,
a President for Epistles* (1617). Writing a letter to an absent friend, a
man might wish to say, "Make my mind happy to behold which mine
eies cannot, I meane your selfe in your Letters, for they are the liue-
liest Characters of that Figure which we adore with most earnest-
nesse, and though the words be blacke, yet is their sound so cleere &
mysticall, that they straine the brightnesse which contains them, &
make the [e]are couet no noyse but their repetition" (G1). The same
metonymy of hand and voice is celebrated in a commonplace book
of the 1630s (now Folger MS V.a.345), in a poem entitled "Vocis. /
Speeches, truth instrument, ye Pen":

When the rich art of writing first was borne
Pens spake and made a miracle a scorne,
Speech was a wonder vilifyed, when men
Spake wthout voices, the deputed pen
Did ye tongue['s] office, and ye ey turnd eare
To ye dead voice and p[ro]perly did heare,
It maz'd ye world, the Interp^{r}ter of thought
Should have a nother channel than ye throate.

(99)

Such slips of, with, and between tongue and pen are ultimately informed by the double meaning of the Latin *vox*: both "voice" and,
metonymically, the "word" that voice utters.

David Brown is altogether typical of early modern authorities
in accepting not only the synonymy of voice and word but the uncomplicated relationship both enjoy with conceptions in the mind.
Brown's *The New Invention, Intituled, Calligraphia: Or, The Arte of Faire
Writing* (1622) is less a manual on practicalities of penmanship than
a philosophy of writing. It was, perhaps, Brown's analytical cast of

mind, rather than specimens of his work, that drew the attention and patronage of James VI and I, who first encountered his fellow Scotsman during a sojourn at Holyrood House in Edinburgh. If the term had not already been appropriated by linguists, Brown says, he would have styled his work a "grammar" of writing, since writing, like language, can be reduced to a set of fundamental principles (1622: C8ᵛ). True to his analytical method, Brown distinguishes five "causes" of writing, whereof the efficient cause is the writer, the material cause ink, the instrumental cause the pen as guided by "Members of the Writer his Bodie, as the Hand, Thombe, and Fingers" (1622: ¶¶¶8ᵛ), and the formal cause "the externall shape of the Letter, whether it be perfectly or imperfectly proportioned" (1622: ¶¶¶¶1ᵛ). The final cause—and for our purposes this is the telling distinction—remains

> the former signifying of articulate voyce, whereby the thoughts of the minde are interpreted, and the demonstrating of the minde without the voyce; for as *Aristotle* teacheth, Writs or Letters are the Symboles of Voyces or Wordes, (howsoever it may bee thought that the Voyce beeing invisible cannot bee represented by anie externall Signe) the Voyces Symboles of the Conceptions of the Minde, and the Conceptions of the Minde, Images of thinges which bee outwith the Minde: and that both of divine Writs and Humane. (1622: ¶¶¶¶1ᵛ–¶¶¶¶2)

The "elements" of writing, which Brown enumerates as (1) form, (2) knowledge, (3) name, and (4) sound, all function as symbols or signs of voice.

Brown insists on the physicality of sound. Syllables are syllables, words are words, sentences are sentences, not because they possess any distinguishing features in and of themselves, but because they match speech sounds:

> A written sillable, is rather called a sillable; because it signifieth a great part of the voyce . . . than for anie respect eyther to the matter or forme of the letters themselves whereof it is composed. A written Worde, is rather called a Worde, because it representeth a greater part of the voyce, which beeing uttered, beateth the Aire; than for anie regard eyther to the matter or forme of the syllables, or letters, which bee included therein. And a written sentence, is named a sentence, rather because it signifieth the greatest or longest sound, or part of the voyce, (that is, the perfect meaning, and value of the wordes, one or moe, which it doeth comprehende) than for the matter, or composition of the words themselves, as they bee written.

The movement from single letters to words to complete sentences is a movement toward what can only be called *presence:* "the first

whereof is viue, the second viuer, and the thirde moste viue of all"
(1622: D2–D2ᵛ).

Reading and writing may have been taught as separate technologies in early modern schools—first reading, then writing—but
both technologies started just where Cicero and Quintilian specified:
with the spoken word. Early modern authorities on education took
Cicero and Quintilian at their word in advising schoolmasters to have
their pupils sound words aloud as they wrote them down, in just the
way that they would sound words aloud as they read them from an
already written text. John Brinsley in *Ludus Literarius: Or, The Grammar Schoole* (1612), for example, advises the teacher to read out a sentence in English, taking care to "utter each word leasurely and treata-
bly; pronouncing every part of it, so as every one may write both as
fast as you speake, and also faire and true together." While the
teacher is speaking, the students should not only write what they
hear but sound out the phonemes themselves: "as they are writing,
cause everie one in order to spell out his 2. or 3. words together,
speaking up, that all his fellowes may heare, & may goe on in writing
as fast as he spels and you speake." After the students have translated
the sentence into Latin, the process is reversed: each student in turn
reads out his translation while the others take dictation. As the final
step, the teacher should challenge the students to memorize the
translation, "to trie which of them can repeate the soonest without
booke, that which they have made" (1612: 151–152). The effect of such
exercises, for the writer at least, would be to imbue his hand with the
sound of his own voice.

One difficulty with both the structuralist and the poststructuralist models of orthography is the way they reify both "writing" and
"speech." "Writing," after all, is an abstraction, the inductive result
of grouping together all sorts of occasions, materials, and techniques
whereby people make graphic marks on a planar surface. If we step
out from under the category of "writing" and look around instead at
individual *acts* of writing, we can distinguish not one writing technology in early modern England but several. Instead of writing we find
writing*s*. Several early modern manuals of penmanship suggest as
much. In *The Writing Schoolemaster: Conteining three Bookes in one*
(1590), for example, Peter Bales devotes separate "books" to "swift
writing," "true writing," and "fair writing." The three technologies
are distinguished from each other by their differing relationships to
speech. Swift writing, or "brachygraphy," is a kind of shorthand, with
which "a man may take a whole sermon or other speech whatsoever,
verbatim, being delivered treatablie" (1590: C2). True writing, or orthography, is a translation of the transcribed speech sounds of
brachygraphy into the graphic system "as it is now generally printed,

used, and allowed, of the best & most learned Writers" (1590: title page). In just the terms Mulcaster proposes, Bales's orthography is sound disciplined by reason and custom. Finally, fair writing, or calligraphy, is writing turned into an art form, writing that calls attention to itself as a medium.

Bales's brachygraphy, the subject of a separate volume published a few years after his "three books in one," uses the printed page to lay out the ground plan for a "house of memory." Single letters with added points, jots, and tittles stand for commonly used words beginning with that letter. By memorizing these cues—Bales recommends doing so in groups of twelve—a student theoretically has the entire English language literally at his finger tips. For words that are not in Bales's lists a student can rely on synonyms or else make up his own symbols. What Bales proposes, in effect, is a continuum that stretches from sound-dominated graphemes at one end to sight-dominated graphemes at the other. Swift writing remains closely connected to speech; fair writing fetishizes the signifier over the signified. In the middle is ordinary orthography. Like Mulcaster, Bales endows the middle with the power to render sound, however mediated it may be by abstract rules and graphic conventions. "True" writing is not just true to rule but true to sound. Harold Love would complicate Bales's middle range by distinguishing italic hand, closer to the pole of fair writing, from secretary hand, closer to the pole of swift writing. Contrived originally as a medium for copying classical texts, demanding painstaking adherence to printed models, proscribing touches of individuality, italic hand does seem to have muted a sense of the writer's presence—a situation Goldberg claims to be true for *all* forms of early modern handwriting. Secretary hand, by contrast, fostered individual difference, including, Erasmus and others suggest, a sense of individual voice (Love 1993: 109). There is more to early modern handwriting than meets the eye.

The culture of early modern England, with its complex reciprocities between orality and literacy, recognized a variety of ways in which speech could be given visible form, not only in writing but in other graphic signs. Learning to read the signs of speech in early modern texts underscores the point made by Jack Goody, Michel de Certeau, and Henri-Jean Martin: orality and literacy, far from being fixed entities, are systems of communication that exist in complicated, changing, culturally specific relationships to each other (Goody 1977: 36–51; de Certeau 1984: 133; Martin 1994: 87). In Goody's summation, each graphic system has its own logic and dynamism (1977: 76–78). The degree to which sound is registered in a given graphic system is very much a cultural variable. Although Garrett Stewart argues that every written text in English is a "phonotext," playing out deconstructive tensions between the fixities of printed words and the potentialities of voiced sounds, Henri-Jean Martin's survey of writing systems in *The History and Power of Writing* includes several examples from antiquity in which the aim was not to represent spoken discourse but to commit to memory discrete units of information, which a scribe was free to configure in ways quite different from speech (Stewart 1990: 35–142; Martin 1994: 87–88).

Written texts in early modern English stand in sharp contrast to such systems. In Goody's terms, the graphic system of early modern English may find its logic in Latin grammar but its dynamism comes from speech. What we need to look for are not morphemes, minimal units of semantic meaning, but *graphemes*, minimal units of visual meaning. Styles of hand, typefaces, illustrations, spaces on the page, the physical medium on which these marks are made—all of these things can signify in a communication system that gives primacy to speech. Graphemes are signs: they point toward something that exists before and beyond the sign itself. Writing "members" speech into graphemes; the viewer of the signs *re*members. Marks on the page become registers of bodily experience. Among the graphic systems, in early modern England at least, there seems to be an inverse relationship between graphic information and somatic content: the less graphic information, the greater somatic content. A survey of the media in early modern England might begin with those in which the graphic signs are minimal, the somatic content extensive, and end with those in which the graphic signs are extensive, the somatic content minimal.

Graphemes mediate between sound-in-the-body and sound-on-the-page. The common demoninator in this transaction is *body:* paper and ink as material entities stand in for muscles and air as material entities. The paper functions as a kind of *membrane*, or skin, in just way Jonson suggests in his anatomy of speech: as visual evidence of an acoustic event. (The Latin *membrana* refers generally to skin but specifically to an animal skin that has been prepared as a writing surface—that is, to parchment.) Stature, skin, and articulated structure are legible counterparts to frequency, duration, noise, timbre, inharmonics, vibrato, timing and rhythm, and radiation/propagation. Pictures, with or without snippets of text, can stand as signs of sound. Take for example the title page to the 1615 edition of *The Spanish Tragedy* (fig. 4.6). Within a single frame the woodcut conflates three separate moments from 2.4 and 2.5. Reading from left to right, Hieronimo is shown discovering his son Horatio hanged in a bower. "Alas it is my son Horatio," says the scroll, paraphrasing 2.5.14. To the right is depicted an earlier moment when Horatio's lover Bel-imperia is being abducted by her brother Lorenzo and his confederate Balthazar

The Spanish Tragedie:

OR,

Hieronimo is mad againe.

Containing the lamentable end of *Don Horatio*, and
Belimperia; with the pittifull death of *Hieronimo*.

Newly corrected, amended, and enlarged with new
Additions of the *Painters* part, and others, as
it hath of late been diuers times acted.

LONDON,
Printed by W. White, for I. White and T. Langley,
and are to be sold at their Shop ouer against the
Sarazens head without New-gate. 1615.

Figure 4.6. Title page to Thomas Kyd, *The Spanish Tragedie* (1615). Reproduced by permission of the Folger Shakespeare Library, Washington, D.C.

just after they have hung Horatio. "Murder, helpe Hieronimo," cries Bel-imperia. "Stop her mouth," commands a black-faced Lorenzo, echoing 2.4.62–63. The same design—indeed, the woodcut appears to be the very same—was used by another printer when he issued the ballad of "The Spanish Tragedy" five to ten years later (see fig. 4.5). Aside from Middleton and Dekker's *The Roaring Girl* (printed in 1611), which carries one of Moll's lines outside the frame that sets off Moll herself, R. A. Foakes counts seven other printed scripts that are illustrated with speeches on scrolls: Greene's *Tu quoque* (printed in 1614), Middleton's *A Game at Chess* (printed in 1624), Greene's *Friar Bacon and Friar Bungay* (printed in 1630), William Sampson's *The Vow Breaker* (printed in 1636), and Dekker, Rowley, and Ford's *The Witch of Edmonton* (printed in 1658). The fact that two plays in Latin, George Ruggle's *Ignoramus* (printed in 1630) and Edward Forsett's *Pedantius* (printed in 1631) also carry illustrations with speeches in scroll work (albeit in engravings, not woodcuts) suggests that the convention was not beneath academic contempt (Foakes 1985: 87–147).

The mnemonic function of these scrolls is apparent in every case: what is quoted is only a tag that calls to mind a longer speech. That is to say, speech is *dis*membered in order to be *re*membered. Cued by the scroll, an absent voice is called into presence. The most common space for such devices in sixteenth- and seventeenth-century print is not the title pages to playscripts but broadside ballads. The fact that "The Spanish Tragedy" woodcut seems to have appeared first on the 1615 quarto of Kyd's play indicates the ready communication between plays and ballads in early modern culture (Smith 1991b: 127–144). In both cases, the printed words point off the page and into the air. Following that prompt, we should perhaps pay particular attention to title-page advisories that a play is printed "as it hath of late been divers times acted." In the case of the 1615 quarto of *The Spanish Tragedy*, twentieth-century editors have taken that phrase to mean that the text incorporates recent additions to Kyd's original play. In common with all such notices, however, it may be setting up the entire script as a kind of scroll, as a shorthand transcript of spoken words that helps a reader literally to *re*member the play as he or she may have heard it in performance, as an experience of sounding bodies moving in space. Certainly that was the case for the actors putting on the play. Like the psalm-singers in Fella's drawing, each performer saw only his own part and no one else's. Vis-à-vis the other performers, an actor's craft, like a countertenor's, was piecework of a sort. In such circumstances, no one person sees the whole: the whole exists only as a sounded performance, an ensemble of all the individual parts. Full remembering happens only in performance.

Pictures without scrolls might also function as graphemes of sound. Foakes, for example, catalogs 26 printed scripts that contain speechless images—nearly four times as many as images with scrolls. Many of these mute pictures do carry the protagonist's names, implicitly connecting each protagonist with the speeches inside in a kind of graphic dumb show. The same is true for printed ballads. Broadsides that feature wordless images of the first-person singer and his interlocutors far outnumber the printed sheets that assign those images particular speeches.

Along the continuum from minimal graphemes to replete graphemes, handwriting occupies an intermediate position. It transcribes sounds as letters, but it has the capacity to do so in a form quite specific to the writer's own body. At the same time, it can render speech in a form quite alien to the writer's body. For the deliberate, calculated hand Bales calls "fair writing" or "calligraphy" David Brown proposes a spatially precise term: "*set* writ." *There* it is, he implies, a text and a technique waiting to be copied. The objective quality of set writ finds its here-and-now counterpart in "*common* writ" or "*current* writ," set writ that has been adapted for speed. "Currency," in Brown's scheme, is a matter both of space (the hand moves rapidly across the page) and time (thought and writing happen simultaneously). As a system of graphic signs, print exceeds even the most elegant and singular set writ in its *there*ness, its separation from the speaking body.

Handwritten letters are drawn on a surface; printed letters are cut and impressed. Force and energy in the two technologies are deployed in quite different directions. In printing, the force is directed onto and *into* the paper—a quality that has been lost in the photo-offset technology of the late twentieth century. With sixteenth- and seventeenth-century books and broadsides it is possible to feel the text with one's fingertips as well as to see the text with one's eyes. In handwriting, on the other hand, force is directed not so much onto the paper as into the space above the paper, as the writing hand moves onward in time and space. Brown catches this effect in his argument that writing is superior to printing. Partly that superiority is due to writing's logical priority (printing could hardly exist without writing), partly to its low-tech efficiency (it requires no second person, no artificer, who stands between the writer and the record of his thoughts), and partly to its unique capacity for registering thought-in-time. In this sense, writing can be defined as "a present framing and expressing of one Letter after another, to signifie the articulate voyce of the Tongue, whereby the thoughtes of the minde are expressed to those who bee present, and to interprete the minde, (without use of the tongue) unto those who bee absent." Hence writing operates in three time frames at once: it remembers the past, it captures the process of thought in the present, and it opens out into the future. Printing—or at least the *act* of printing—has always the quality of *past*ness about it:

> Now although *Printing* may say, *Framing* and *Expressing*, yet it may not say, *Present framing*, and *expressing*: that is, both to frame and set downe Letters, and therewith to expresse wordes and sentences at everie instant, without the ayde of other Artificers: for the worde *present* is proper onelie to Writing, because the action thereof is readie and easie, to bee prosecuted at everie occasion, without the helpe of such secundarie meanes.
>
> Againe, *Printing* may well say *Framing* and *Expressing* of Leeters [*sic*] in the plurall number, but it may not say, *of one Letter after another* advisedlie in the singular: because it consisteth in stamping, or imprinting of manie Letters, (right or wrong) with one impression: for they may well be set in order severallie, but the Impression must be together. (1622: ¶¶¶2ᵛ–3)

Printing, in Brown's view, is a technology infinitely less sensitive than writing is to the unfolding of events in time.

Hence also the relative insensitivity of print to voice and sound.

As a seventh site of resonance between sounding voice and graphic media, print stands at the farthest remove from the speaking body. "Houses of memory" in space and circular designs in time exist only within the speaker's mind, musical notation depends on somatic cues, scrolls provide snippets that call up longer speeches stored in the reader's memory, visual images position the viewer in aural situations, handwriting captures the spatial and temporal immediacy of speech. In comparison with all these, a printed text exists relatively independent of its maker. The move from memory to print is a move, by stages, outside the speaker's body, toward another body, toward an object that does not need the speaker to make sense.

If the relationship of hand to writing is a matter of synecdoche and metonymy, the relationship of the human body to typography is a matter of metaphor. We can see the situation graphically in two guises: in the conventional names for the "body" of lead type and in the anatomical principles that were imagined to inform the design of Roman letters. If the pen functions as an extension of the human body, the press creates a body apart from the human body. Joseph Moxon's *Mechanick Exercises on the Whole Art of Printing* (1683–84) details the multiple ways in which this phenomenon happens. In early modern print technology, "body" is a property of punches (which possess a "face" and a "shank"), of moulds (which possess a "body" and a "mouth-piece"), of registers (which possess "shoulders," a "neck," and a "cheek"), of matrices (which possess a "face"), of composing sticks (which possess a "head," a "back," and "cheeks"), and of the printing press itself (which possesses a "head," a "cap," "feet," and "cheeks"), as well as of the printed letter (which possesses both "face" and "body") (Moxon 1958: 19–21, 41–42, 99–106, 135–150). At every step in the process that transforms a manuscript into a printed text, a body of some sort interposes itself between the act of speaking and the act of reading. Impressed onto, and *into*, a sheet of pounded rags, letters tend to lose their character as marks of sound in time and become instead objects in their own right.

With respect to typefaces, Geoffroy Tory's *Champ Fleury* is only the most fanciful instance of a phenomenology of the printed word that also includes Dürer's *Of the Just Shaping of Letters* and Luca Paccioli's *Divina Proportione*, said to be derived from designs by Leonardo da Vinci. In Tory's scheme, all the Roman letters are reducible to variations on the shapes I and ○, which in turn derive from the proportions of the human body (figs. 4.7 and 4.8). As Tory observes with respect to ○, "The man in this figure, with feet and hands extended to equal distances, & the ○, meet in the square, in the circle, and in the centre, which betokens the perfection of the human body & the said ○, since the circle is the most perfect of all figures & the most

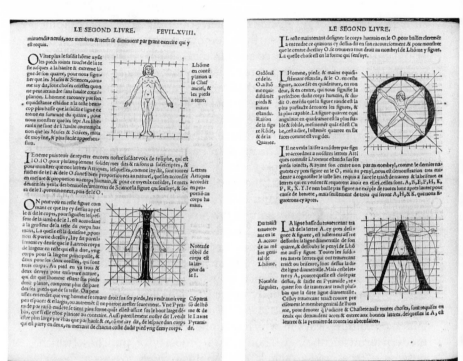

Figure 4.7. (*left*) The letter I. From Geoffroy Tory, *Champ Fleury* (1529), BL shelf mark C.31.K. Reproduced by permission of The British Library, London.

Figure 4.8. (*right*) The letter O. From Geoffroy Tory, *Champ Fleury* (1529), BL shelf mark C.31.K. Reproduced by permission of The British Library, London.

comprehensive. The rectangular figure is the most stable and solid, especially when it is a cube, that is to say, having six faces, like dice." To a properly trained eye, the precise center differs between I and ○. With the spread figure, the navel forms the center, with the erect figure the groin. In that difference there is both graphic and ethical significance. It is at groin-level, Tory observes, that the cross-strokes should be made in A, B, E, F, H, K, P, R, X, and Y, "to signify that Modesty and Chastity are required, before all else, in those who seek acquaintance with well-shaped letters" (1927: 48). In such speculations we find ourselves at a far cry indeed from [o:]. "*Language* most shewes a man: speake that I may see thee": in type technology Jonson's figurative command becomes literal fact.

Gauging the relationship between handwriting and movable type, historians tend to speak in almost military terms. Print is said to "triumph" over script; the newer medium is said to "conquer" the older.

In oblique ways as well as direct, early modern witnesses suggest a more complicated relationship. Erasmus argues for the authenticity of a well-cultivated hand, but he does so in print, in a text first published by the Aldine Press. Billingsley and Brown affirm in no uncertain terms that writing claims precedence over print since print is copied from writing, and yet both writing masters depend on print to publicize the very idea. Bales is more accommodating: he advises that, in learning to write, a sudent should take a printed book as an exemplar. Most complicated of all is Mulcaster's observation that sound alone cannot dictate orthography, otherwise we would occupy a lawless state "wherein everie mans brain was everie mans book, and everie privat conceit a particular print" (1582: 104). In Mulcaster's conceit, book-making technology has been thoroughly acculturated to orality, if not orality to book-making technology. As de Certeau observes, in the polarity of "oral" and "literate" one term is inevitably privileged over the other (1984: 132–133). In the case of Mulcaster's conceit, the privilege belongs to orality. Sound-based communication remains the reality; the new medium of print, scarcely a hundred years old, serves that reality as a metaphor. Just such a disparity in conceptual force helps to explain why, under certain circumstances, black-letter type, like secretary hand, can signify vernacular speech. Before the "triumph" of print, it was type that served speech, not vice versa.

Evidence that readers responded to oral cues can be found not only in teaching manuals but in the marks readers made in books. Coote's *The English Schoole-Maister* is absolutely typical of early modern pedagogy in the way it presents learning to read as an oral activity. First a student learns to recognize and to call aloud phonemes, then single syllables, finally polysyllabic words, first of two syllables and then of three (1596: 30–32). Brown in his *New Invention* insists that a subvocalization of sounds goes on, even when a practiced reader sits silently with a book:

> albeit Reading bee often times used without uttering of the Voyce, and
> therefore may bee thought to enervate, and cut away, the Pipes of the
> Sound, signified by these Symbols, or, at the least, attributed unto
> them: yet it is otherwayes: for such kinde of Reading proceedeth rather
> of some contracted habite, or custome of the sound of Letters, learned
> of before through oft reading; than of the speculation, or inspection, of
> their Forme, which then is seene. Otherwise, wee could no more reade
> without uttering of the written wordes which wee see, than expresse
> them without a viue sight thereof: I meane such as wee haue not re-
> cent in our memorie. (1622: D3–D3^v)

Marks made by readers attest the truth of Brown's remarks. The majority of such marks—in the form of marginal notes, asterisks, quotation marks down the left margin, sometimes a hand with a pointing finger—serve to set apart memorable passages from the rest of the text (Kintgen 1990: 1–18). In effect, the reader's hand "members" the text, divides it into parts, so that the parts can be *re*membered on the reader's own terms.

"Sentences," the early modern term for these membered and remembered texts, catches the oral basis of the whole exercise. A reader might even transfer remembered passages into a commonplace book, but he or she would do so, not to fix the text like a flower pressed between the leaves of a book, but to keep the text even handier for use in the reader's own conversation (Crane 1993: passim). The full title of the first of John Florio's Italian-English language guides leaves no doubt that "sentences" belong not on the page, but in the reader's mouth: *Florio His first Fruites: which yeelde familiar speech, merie Prouerbes, wittie Sentences, and golden sayings.* Thus, in the chapter devoted to "Diuers sentences diuine and profane" Florio's interlocutors trade such speeches as "Three things neuer want commendations, Good wine when it is drunken, a wise Sentence when it is spoken, and a good man in aduersitie" (1578: E3). When one of the speakers marvels that the other knows so many sentences, his companion replies, "Sir, I wyl tel you, I haue readde them often, and so I keepe them in memory, for when a man wil keepe a thing in memory, let him reade it often" (1578: G2). But the "fruits" of stored-up reading are in the speaking. Quotation marks in early modern practice distinguish not so much what a speaker or a writer has said as what the *reader* might say (de Grazia 1991: 57–71). From pictures to handwriting to print the eye surveys its phenomenological dominion. In varying degrees, however, it must share that dominion with other parts of the body, with the hand and the ear. Sight can know things that are *there,* but it needs the entire body to know them *here,* to know them through and through.

The move from houses of memory to musical notation to scrolls to illustrations to handwriting to print is a gradual move out of the sounding, speaking body toward an artificial body. Even with print, however, the artificial body never quite gains the independence that it has for us, products as we are of a culture that communicates primarily through electrical impulses that are turned into print or into facsimile images. In hindsight, it is easy for us to talk about the "triumph" of printing in early modern Europe. What we are apt to miss is the resistance of voice to the new medium. In a culture that still gave precedence to voice—in legal practice, in rhetorical theory, in art made out of words, in the transactions of daily life—we should be looking, not for evidence of the hegemony of type technology, but for all the ways in which that newly discovered resource was colonized by regimes of oral communication. Between voice and type, ecological theorists like Niklas Luhmann would have no difficulty in identifying which was the dominant system of communication in early modern England. If one system "resonated" in another, it is the voice-based system that, for the time being at least, was doing the louder sounding.

Such considerations should prompt us to think again about the ways in which voice is signed in early modern culture. Whatever form they take—finger positions on the neck of a lute, musical notes on a staff, tags of speeches written out on scrolls, *sententiae* copied down in secretary hand, a playtext printed in quarto—graphemes are signs: they point toward something that is absent. Since Saussure, we have taken those marks of sound and speech to be *symbols,* signs with only an arbitrary connection with the thing toward which they point. A survey of early modern media, all of them coupled in various ways with the sounding human body, suggests that graphemes may have functioned instead as *indices,* as signs with a natural or metonymical connection between the sign and what it represents. The signs of sound all possess a body. In early modern culture, those artificial bodies bear direct and intimate relationships with the sounding body. The speaker himself disposes the parts of a speech within an internal "house of memory." A musical performer carries within her mind and muscles the somatic memory of where to place her fingers, how to lengthen or shorten her vocal cords. With or without a scroll of speech, the body in a woodcut stands in for the speaker's body. Writing at the speed of hearing, a practitioner of Bales's "brachygraphy" uses his own body, his hand, to record what he hears. A reader takes in hand a printed book, but she turns the fixed body of type into a living, sounding thing by marking "sentences," copying them down perhaps in a commonplace book and making them part of her own conversation. All of these techniques are ways of remembering, and all of them insinuate the human body. We are adept at reading graphemes as symbols of semantic concepts; what we need are ways of reading graphemes as indices of somatic experience.

5 Some Propositions Concerning ○

○ Birds sound in chirps. Dogs sound in barks. *Homines sapienses* sound in ○.

○ ○ measures space. ○ calibrates time. ○ circumscribes the horizon of sound.

○ As an organization of lungs, muscle, larynx, mouth, tongue, teeth, lips, ears, and nerve endings, ○ is a constant across time. ○ is biological materialism.

○ As a set of protocols for crying out, speaking, singing, remaining silent, and listening, ○ is a variable across time. ○ is cultural determinism.

○ As the experience of crying out, speaking, singing, remaining silent, and listening, ○ can be known only in terms of ○. ○ can be turned into an object and studied from the outside, but the thing thus objectified and studied is not the same thing as ○. ○ is not quintessential knowledge; it is quintesensual experience.

○ ○ is both signified and signifier.

○ ○ belongs to the Imaginary before it is transformed by the Symbolic, and it never loses that Imaginary trace.

○ If S/Z is *plaisir du texte*, ○ is *plaisir du corps*.

○ A well-wrought urn is no more present than ○. A "site" or "space" is no more present than ○. A "text" is no more present than ○.

○ If ○ is absence, it is *present* absence. As I experience them, presence and absence are not binary opposites but relative conditions along a continuum. The effects of presence and absence vary according to media of communication, to persons, to social circumstances, to the disposition of bodies, to time.

○ ○ centers the listening subject within the horizon of hearing. ○ centers the speaking subject within the range of voicing. ○ centers speaking and listening subjects within horizons of place, time, and culture.

○ ○ centers speaking and listening subjects who may be situated quite otherwise in writing.

○ The cultural horizons of ○ are not concentric.

○ ○ directs me to listen otherwise.

PART
TWO
Within

6

Games, Gambols, Gests,
Jests, Jibes, Jigs

"For O, for O, the hobby-horse is forgot"
—refrain from a lost ballad, quoted in *Hamlet* (3.2.129), *Love's Labor's Lost*
(3.1.27–28), and elsewhere in commercial stagescripts

In the beginning there was [o:]. Ulululululululation.
Then there was [o:] [o:]. A pause for breath.
Then [o:] [o:] [o:]. Two pauses for breath. There began rhythm.
There began syntax. There began music. There began dance. There
began the turn, turn, return to [o:]. The round of May games at Wells,
Somerset, in 1607 traced the full circle of sound, from shouts to
speech, from speech to music, from music to baaing, from baaing to
whoops and hollers. We know about the festivities at Wells in much
greater detail than most such occasions because they came to the
attention of the court of Star Chamber (Stokes 1996, 1: 261–358, 2:
709–728; Underdown 1985: 55–56; Sisson 1936: 157–185). St. Cuthbert's Church was in need of major repairs, and the parishioners
landed upon a way of raising money that had been common all over
England before the Reformation: they staged a church ale (Hutton
1996: 244–261; Hutton 1994: 70–71, 100–101). Instead of lasting only
a day or two, however, the merrymaking at Wells extended over a
period of eight weeks, from May Day through Ascension Day, Whit
Sunday, Trinity Sunday, and St. John's Day, all the way to June 25. The
churchwardens, wary of Puritan opposition to such goings-on, were
careful to secure the permission of the Dean of Wells Cathedral. Instead of simply offering fresh-brewed ale, cakes, cream, and roasted
meat, with perhaps a piper and some dancing, the Wells festivities
included virtually every traditional recreation known to the denizens
of early modern England: a May lord and his court, morris dancing,
Noah and the ark, the Pinder of Wakefield, St. George and the
dragon, and Robin Hood and his merry men, not to mention "skimmington" rides against townsmen who were vocal in their opposition
to the sports. Every verdery (or tax district) of the city, each centered
on a different street and on a different trade, contributed its traditional show or shows. The neighbors, some of them disguised to fit
the fiction, would parade to the churchyard for a dinner and then
parade home again, while people from elsewhere in town and from
the countryside round about watched and cheered. It was the deci-
sion of two of the verderies to turn their shows into mockeries of
Puritan objectors that landed the participants in Star Chamber. All in
all, the Wells shows illustrate Ronald Hutton's point about the fluidity

i festive practices in early modern England: outdoor celebrations might occur at any time from May Day to midsummer, and they might include any number of diversions, according to local custom and the resources available in a given year (Hutton 1996: 244–245).

In addition to a sense of merriment and a celebration of community, what made the Wells shows all of a piece was sound. One of the unrepentant participants, William Gamage, wrote the whole thing up in a ballad that epitomizes the eight-week affair as a procession of pageants—a procession of sounds:

> drumes fyfes and trompetts did sowne apace
> the countrye held at no disgrace
> vnto our towne to make resorte
> to heare the Roaring canon shotte. . . .
>   (Sisson 1936: 179)

Three kinds of sound in particular sent constable John Hole and his friends into a fine frenzy: gunshots, trumpet blasts, and drumming. John Hole's complaint before Star Chamber charges that the revelers were "armed with vnlawfull weapons and drumes & then & there acted not only many disordered Maygames[,] Morice daunces[,] long Daunces[,] men in weamens apparall[,] new deuised lordes and ladyes[,] and Churchales, but further acted very prophane & vnseemely showes & pastimes" (Stokes 1996, 1: 262). (These latter shows and pastimes were presumably the skimmington rides that took the mick out of Hole and his friends.) "Thundring peeces" punctuated the mercers' procession—not only their mock battle between St. George and the dragon but their show of Actaeon and Diana (1: 352, 2: 713). Two trumpeters blasted away as the revelers, on one occasion, prevented Hole from giving his charge to the night watch. The court questioned one witness whether he and his cronies did, "makinge a great noyse then and there[,] come neere vnto the Complainant and sounded trumpettes and hallowed and showted" so that Hole "was thereby soe disturbed soe that hee could not quietly chardge the said watch" (1: 296). The witness confessed that he ordered the two trumpeters to play but denied that he did so in order to interfere with Hole's charges to the watch (1: 297). According to John Yarde, another complainant, the merrymakers then tooted and shouted their way through the streets to the bishop's palace and continued to make noise for an hour (1: 342).

It was drumming, however, that gave the festivities their dominant pulse. And for good reason: as Barnabe Rich attests, drumbeats were among the loudest sounds to be heard in early modern England: "A Drummer is the pride of noyse, *for* he puts downe all but thunder" (1619: 57) (see fig. 3.1). Virtually every event in the eight weeks of revelry was accompanied by one or more drummers. Drum-beats called the revelers together, drumbeats quickened their march-ing and dancing, drumbeats drowned out the voices of opposition. Before dawn on May 1, a drummer accompanied twenty men, women, and children as they paraded through the streets and set up a maypole in the High Street (Stokes 1996, 1: 298–300). For street dancing on Sunday, May 3, the drumming began as early as four o'clock in the morning. When Hole tried to disperse the crowd, he was, quite literally, drummed out of countenance: the dancers, Hole told His Majesty's representatives, "compassed your Subiect (with) drumes (beaten in disgrace) of your Subiect" (1996, 1: 262). When Yarde's wife complained that the maypole was a pagan idol, which she refused to pass on her way to church, an armed band of men paraded through the streets with an Old Testament idol, a "sparked" (i.e., spotted) calf made of wood—all to the accompaniment of fifes, trumpets, and drums. Outside the Yardes' house and elsewhere in the town the men shot at the graven image while one dressed in "Sat-ire Skynns" cried "ba, like a calf" (1: 351). Drums and shot scored the childrens' show on Ascension Day. "Three greate Drvmes" sum-moned as many as a thousand spectators to Robin Hood's show on Trinity Sunday (1: 277). Drums cleared the way for the skimmington rides that made public spectacles of Hole, Yarde, and other Sabba-tarians on June 18 and 25 (1: 308, 317).

The dinning drums through the days and nights of May and June is attested by two of the drummers themselves. Thomas Petters, jour-neyman shoemaker, told the court that "he did vse to play on the drvmm in the monethes of may & June . . . in the streete of wells & sometymes abroad in the feild to fetche home May"—but never dur-ing times of divine service. Sometimes, Petters added, there were as many as three drummers going at once (1: 301–302). John Rodway, barber, likewise acknowledged that he "did divers tymes sound his drume . . . when divers people & companies were present when as they were feching in bowes & may & at such like pastime" (1: 307). Both drummers deny that anyone hired them to play. They did it for pure pleasure. Petter, in his own words, "did so plaie on the drvm & keepe companie to make sporte of his owne minde & forwardnes" (1: 302).

As a veritable anthology of early modern folk revelry, the May games at Wells in 1607 run the gamut from [o:] to [o:]. They find their inspiration outside speech, they pass briefly through language in Gamage's ballad and the Star Chamber papers, they pass beyond language into singing, trumpeting, and fifing, they turn man-made rhythms into ambient thunder through the pounding of drums, they devolve into shouts of hurrah. Hole's several confrontations with the

revelers are paradigmatic. About 6:00 P.M. on Sunday, May 17, he tried
to stop thirty to forty morris dancers who were returning from an ale
in nearby Croscombe. According to one witness, at least a hundred spectators had assembled, some of them "hoopinge laughing &
sporting in merriment as is vsed to be done in such like games &
sportes" (1: 280–281). Hole's main strategy for stopping the dancing
was to try (unsuccesfully, as it fell out) to arrest the taborer whose tap,
tap, tap was speeding the dancers on their way through the streets. It
was later the same evening that Hole's charge to the watch was
drowned out by trumpet blasts and jeers from the crowd. Two weeks
earlier, at the very beginning of the revels, drumming had stifled his
attempt to read the mob the riot act, in the form of a royal proclamation against sports on Sundays. (The question was whether this in-
junction applied all day Sunday or only during times of divine service. The mayor ruled the latter.) In all of Hole's confrontations,
speech tried to silence noise—noise that began in shouts and trailed
off into blasts of brass horns, into gunfire, into cheers. Speech, that is
to say, tried to claim the whole of ○. But to no avail: the evenings of
May and June were filled with morris dancing, with stylized combats
between heroes of legend, with charivari.

For such performances the vocabulary of literary criticism lacks
a term, in large part because the events themselves lack verbal substance. They cross generic boundaries: neither song nor dance nor
drama, they are at once none of the above and yet all of the above.
Middle and early modern English describes such events as "games,"
as varieties of play as opposed to work, or as "gests," as "deeds" or
"exploits." Implicit in both "game" and "gest" is a strong sense of
gesture, of bodily movement. The essential elements in such performances are, indeed, body and sound. When dancers dance, when
fighters fight, when riders ride in procession to a victim's house, bodies move in rhythm with sound. But the nature of that sound need
not be verbal. Therein lies the suspiciousness of games and gests to
Puritan detractors like Hole—and the neglect of games and gests by
textual critics. As different as they may be in many respects, morris
dancing, combats by Robin Hood, skimmington rides, and jigs share
one thing in common: they dance the circumference of sound from
[o:] to speech to music to ambient sound, before they revert to [o:].
"People a Maying" burst into *The Two Noble Kinsmen* as "Noise and
hallowing" (Q 1634: S. D. before 3.1.1). Always threatening to turn
into noise, gests challenge the authority of language.

Even the cultivated music of early modern England, theorized
as an aural translation of the music of the spheres, flirts with this
phenomenon. Thomas Morley's madrigal "Ho who comes here all
along with bagpiping and drumming?" offers a particularly robust

example (musical quotation 6.1). Carefully controlled in Morley's madrigal, "hey nonny nonny" is given rein to run riot in popular forms of music. The cultivated dances of early modern England—the almain, the basse dance, the brawl, the cinquepace, the galliard, the pavane, the volta—attempted to impose similar controls on the kicking legs and flailing arms of folk dances. Skiles Howard contrasts country dancing with courtly dancing on a number of fronts: physical space, the relation of dancers to each other, the orientation of the dancing body. Where country dancing took place outdoors, in the round, courtly dancing took place inside, in rectangular rooms; where country dancing joined dancers together in lines and rings, courtly dancing isolated them as couples and as individuals; where country dancing encouraged freewheeling movement of arms and legs, courtly dancing trained the body upward. In effect, courtly dancing becomes a discourse in its own right, a means of disciplining bodies and subjecting them to the dominant ideology (Howard 1998:

Musical quotation 6.1

(Morley 1962: no. 18)

10–13). One remembers Queen Elizabeth at Kenilworth in 1575, so intent on "delectable dauncing" indoors that she neglected to observe the country sports being staged for her amusement outside.

At the philosophical level, writers like Sir John Davies in *Orchestra: A Poem of Dancing* could "tame" rhythmic bodily movement via comparisons with the cosmic dance of the sun, moon, and planets in the same way that musical theorists could "tame" singing and viol-playing via comparisons with the music of the spheres. The cycle of sound, with its movement from [oː] to speech to music to ambient sound to [oː], is certainly open to such a reading. It all depends on how one chooses to read mankind's divided nature. An Apollonian view would interpret movement around the circle as one toward transcendence: from bodily [oː] to speech to music to music of the spheres to Omega. A Dionysian view, on the other hand, would interpret the movement as a return to bodily [oː]. Both views seem to be present in the contrast between masque and antimasque in performances like Jonson's *Pleasure Reconciled with Virtue*. Cultural exchange may move in both directions—Ronald Hutton presents persuasive evidence that morris dancing migrated from court to country, while the cult of Robin Hood migrated from country to court—but the regimes of sound and body movement in the two realms remain distinct (Hutton 1996: 262–276).

Compared to other manifestations of the O-factor, gests present the greatest challenge to documentation and interpretation. Their relationship to written language is far more tenuous than that of ballads, sermons, or stage plays. People learned how to perform gests not from manuscripts or books, but from each other. Such textualizations as exist are basically of four kinds. Most prominent are hostile accounts: attacks from the likes of Philip Stubbes or court records like the Star Chamber investigation into the May games at Wells. Antiquarians like John Stowe, William Camden, and John Aubrey have left a different kind of record. They view gests as historical residue, interesting as evidence of national identity (in the case of Stow's *Survey of London* and Camden's *Britannia*) or as survivals of ancient practices (in the case of Aubrey's *Remains of Gentilism and Judaism*). Correctors of "popular errors" constitute a third source. Sir Thomas Browne in *Pseudodoxia Epidemica* (1646) and Henry Bourne in *Antiquitates Vulgares* (1725) record folk practices only to judge them according to criteria that are either rational and scientific (Browne) or moral and social (Bourne). Most suspect of all, perhaps, are pastoralizers, professional writers who deal with country sports as commodities for readers to buy and consume or as pawns in "the politics of mirth."

Signs of the gentrification of rural customs are present even in the earliest textualizations like the 1569 broadside ballad "Good Fellows Must Go Learn to Dance" (British Library, Huth Collection 66) and the late 1580s manuscript transcription of "The Merry Life of the Countryman" (Bodleian Library, MS Rawl. poet. 185). The latter declares its political vantage point in the first stanza:

> A Prince dothe sit a slippery seate,
>   and beares a carefull minde:
> the Nobles, which in silkes doe iet,
>   do litle pleasure finde.
> Our safegard and safetie, with many great matters,
>   they scan;
> and non liues merrier, in my mynde,
>   than dothe the plaine countryman.
>
>   (Clark 1907: 361)

After a century of controversy over maypoles and Whitsun ales, the country "fancies" printed in Interregnum anthologies like *Recreation for Ingenious Head-peeces. Or, A Pleasant Grove for their Wits to walke in* (1650) are even more untrustworthy as records of actual events. In all four kinds of textualization, gests are something *observed*, not something *done*. Gests themselves lack a verbal text, even as they are surrounded by verbal texts. They exist at the point, illegible in itself, where a range of verbal texts—ethical, antiquarian, scholarly, literary—happen to converge. Whether a writer is attacking gests or celebrating them, graphic transcripts register, in one way or another, an awareness of the impotence of words: written texts gesture toward events that happen through means other than writing, in a place other than here: out in the country, up in the north, inside the body of the performer, within the ○.

MORRIS, MOORS, *MORES*

Poor Phillip Stubbes. His *Anatomie of Abuses* (1583), intended to castigate the wanton pastimes of early modern England, has unwittingly provided a visual record of practices that otherwise would have disappeared from memory. The same eye that imagined spectators leaving the theater to go home and play the sodomites, or worse, sees in morris dancers the very image of devils incarnate. Stubbes's morris-men are said to be the servants of a Lord of Misrule, the kind of figure who might preside over May Day festivities or a church ale or a Midsummer watch. Decked out with scarves, ribbons, laces, gold rings, precious stones, jewels, bells on their legs, and handkerchiefs in their hands and around their necks—handkerchiefs "borrowed for the most parte of their pretie Mopsies & loouing Besses, for bussing them in the dark"—the dancers make their way to the churchyard.

What offends Stubbes is not only the riot of color he sees but the pandemonium of sounds he hears:

> Thus al things set in order, then haue they their Hobby-horses, dragons & other Antiques, togither with their baudie Pipers and thundering Drummers to strike vp the deuils daunce withall. [T]hen, marche these heathen company towards the Church and Church-yard, their pipers pipeing, their drummers thundring, their stumps dauncing, their bels iyngling, their handkercheifs swinging about their heds like madmen, their hobbie horses and other monsters skirmishing amongst the route. . . .

The commotion, at least in Stubbes's heated imagination, bursts into the church during divine service, making "such a confuse noise [sic], that no man can hear his own voice." The parishioners don't seem to mind: they "look," "stare," "laugh," "fleer," and jump up on the pews to get a better view (1877–79: 147). Stubbes, for his part, is a man who wants to hear his own voice. The parishioners and the dancers, for their part, take the "confuse noise"—or what Stubbes *hears* as confused noise—to be the essence of the event: a synaesthetic synchrony of pipes, drums, bells, hands, arms, legs, and feet.

Other early modern witnesses confirm Stubbes's testimony that morris dancing is as much about sound as it is about movement. The "May-game" Henry Machyn witnessed in London on June 24, 1559—the first year of Elizabeth's reign—included the same jumble of festive elements as the Wells games of 1607, with the same rogation round the town: a giant, the nine worthies "with spechys," "a goodly pagant with a quen c . . . [manuscript damaged] and dyvers odur, with spechys," St. George and the dragon, "the mores dansse," Robin Hood, Little John, Maid Marian, and Friar Tuck with "spechys rond a-bowt London." Speeches there may have been, but Machyn also notes that the whole affair was accompanied by "drumes and gunes." The shows were transported to Greenwich the next day and played before the queen and council (Machyn 1848: 201). Shouts and the sounds of a tabor draw the anonymous author of "The Puisnes Walks About London" to a morris dance in Fleet Street (Garry 1983: 222). In *Plaine Percevall the Peace-Maker of England* (1590?) Richard Harvey finds in morris dancing a metaphor for the distracting hubbub that keeps people from being able to make out the real issues in the Marprelate pamphlet-wars:

> If *Menippus*, or the *Man in the Moone*, be so quick-sightd, that he beholds, these bitter sweete Iests, these railing outcries: this shouing at *Prelats* to cast them downe, and heauing at *Martin* to hang him vp for *Martilmas biefe*: what would he imagine otherwise, then as that stranger, which seeing a *Quintessence* (beside the foole & the Maid Mar-

ian) of all the picked yoouth, straind out of an whole endship, foot-
ing the Morris about a *May pole*. And he, not hearing the crie of the
hounds, for the barking of dogs, (that is to say) the minstrelsie for the
fidling, the tune for the sound, nor the pipe for the noise of the tabor,
bluntly demaunded, if they were not all beside themselues, that they
so lipd and skipd without an occasion. (1590: 8–9)

The morris is as much a matter of sounds piercing the air as of bodies
moving in space.

The varieties of sound writers most often associate with the morris are four: the jingling of bells, the shrilling of pipes, the rasping of
fiddles, and the dinning of drums. Each of these sounds has its own
distinctive properties, each its own cultural meanings. Together, they
serve to "place" morris dancing quite precisely along the curve of the
○: beyond speech, beyond music, well into the range of naturally
occuring ambient sounds that ultimately encompass the primal human [o:]. Langham may describe the morris dance at Kenilworth in
1575 as "according too the auncient manner, six daunserz, Mawdmar-
ion, and the fool" (1983: 50), but most modern historians of dance
agree that morris dancing, both in England and on the continent of
Europe, had its origins fairly recent to Langham's time, in court
dances of the fifteenth century. It is probably no accident, Ronald
Hutton suggests, that the earliest reference to the popular morris in
England occurs in 1507 at Kingston-upon-Thames, in the neighbor-
hood of Richmond Palace and Hampton Court (Hutton 1996: 264–
265).

An early sixteenth-century version of one of these courtly "mor-
isques" in France is described by Jehan Tabourot in the dance manual
Orchésographie (1588) that he published under the name of Thoinot
Arbeau:

> In my young days, at supper-time in good society, I have seen a
> daubed and blackened little boy, his forehead bound with a white or
> yellow scarf, who, with bells on his legs, danced the *Morisques* and,
> walking the length of the room, made a kind of passage. Then, retrac-
> ing his steps, he returned to the place where he began and made an-
> other new passage, and continued thus, making various passages very
> agreeable to the onlookers. (1925: 148)

Whatever may have changed about the morris as it migrated from
court to country, it retained three elements in Tabourot's description:
scarves, blackface, and bells. Indeed, Tabourot sees bells as the distinguishing feature of morris dancing and ingeniously tries to connect
it with ancient Roman bell-dances mentioned in Macrobius and Vir-
gil. One measure of the morris's rise to popularity in England is the
fact that in 1567–68 alone 10,000 morris bells came into the country

through the Port of London (Hutton 1996: 268). Literary evidence at least suggests that all this ringing happened at different pitches—a refinement that is not characteristic of morris dancing today. As the country clowns in Dekker, Ford, and Rowley's *The Witch of Edmonton* (1621) prepare for a dance, they make a list of just what they will need. "A new head for the Tabor, and silver tipping for the Pipe," Cuddy Banks reminds his "Morrice-mates." "Remember that, and forget not five lesh of new Bells." Crooked Lane in London is the place to go for those, one of his mates volunteers. The bells Cuddy and his friends have in mind are graded according to pitch. Cuddy wants all trebles:

> [CUDDY]: Trebles: buy me Trebles, all Trebles: for our purpose is to be in the Altitudes.
>
> 2[ND CLOWN]: All Trebles? not a Mean?
>
> CLOW[N]: Not one: The Morrice is so cast, we'll have neither Mean nor Base in our company. . . . (2.1.37–45, 3.1.51 in Dekker 1953–61, 3: 507, 521)

When they perform their morris later in the play, a fiddler comes on with the dancers and arranges them in "Morrice-ray," according to the pitch of their bells: "fore-Bell" and "second Bell" here, "Tenor" and "great Bell" there (3.4.14–15). The effect would have been one of percussive sound across a wide range of frequencies.

Pipes are capable of playing tunes, and for this reason Richard Harvey contrasts the pipe with the tabor, but at least one witness suggests that the pipe's provision of "vertical" rhythm was just as important as "horizontal" melody. Pipe and tabor were, after all, played by one and the same person, who constituted a kind of one-man band: striking the tabor with one hand, the performer fingered the pipe with the other. The sound produced by the pipe was a high-pitched whistle (Munrow 1976: 13–14). In his character-book *Whimzies* (1631) Richard Brathwait casts a piper as a "drone," in two respects: he lives off other peoples' largess, and he produces the same sounds over and over again. However repetitive his tunes, the piper provides the toe-tapping rhythm necessary for dancing: "In Wakes, and rush-bearings he turnes flat *rorer*. Yet the Youths without him can keep no true *measure*. His head, pipe, and leg hold one consort. He cannot for his hanging fit himselfe to any *tune*, but his active foote or great toe will keepe time." The authorities may have tried to outlaw morris dancing by outlawing itinerant musicians, but the piper has cleverly adapted his art to the shift in the wind and has given up melody entirely: "Since hee was enacted *Rogue* by Parliament, hee ha's got hold of a shamelesse tunelesse Shalme to bee his consort, that the statute might take lesse hold of his single quality" (1631: L11–L11ᵛ).

The fiddle, too, served its rhythmic turn—with distinctly sinister overtones. In Harvey's metaphor, "sound" is to "tune" as barking is to crying, as "noise" is to piping, as fiddling is to "minstrelsie." That is to say, fiddling is a form of noisemaking, of percussion. Of all the instruments in the morris, it seems to have been most closely associated with the Devil. The country clowns' dance in *The Witch of Edmonton* is stymied when Mother Sawyer's familiar, Dog, jinxes Sawgut's fiddle. The dance can proceed only when Dog takes the fiddle in hand and plays the morris himself (3.4.24–49). None other than Satan himself figures as captain of the morris in one of the dreams attributed to Sara Williams in Samuel Harsnett's *A Declaration of egregious Popish Impostures* (1603). As in *The Witch of Edmonton*, it is the fiddler who takes the lead: "*Frateretto, Fliberdigibbet, Hoberdidance, Tocobatto* were foure deuils of the round, or Morrice, whom *Sara* in her fits, tuned together, in measure and sweet cadence. And least you should conceiue, that the deuils had no musicke in hell, especially that they would goe a maying, without theyr musicke, the Fidler comes in with his Taber, & Pipe, and a whole Morice after him, with motly visards for theyr better grace" (1603: 49).

Bells, pipe, and fiddle contribute their parts, but the drum is the *sine qua non* of morris dancing. In Barten Holyday's play *Technogamia*, acted at Christ Church, Oxford, in 1617, the dancers are all ready to go but lack a taborer. "No matter," says one, "since he does not come, wee'll sing, and so make musicke to our selues. Who can tune the Morrice best?" At that moment, the drummer arrives (Garry 1983: 220–221). What's needed, the dancers imply, is not so much a melody as a marker of rhythm. "Tuning" the morris is sounding out the beat. When the human voice joins the morris, it does so as a percussive instrument. Who needs a tune? What is required is [o:] [o:] [o:]. To the tabor's tap tap tap, the pipe's pip pip pip, and the gut-string's grate grate grate might possibly have been added the clap clap clap of clubs, as the morris dancers faced one another and thwacked the staffs they carried. That, at least, is a feature of modern morris dancing. With or without staffs, morris dancing was perceived—at least by people who bothered to think about such things—to have had its origins in "Moorish dancing." "[L]ooke right and straight / Vpon this mighty Morr—of mickle weight," intones the Schoolmaster who introduces the morris dancers in *The Two Noble Kinsmen*. "Is—now comes in, which being glewd together / Makes Morris" (Q 1634: 3.5.119–122). The pedant's etymology may find visual confirmation in the blackface that sometimes (but not always) constituted part of morris dancers' regalia. Whether or not the courtly dances of the late fifteenth century had any connection with actual Moors—celebrating the expulsion of the Moors from Spain or reenacting battles with the

Moors in the Crusades—the morris of the sixteenth and seventeenth centuries seems to have taken shape as choreographed combat (Hutton 1996: 262–263; Baskervill 1929: 352–357). Tabourot classifies dances, in part, by the instruments used to accompany them. Drums and pipes he includes among instruments of war (1925: 25). And war brings noise.

One speaks of the dancers, in modern times at least, as a "morris *side.*" In the sixteenth century two relevant senses of "side" were current: one having to do with military combat (OED, "side" 20a) and another with choral-singing (19d). Sport came later, at the end of the seventeenth century (20b). One highly conspicuous source of combat often accompanied morris dancing in the form of hobby-horses, men wearing horses' bodies about their middles. In illustrations they look very much like two-legged centaurs. In most circumstances hobby-horses meant trouble—trouble that was very much part of the dance. Stubbes describes hobby-horses as "skirmishing" among the crowd of dancers. In *Technogamia* a hobby-horse arrives with the taborer and proceeds to knock down the other morris dancers. "O my arme, my arme!" groans one. "O my shinne!" exclaims another. "I haue hurt my breast," says a third. "O the side of my face!" complains a fourth. The dancer impersonating Maid Marian pleads with the hobby-horse to "dance the morrice quietly with us"—to no avail (Holyday 1942: 83). It is the hobby-horse's reputation for brawling that prompts Bomby in Fletcher's *Women Pleased* (1613?) to give it up, on his wife's advice: "This Hobby-horse sincerity we liv'd in / War, and the sword of slaughter: I renounce it, / And put the Beast off. . . ." "Will you daunce no more Neighbour?" asks one of his morris-mates. "Shall the Hobby-horse be forgot then?" asks another (4.1.155–170 in Beaumont and Fletcher 1966–1996, 5: 500). Sometimes stylized combat could become the real thing. John Aubrey describes the custom at Long Newnton, Wiltshire, of marking Trinity Sunday by having a maid of the town drape a garland of flowers around the neck of a bachelor from another parish, while claiming a monetary payment and granting three kisses (1972: 193). In June 1641 men carrying the garland were met by a gang of men from nearby Malmesbury, led by one "with a hobby-horse, and bells on his legs." "Win it and wear it," challenged one of the Malmesbury morris men, "come three score of you, you are but boys to we." A bloody melee ensued (Underdown 1985: 96). One wonders if it was not hobby-horses that inspired "horseplay" as a term for knockabout among the boys. On this occasion at least, the regular rhythms of bells, pipes, fiddles, and drums dissolved into noise.

For all its possible origins in narratives of combat, what emerges from sixteenth- and seventeenth-century allusions to the morris is less a story line than a repertory of body movements. The precise configuration of the dancers vis-à-vis each other is not known, but the movements of the dancers are characterized in terms of jumps and jerks. In *Henry VI, Part Two,* Jack Cade in the midst of battle is said to "capre vpright, like a wilde Morisco, / Shaking the bloody Darts, as he his Bells" (F1623: 3.1.365–366). "At it now Thomas lustily," Turnop invites his fellows in Anthony Munday's *John a Kent and John a Cumber* (1596), "and let vs ierk it ouer the greene" (1923: 42). The pattern of jumps and jerks might assume one of two basic forms: a line or a circle. A morris "side" suggests a line, but numerous witnesses seem to envision the morris as a ring-dance. In Nashe's *Summer's Last Will and Testament* (1592) the spokesman for Spring brings on a morris troop with a hobby-horse. The running commands he gives them while the dance is in progress conjure up a circular space: "About, about, liuely, put your horse to it, reyne him harder, ierke him with your wand, sit fast, sit fast, man; foole, hold vp your bable there. . . . So, so, so; trot the ring twise ouer, and away" (1904–1910, 3: 239–240). Rafe invokes Spring in *The Knight of the Burning Pestle* (1607) by invoking dance: "The Morrice *rings* while the Hobby-horse doth *foote* it *feat*eously," all puns very much intended (4.Int.38 in Beaumont and Fletcher 1966–1996, 1: 76).

The most primitive form of dance is, in fact, a circle. At its center, originally, was a physical object—a maypole, for example (Howard 1998: 8). At its center, vestigially, is a *phenomenal* object: the experience of the dance. The sounds of the morris—bells, pipes, fiddles, drums, clubs, and cries—serve to conjure up that circular space and to keep it alive with sound and movement. It is "the painted Pipes of worldly pleasures" that Fletcher's Bomby gives up along with his hobby-horse (4.1.162). From rogation rites we know that processional forms of movement can serve to bind a community together. At Wells in 1607, for example, one witness describes how the lord and lady of May led off a dance that moved from one end of the town to the other: about eight o'clock on Sunday evening, the third of May, "divers menn & their wives & others of good sorte in the saied City did to delight themselves procure minstrelle. to play before them & did dawnce or goe in a rownde & one after another in the streetes of wells" (Stokes 1996, 1: 301). On a smaller scale, a line dance may arrange the dancers into "sides," but in moving about the dancing space the participants create a metaphorical circle if not a physical one, setting themselves and their joint experience apart from the larger world. So it is with "morris-mates." As a combination of combat and communion, morris dancing enacts the dynamics of male identity-formation in early modern England, a process I have elsewhere styled "the myth of combatants and comrades" (1991a: 31–77). Affiliation is affirmed through antagonism; bodies move in opposition yet in synchrony. That ritual act takes place beyond language, at a point where music, ambient sound, and the primal [oː] ring out together.

FOOLING AROUND

If there was a plot to the courtly "Moorish" dances of the late fifteenth and early sixteenth centuries, it was combat among knightly dancers for the favor of a fair lady. During the Christmas revels of Henry VIII's court at Richmond Palace in 1515, "moresks" were performed by five dancers and a Fool, all of them displaying their terpsichorean capabilities, like so many cocks, for the approval of two ladies impersonating Venus and Beauty (Hutton 1996: 264). In many popular versions of this rite, the Lady has become the "Maid Marian" of Robin Hood lore (Wiles 1981: 20–24). In most cases it is the Fool who wins her hand. Indeed, the Fool and Maid Marian make an archetypal pair, as an epistle in Nicholas Breton's *Poste with a Packet of Mad Letters* (1605) attests: "for your steeple tyre it is like the gaude of a *Maide-Mairan*, so that had you a foole by the hand, you might walke where you would in a Morice dance" (1605: D2ᵛ). The Fool succeeds where others fail, for two very good reasons. First, he is the best dancer. Second, because he is the best dancer, he must also be the best lover.

One of the stage jigs preserved among Edward Alleyn's papers shows how the plot of the morris could be turned into a commercial property without losing any of its sexual energy. "The Wooing of Nan" stages the competition among Rowland (the hero of many jigs), Pierce the farmer's son, and an unnamed Gentleman for the hand of Nan. The tune is not specified, but each of the suitors gets a stanza or two and a turn on the floor with Nan. Enter the Fool. The tune changes, and the contest is over.

GEN[TLEMAN]
    whie how now sweet nany I hope you doe Iest
WE[NCH]
    no by my troth sir I love the foole best
    & if you be Ielous god give you god night
    I feare you are a gelding you caper so light   *Exnt.*
GEN[TLEMAN]
    I thought she had but Iested & ment but to fable
    but now I doe see she hath play[d] wᵗʰ his bable
    I wishe all my frends by me to take heed
    that a foole com not neere you when you mene to spee[d].
    (Baskervill 1929: 436)

Nashe in his preface to the 1591 printing of Sidney's *Astrophel and Stella* similarly links dancing around and *fooling* around: "my stile is somewhat heauie gated, and cannot daunce trip and goe so liuely, with oh my loue, ah my loue, all my loues gone, as other Sheepheards that haue beene fooles in the Morris time out of minde" (Nashe 1904–1910, 3: 332). Although the popular morris is usually portrayed as a group dance, it provided occasion for bravura turns by single dancers (Baskervill 1929: 355–356). The expectation that the Fool would be the cleverest dancer helps to explain why Will Kemp could cap his career as a stage clown by dancing the morris from London to Norwich. Kemp presumably had lots of experience not only in roles like Launce and Dogberry but as the lead dancer in jigs like "The Wooing of Nan."

Popular versions of the Fool and the Lady, in the form of plays acted on Plough Monday, were noted down from oral tradition in the late eighteenth and early nineteenth centuries. The text collected at Revesby, Lincolnshire, in 1779 may actually have been written by Sir Joseph Banks's sister Sara for a "rustick" performance before house guests, but it revels, most unladylike, in the same bawdy as "The Wooing of Nan" two centuries earlier. A variety of suitors dance for the hand of Cicely. Ralph the Fool ("I have a tool / Will make a maid to play") bests them all.

CICELY
    Raf, what has thou to pleasure me?
FOOL
    Why, this, my dear, I will give the[e],
    And all I have it shall be thine.
CICELY
    Kind sir, I thank you heartelly. (Chambers 1933: 113, 118)

About the identity of the Fool's "*this*" there can be no doubt. Less certain is the connection between the eighteenth- and nineteenth-century texts and the Fool and Maid Marian of the sixteenth-century morris. Plough Monday plays may represent a continuous oral tradition, or they may be an innovation of the eighteenth century—a festive addition to the much older custom of parading a ceremonial plough through the streets on the first Monday after Twelfth Night (Hutton 1996: 129–131). In either case, the Fool and the Lady dance at the center of the circle.

Taken all together, these witnesses demonstrate how the plot of the Fool and the Lady could be realized across a wide range of sounds: as mimed action while fiddle, pipe, and tabor keep time, as a sung and danced ballad in the form of a stage jig, as a rhymed text that calls dancers into the dance. In every case, words are less important than body movement. Speech functions largely to provide dance cues. The purpose of the sounds—words included—is to send the dancers' bodies capering into space. The ditty sung by the Fool and the Lady at the end of the Revesby play catches that momentum:

> We will have a jovel weding, the fiddle shall merrily play.
>   ri forlaurel laddy ri forlaurel lay.
>
> (Chambers 1933: 97)

In all these performances, sense gives way to nonsense, speech to music, music to ambient sound.

In the popular morris, on the stage of the Rose, on village greens on Plough Monday, the Lady was no lady at all but a man in women's clothes. Urging on the English in their assaults of Ireland, Barnabe Rich exclaims, "Let them shew themselues to bee like men, that doe now shew themselues like women, to looke like Maid-marrian in a Morris-dance, fitter for a Sempsters shop, then to fight for a Countrey" (1619: 22). Nashe gleefully seizes on the custom of cross-dressing when he casts prominent Puritans in the roles of morris dancers in *Pasquil and Morforius* (1589). Martin Marprelate plays Maid Marian, Dame Margaret Lawson gives him his gear, the militant Calvinist Giles Wiggenton cavorts as the Fool, and William Paget, the staunchly Protestant son of a Recusant father, clears the way:

> *Martin* himselfe is the Mayd-marian, trimlie drest vppe in a cast Gowne, and Kercher of Dame *Lawsons*, his face handsomelie muffled with a Diaper-napkin to couer his beard, and a great Nosegay in his hande, of the principalest flowers I could gather out of all hys works. *Wiggenton* daunces round about him in a Cotten-coate, to court him with a Leatherne pudding, and woodden Ladle. *Paget* marshalleth the way, with a couple of great clubbes, one in his foote, another in his head. . . . (1904–1910, 1: 83)

The amorous byplay expected between the Fool with his leather ballocks and wooden cock and Maid Marian with her beard barely concealed is confirmed by the assigning of parts in *John a Kent and John a Cumber*. Turnop and his mates want to cheer up John a Kent with a morris dance. "[L]et mayde Marian haue the f[irst] flurt at him, to set an edge on our stomacks," says Turnop, "and let me alone in faith to ierke it after her." Spurling approves the idea: "Now by my troth well aduisde good neighbour Turnop, Ile turne her to him if he were a farre better man then is, too him, too him, touch him roundly." The maid in this instance is to be played by a character labeled "Boy": "what? think ye I am afrayde of him? infaith Sir no" (Munday 1923: 42). John a Kent himself is drafted to play the Fool. In *The Witch of*

Edmonton the male playing Marian may be someone of maturer years: assigning the pitches of the bells to participants in the dance, the fiddler gives to Maid Marian not the treble bell, the second bell, or the tenor bell but "the great Bell"—the bell with the deepest pitch (3.4.15 in Dekker 1953–1961, 3: 533). A Fool with outsized genitals, a Lady with a beard: the possibilities for sexual jibes are very much part of the occasion at hand. In *John a Kent and John a Cumber* the boy playing Marian is explicitly instructed to "flurt" with John a Kent, to "touch him roundly." In the Revesby play the Fool makes his entrance from behind—in more ways than one. A dancer called Pickle Herring has almost won Cicely's favors.

P[ICKLE] H[ERRING]
    Nay, then, sweet Ciss, ne'er trust me more,
    For I never loved lass before like the[e].

Enter Fool.

FOOL
    No, nor behind, neither.
    Well met, sweet Cis, well over-ta'en! (Chambers 1933: 118)

If such open flirtations with sodomy are possible, it is because the performance relaxes the rules not only of social decorum but of verbal syntax. The essence of the event is not, after all, in words but in nonverbal sound. There is no need to identify with the disguises, no self-reflection about motivation, no worry about consequences. The sequence of jerks and jumps is a game, a mere sug*gest*ion, something done *sub:* gest. Under the aural/syntactical circumstances, all sorts of turns are possible.

ROBIN HOOD AND HIS FOES, BRITONS AND DANES, ST. GEORGE AND THE DRAGON

Thanks to ballads going back to at least the fourteenth century, Robin Hood bursts into the festive ○ fully equipped with all manner of narratives (Knight 1994: 44–97; Hutton 1996: 270–274). Hence, perhaps, the "spechys rond a-bowt London" Machyn heard in 1559. Such speeches as survive, however, point to a sense of Robin Hood that has less to do with telling a story than with playing a game. Two documents suggest what Robin Hood and his men may have sounded like once they had ceased to be participants in a festive procession and had taken on voices of their own: a fragmentary text in a manuscript datable to about 1475 (now Trinity College, Cambridge, MS R.2.64) and "the playe of Robyn Hoode, verye proper to be played on in Maye games" appended by the printer William Copland to his edition of *A Mery Geste of Robyn Hood*, printed about 1560. As Stephen Knight and David Wiles have demonstrated, neither of these texts offers a continuous dramatic narrative. Rather, each presents a series of fictional invitations to the real business at hand: physical trials of combat (Knight 1994: 98–115; Wiles 1981: 31–42). If courtly masques are invitations to dancing, Robin Hood plays are invitations to combat.

Though written out as one continuous text, the late fifteenth-century manuscript seems to offer two such invitations in two different scenarios. In the first, an unnamed spokesman offers to challenge Robin Hood in the presence of a sheriff: "Syr sheryffe for thy sake / Robyn Hode wull y take." The sheriff promises "golde and fee" and the game is underway. The challenger squares off against Robin Hood in a series of trials of prowess. First comes shooting, presumably an archery match:

[CHALLENGER]
> Robyn Hode ffayre and fre / Undre this lynde shote we

[ROBIN HOOD]
> With they shote y wyll / Alle thy lustes to full fyll

Next they take a hand at cleaving a stake, then casting stones, then throwing an axle-tree. In each case Robin wins. Next comes wrestling:

[ROBIN HOOD]
>   Have a foote be fore the
> Syr knyght ye have a falle

[CHALLENGER]
>   And I the Robyn qwyte shall
> Owte on the I blowe myn horne

[ROBIN HOOD]
>   Hit ware better be un borne

In blowing his horn the challenger may only be issuing a metaphorical boast, or he may be summoning his own men to square off against Robin's. (The second scenario in the Cambridge manuscript is a setup for group combat.) Single hero against single hero, or side against side, the climax of the first game seems to be a sword fight:

[ROBIN HOOD]
> Lat us fyght at ottraunce / He that fleth god gyfe hym myschaunce

Once again Robin wins, and the fiction goes so far as to have Robin cutting off the challenger's head, stealing his clothes, and carrying his head away in the victor's handy hood (Dobson and Taylor 1976: 205–207). What seems to be going on in this text is no more than a choreographed version of the games of skill John Stow records from the London of his youth: "On the Holy dayes in Sommer, the youthes of this Citie, haue in the field exercised themselues, in leaping, dauncing, shooting, wrestling, casting of the stone or ball, &c" (1971, 1: 94).

Billed on the title page as "a newe playe for to be played in Maye games very plesaunte and full of pastyme," the speeches appended to Copland's *Gest* likewise offer two separate invitations-to-combat in one continuous text. In both of them Robin Hood acts as choreographer as well as protagonist: he calls the players into presence and tells them what to do:

Now stand ye forth my mery men all,
And harke what I shall say;
Of an adventure I shal you tell,
The which befell this other day.

Robin Hood's opening speech typifies the way in which *listeners* to the "play" are drafted into being *participants* in the play. Robin Hood tells the story of encountering a stout friar and giving him battle, and then proceeds to cast his listeners as combatants in a restaging of the story. *Who* the characters in this story are is less important than what they can *do*. They are defined by how well they can fight. Friar Tuck introduces himself in just such terms:

But am I not a jolly fryer?
For I can shote both farre and nere,
And handle the sworde and buckler,
And this quarter staffe also.

Robin Hood's challenge to the friar sets up the first game. An exchange of verbal insults ends in Robin's demand that the friar carry him across a stream:

Harke, frere, what I say here;
Over this water thou shalt me bere;
The brydge is borne away.

"This water" may provide a fictional setting for the combat, but Friar Tuck's speech after he has wrestled with Robin and thrown him down conjures up a circular space in which the game takes place—a circle that has a "within" and a "without":

Now am I, frere, within, and thou, Robin, without,
To lay the here I have no great doubt.
Now art thou, Robyn, without, and I, frere, within,
Lye ther, knave; chose whether thou wilte sinke or swym.

Bodily combat is succeeded by *aural* combat. Robin asks the friar to let him blow his horn and summon his hound. Taunts the friar:

> Blowe on, ragged knave, without any doubte,
> Untyll bothe thyne eyes starte out.

The same sense of circular space as before is invoked when the friar sees coming, not a hound, but all of Robin's men. The friar, in turn, asks leave "to whistell my fyll." Robin replies:

> Whystell, frere, evyl mote thou fare!
> Untyl bothe thine eyes starte.

The friar's men bring on clubs and staffs for a pitched battle with Robin's men. What ensues may not be an actual fight but a morris dance, since the prize is "a lady free," which Robin offers to Friar Tuck if he will join his band. The friar's response closes the game with an invitation to dancing and (assuming that the lady, if present, is a male in women's clothes) to sodomitical by-play. The friar takes the lady's hand and becomes the Fool of the morris:

> Here is an huckle duckle,
> An inch above the buckle.
> She is a trul of trust,
> To serve a frier at his lust,
> A prycker, a prauncer, a terer of sheses [=sheets?],
> A wagger of ballockes when other men slepes.
> Go home, ye knaves, and lay crabbes in the fyre,
> For my lady and I wil daunce in the myre for veri pure joye.
> (Dobson and Taylor 1976: 210–214)

In all these texts speech is subordinated to action. "What seems to be a textual silence," Stephen Knight argues, "may indeed be the high point of performance: the dramatic character of the plays can itself be suggested by a lack of words" (1994: 101). David Wiles agrees: "the words amount to little more than a scenario or mnemonic providing a framework for improvisation. Within a given locality one would expect the words to have been transmitted orally" (1981: 36–37). The Coventry Hock Tuesday play, with its combat between Britons and Danes, is another example, on a grander scale. So, too, are the so-called "mummers' plays" collected in the eighteenth and nineteenth centuries. The texts vary from place to place, but in all of them a Christian hero (usually St. George) fights one or more evil adversaries (usually a Turkish Knight, sometimes the dragon), is killed, and is resurrected for a final triumph by a miracle-worker (usually The Doctor). The "plot" of the combats remains the same; what changes are the verbal cues. At Netley Abbey, Hampshire, for example, St.

George announced himself as "King George" and the Turkish Knight as "Turkey Snipe" (Brody 1969: 131–135). Written references to these plays are entirely lacking before the eighteenth century, casting serious doubt on their existence before that date (Hutton 1996: 70–80). At the same time, St. George and the dragon did more than parade at Wells in 1607. John Hole describes the event as "a man on horsback in armor with sword & speare representinge St George and the Couterfeite of a dragon with a man within him that carryed the same and they boath represented or Acted the fighte betweene the dragon & St. George" (Stokes 1996, 1: 340). At Norwich the annual celebration of St. George's Day seems to have included skirmishes between "Old Snap" and the locals—along with lots of noise. Orders abolishing the celebration in 1645 specify that there will be "no beating of drums or sounds of trumpets; no snap dragon, or fellows dressed up in fools' coats and caps" (Underdown 1985: 259). An allusion in Fletcher's comedy *The Woman's Prize* (1611) implies that "St. *George* at *Kingston*," epicenter of the morris, featured combat between the dragon and a less-than-valorous St. George,

> Running a foot-back from the furious Dragon,
> That with her angry tayle belabours him
> For being lazie.

A "fearfull dwarfe" bids the champion fly. "Here is no place for living men" (1.3.19–28 in Beaumont and Fletcher 1966–1996, 4: 29). Was there a death-and-resurrection plot to some of these early modern games of St. George and the dragon? The allusions are too sketchy to allow a firm "no" any more than a firm "yes."

St. George of Wells, St. George of Norwich, St. George of Kingston: it is the localization of combat games—their presence in the here-and-now, in *this* city or village, within *this* circle of sound—that helped make them suspect to increasingly centralized authority in early modern England. Depending on local circumstances, Robin Hood games could be supportive of the social structure, as villagers in green garb collected funds for the parish, but they could also be subversive of the social structure, as disaffected men used Robin Hood's outlaw status as a convenient front for rudeness, or even riot. It was not a Protestant but a Catholic, Mary Queen of Scots, who suppressed Robin Hood plays in 1562 (Knight 1994: 111–115; Underdown 1985: 110–111, 135). Shortly thereafter they begin to disappear from local parish accounts, to be succeeded by commercial exploitations of the Robin Hood story in plays like *George a Green the Pinner of Wakefield* (1592), Munday's *The Downfall . . .* and *The Death of Robert Earle of Huntington* (1598–99), and *Looke About You* (1600).

"ROUGH MUSIC"

Among the entertainments at Wells in 1607 it was not morris dancing or Robin Hood or "St. George of Wells" that got the organizers in trouble with His Majesty's government but the townsmen's public ridicule of seven local spoilsports: the constable and capitalist weaver John Hole, a grocer Humphrey Palmer, a pewterer Hugh Meade, a moneylender John Willis, a notary Richard Bowrne, a haberdasher John Yarde, and Yarde's wife Anne. Each of these objectors came in for boisterous attack. In Mistress Yarde's case, it was not only her objection to their maypole that drew the revelers' derision but her supposed adultery with John Hole. Later the Puritanical Mrs. Yarde would be publicly accused of adultery before the church courts (Sisson 1936: 165–167, 172–174). The victims' names were simply too good to be true: in a wonderful chiasmus, *she* played the "yard" or penis to *his* "hole." The joke was not lost on the revelers, who set up a portrait of Mistress Yarde as the backboard for the bowling game of "Nine Holes." The course was (what else?) a *yard* long. The game became a public spectacle when the revelers paraded their device through the streets. The ringleader (apparently Gamage the balladwriter) was accused of trundling a ball into the hole while crying out, "I hole." People in the crowd are said to have replied, "you hol not for a Crowne, or for vjd" and "he holes yt not within a yard" and "yf you will hole yt, yt must be in a Meade" (Stokes 1996, 1: 278–279, 287). Mistress Yarde happens to have dwelt at "the sign of the Crown," which does triple duty for the revelers as a coin bowlers might wager and as visible proof of Master Yarde's cuckoldom (G. Williams 1994, 1: 337).

Public humiliation came to the other objectors in the form of a skimmington ride organized by the tanners, chandlers, and butchers of Southover verdery. A cross-dressed man on horseback lampooned the clothier John Hole, two men riding on one horse and flinging grain to the crowd represented the grocer, two others riding face to face with book, desk, inkwell, and a bag full of counters impersonated the moneylender and the notary, another decked out in old hats put in an appearance for the haberdasher, while a seventh rode as the pewterer, complete with saucers, dishes, and a hammer that he used to make a great racket. It is the impersonator or effigy in such cases—the "skimmington"—that gives the procession its usual seventeenth-century name.

Skimmington ride, charivari, rough music: by whatever name, processions like those at Wells in 1607 drew from the same repertory of festive practices that celebrated communal holidays. Rides against shrewish wives, hapless cuckolds, and ill-matched brides and grooms could take place at any time of the year, whenever occasion arose, but Martin Ingram points out how many of them seem to have occurred during and near important holidays. It is tempting, he concludes, to see them as "part and parcel of the festivities" (1984: 96). *Noise* was as much a part of a skimmington ride as visual disorder. Cries and shouts, the baaing of the painted calf, the sounds of shots at the image, the clink of the notary's counters, the clattering of the pewterer's hammer are altogether typical of the cacophony that in later times earned such processions the name "rough music." Ingram cites an instance at Quemerford, Wiltshire, in 1618, in which a suspected cuckold and his wife were visited by a party of men estimated by the victims at three or four hundred, some of them armed like soliders. A drummer led the way. An impersonator of the cuckold, crowned with horns and wearing a smock, rode along. When they reached the victims' house, "the gunners shot off their pieces, pipes and horns were sounded, together with lowbells and other smaller bells which the company had amongst them, and rams' horns and bucks' horns, carried upon forks, were then and there lifted up and shown. . . ." The crowd threw stones at the windows, forced their way in, fetched the offending wife from the house, and threw her into a wet hole, where she was trampled, beaten, and covered with mud (Ingram 1984: 82).

Stage representations of charivari are scored with the same raucous sounds. There is a skimmington-like moment in *Epicoene* (1609), for example, when Morose is plagued with *"Musique of all sorts."* "O, a plot, a plot, a plot, a plot, a plot vpon me!" Morose exclaims. "This day, I shall be their anvile to worke on, they will grate me asunder. 'Tis worse than the noyse of a saw" (3.7.1–13 in Jonson 1925–63, 5: 217). In Thomas Heywood and Richard Brome's *The Late Lancashire Witches* (1634), a bridegroom whose yard has been bewitched, making him unable to consummate his marriage, is visited by a derisive mob: *"Enter drum (beating before) a Skimington, and his wife on a horse; Divers country rusticks."* In this case, the feckless groom and his fuckless bride pull the impersonators from their horse and beat them—but with a curious gender inversion: *he* beats the stand-in for the bride, and *she* the stand-in for the groom. *"Drum beats alar. horse comes away."* Fighting ensues among all onstage, reaching a climax when the bride, Parnel, tussels with the mocker of her husband: *"(make a ring Par. and Skim. fight."* "Beat drum alarum," yells one of the bystanders (1634: H4ᵛ). On the South Bank, as by the village pond, it is rhythmic cacophony—bangs, blares, booms, throbs, thuds, thumps, clamor, clatter, catcalls—that excites bodies into action and puts the *ride* into a skimmington.

On occasion, the ride can assume the form of a dance. Noises are regulated into a kind of music; body movements, into a kind of choreography. In Fletcher's *The Woman's Prize* (1611) a charivari is staged, most unusually, by a group of women—a city wife, a country wife, and three maids—against the wily Italian suitor for the favors of two "masculine" sisters. The women's triumphant retreat from the confrontation is accompanied by "all the pans i'th Town / Beating before 'em," but the confrontation itself is preceded by fiddle-music, a dance by the women "with their coats tuckt up to their bare breeches," and a drinking song to the health of "the woman that bears the sway / And wears the breeches" (3.2.1–2, 2.6.36–57 in Beaumont and Fletcher 1966–1996, 4: 63, 54). When the bride and bridegroom try to dance at their wedding in *The Late Lancashire Witches* the bewitched musicians, like those in *Epicoene*, play "*Every one a severall tune*" (F4). *Horn*-music, the blares of a sow-gelder's [*sic*] horn, greets the old suitor of a young bride in Brome's *The English-Moor, or the Mock-Marriage* (1637). "How now! O horrible!" exclaims the victim. "What hidious noise is this," asks his companion. Maskers have arrived, the servant explains.

> Their tuning their pipes,
> And swear they'll gi'ye a willy nilly dance
> Before you go to bed, tho' you stole your Marriage.

What ensues is an anti-masque of horned beasts: a stag, a ram, a goat, and an ox "*to musick of Cornets and Violins*" (Brome 1873, 2: 14–16).

"Untune that string, and hark what discord follows": in all its manifestations "rough music" is a conscious disordering of the soundscape, a substitution of *noise* for concord. Noise has two possible definitions, one physical and one psychological or social. In terms of frequency and amplitude, noise is a combination of random frequencies across a wide spectrum, at a high amplitude. The banging of pots and pans offers a prime example. In psychological terms, that registers as confused pitch and loud volume. What counts as confused and loud is subject, of course, to social definition: noise is what a community deems to be noise. Both definitions of noise, the physical and the psychological/social, apply to "rough music." Ingram hears in the cacophony of skimmington rides a calculated contrast to the harmony expected in social relations (1984: 98). The banging of the pewterer's wares, the baaing of the painted calf, the blasts of gunshot, the beating of drums: these are audible signs of social discord. Charivari can be interpreted, in this light, as a conservative gesture. But rhythmic percussive sounds have their own subversive power. Drumbeats stir the blood; the hypnotic recurrence of [oː] [oː] [oː] and [oː] [oː] [oː] and [oː] [oː] [oː] drives on, on, on, and on.

THE JOG IN THE JIG

Court records suggest that dancelike skimmingtons in stage plays had real-life counterparts in the towns and villages of early modern England. Two such instances have been described by C. J. Sisson (1936). "Michael and Frances, To the tune of *filiday fllouts mee*" came to the attention of the Council of the North in 1602, when Edward Meynell, squire of Hawnby, Yorkshire, used the ballad to ridicule a neighbor, Michael Steel, by suggesting that Steel was having an affair with his maid servant, Frances Thornton. "Michael and Frances" had first been performed by the devisers for the amusement of themselves and their friends during the Christmas holidays, but shortly, Steel complained, it had been taken up by professional players, "who by practyce and procurement haue at the ending of their playes sunge the same as a Jygg to the great sclandall" of said Steel (Sisson 1936: 130). In the jig Michael and Frances sing and dance their way into bed behind the back of Michael's old, complaining wife.

Another piece of skimmington-work, "ffooles fortune to the tune of A: B: C:," was brought to the attention of Star Chamber in 1622 by the family of an heiress in Shropshire, outraged that a disappointed suitor had turned her into the promiscuous wench of "Jock and Jenny" ballads (Sisson 1936: 148–156; Somerset 1994, 1: 23–33). Humphrey Elliott, disappointed in his own attempts to woo Elizabeth Ridge and lay hold of her fortune, planted a confederate, Edward Hinkes, as a servant in Elizabeth's household with instructions that Hinkes should press Elliott's suit. When even that stratagem failed, Elliott and his friends wrote "dyverse scandaluse & lybelluse and infamus verses in nature & forme of a plaie dyaloggwise wherin it was devysed that one of the actors should bee apparelled in womens apparell & bolstered & sett forth as though shee were great with Child & should appsonatt the said Elizabeth vnder the name of Ienney & one other of the said actors should apparsonate the said Edward Hinkes vnder the name of Iockey." In the ballad Jenny delivers her bastard; Jockey (who hails from the North) gets off scot-free. The tune changes to "the new masque" as Jockey jumps into the plot—and out of it—for the last time:

> the proverb it proues still true for mee
> (wheras all doe saie its better) to be
> a foole then to bee a rich mans Child
> for fooles they haue fortune the wyse are beguild
> for Iockey still goes free goes free
> for Iockey still goes free. . . .

These verses, the court was told, Elliott and his confederates "did malliciusly scandallusly and lybellusly repeat plaie & acte in the presence of a great multytude" all over Shropshire (Somerset 1994, 1: 26, 33).

Though popular in origin, jigs were taken up by professional ballad-writers and professional entertainers, and it is in that commodified form that most of the extant texts have been preserved (Baskervill 1929: 373–605; Sisson 1936: 125–129). By the 1590s many witnesses attest that jigs were serving as conclusions to stage plays in the public playhouses of London. The term "jig" in these references probably refers to any number of devices: a one-person song like "When that I was but a little tiny boy" with which Feste concludes *Twelfth Night;* a song in dialogue like the printed broadside "Frauncis new Iigge, betweene Frauncis a Gentleman, and Richard a Farmer," also known as "Attowell's Jig" after a player in Lord Strange's company; dancing without a dramatic scenario, such as the Epilogue offers to perform in *Henry IV, Part Two;* or dancing *with* a dramatic scenario, as in the numerous adventures of Rowland, left behind as printed broadsides inspired by English players on tour in Germany (Sisson 1936: 125–126; Baskervill 1929: 491–605). It is not clear just which kind of "jig" Thomas Platter witnessed after a performance of *Julius Caesar* at the Globe in 1599: "On September 21st after lunch, about two o'clock, I and my party crossed the water, and there in the house with the thatched roof witnessed an excellent performance of the tragedy of the first Emperor Julius Caesar with a cast of some fifteen people; when the play was over, they danced very marvellously and gracefully together as is their wont, two dressed as men and two as women" (1937: 166). By 1599, however, jigs were on their way to becoming *declassé.* Kemp left Chamberlain's Men the very year Platter witnessed the jig at the end of *Julius Caesar,* and the troupe seems to have left jigs to their competitors north of the City. When the Middlesex magistrates issued order for the suppression of "lewde Jigges" in 1612, they named only the Fortune, the Curtain, and the Red Bull (Gurr 1992: 175–176). Sisson argues that it was topical satire, a form of defamation, that inspired the order (1936: 126). If so, libels took the form of sexual jests. Hole and Yarde at Wells may have been despised as Puritans or as capitalists, but they were attacked in the guise of fornicator and cuckold. Just so, the plots of the jigs collected by Baskervill all turn on sexual adventurism. Seducers in these jigs are not always so successful as the Fool in "The Wooing of Nan" or Jockey in "Fool's Fortune." In the two most famous jigs of the sixteenth century, "Rowland" and "Singing Simpkin," would-be seducers are foiled by the lady's faithful lover or her husband—but not before the audience has had its voyeuristic pleasure.

The pleasure in jigs consists not only in what theatergoers get to

see, but in what they get to *hear*. The singing voices of the protagonists sound out to the same accompaniment as in morris dancing: to
pipe and tabor. Against the vertical tap tap tap of the tabor and the
horizontal passages of the pipe, jigs were played out across one or
more changes of tune, each of them signaling a turn in the plot and
a change in the dance. *"The tune changes"* is as essential to the stage
jig as *"Enter,"* *"Exit,"* and *"He dies"* are to scripts. In "The Wooing of
Nan," for example, wooing is a matter of wowing the wench with
toes, heels, calves, and thighs. Rowland, the chief dancer, is challenged at the peak of his success by Pierce:

[WENCH]
  Whie bony dicky I will no[t] forsake
    my bony rowland for any gold
  if he can daunce as well as perce
    he shall haue my hart in hold
PER[CE]
  why then my harts letts to this geer
    & by dauncing I may wonn
  my nan whose love I hold so deere
    as any creatur vnder sonn[.]

A "Gentleman" succeeds Rowland and Pierce in taking a turn. But
when the Fool enters, the tune changes, and Nan changes *her* tune:
"wellcom sweet hert & wellcom my tony / wellcom my none [*sic*]
true love wellcomm my huny" (Baskervill 1929: 434, 436).

Such jogs in the jig occur even when there may not have been
dancing. "Michael and Frances" goes through six changes of tune in
its devious course toward adultery, "Fool's Fortune" through four. In
"Attowell's Jig" there are four successive tunes: "Walsingham," during which Besse promises the gentleman Francis that she will sleep
with him by sending her farmer husband Richard on a false errand;
"the Jewish dance," in which Richard makes his entry; "The Bugle
Boe," in which Besse connives with Richard and Francis's wife to trick
the errant gentleman into bedding his own wife in Besse's clothes;
and "Go from my window," in which Francis, after an hour of pleasure, discovers the trick and one and all are reconciled (Baskervill
1929: 450–464). Like *Measure for Measure* and *All's Well that Ends Well*,
"Attowell's Jig" lets the audience enjoy the prospect of illicit love and
yet keep a certain moral distance. "Singing Simpkin" (1595), originally danced by Will Kemp, begins with an eruption of tripping feet
and tripping tongues:

WIFE
    Blind *Cupid* hath made my heart for to bleed,
        Fa la, la, la, la, la, la, la, la.

SIMPKIN
But I know a man can help you at need,
    With a fa la, la, la, la, fa, la, la, la, la, la, la.

It ends, at least as a later clown, Robert Cox, revised it for performance in the seventeenth century, with a certain curbing of libidinous
freedom. Rival after rival tries to join the dance. First Bluster the brag-
gart soldier bursts into the circle ("The tune alters"), and Simpkin
hides in a chest. When the Wife's husband ("The Old Man") arrives
on the scene, she plots to use Bluster's aggression as a front for Simpkin's hiding. The Old Man takes pity on Simpkin and goes out to buy
wine. Simpkin exults:

Thy husband being gone my love,
    Wee'l sing, wee'l dance, and laugh,
I am sure he is a good fellow,
    And takes delight to guaff.

But The Old Man—at least in the revised version—has been eaves-
dropping and catches them out. The dance ends in farce as The Old
Man and The Wife turn on Simpkin and *"beat him off."* Simpkin, for
his part, remains saucy to the end. So, too, does The Wife. She never
really repents—"I jested with him, husband, / his knavery to see"—
and there is every indication that Simpkin will, so to speak, rise to
the occasion once again (Baskervill 1929: 444–449).

Like Rowland, Simpkin enjoyed fruitful progeny in the form of
later jigs featuring the same hero in ever new exploits. All the jigs
in the series, even those with ostensibly moral endings, trace out a
movement toward moral license and bodily freedom. Wiles goes so
far as to suggest that, for comedies at least, jigs stand in for the sexual
consummations promised by the romantic ending (Wiles 1994: 56).
The fa-la-la's that set "Singing Simpkin" spinning on its way show
how the licentiousness of jigs is aural as well as moral. *"Jygging vaines
of riming mother wits"* are, as Marlowe implies in his prologue to *Tamburlaine,* more rhyme than reason (Pro.1 in 1: 79). Such an ending may
be natural enough for comedies, but if Thomas Platter is to be believed, even classically-inspired tragedies like *Julius Caesar* ended
with the actors—some of them at least—taking leave of speech. Like
formal epilogues, jigs are designed to set hands clapping and throats
to shouting. According to the Prologue to *The Fair Maid of the Inn*
(1626), *"A Jigg shall be clapt at, and every rime / Prais'd and applauded by
a clamorous chime"* (Pro.9–10 in Beamumont and Fletcher 1966–1996,
10: 559). With respect to the play that has just been concluded, there
are two ways to consider rhyme and chime: most obviously as a re-
nunciation of speech, but also perhaps as an affirmation of speech.
To the extent that they turn blank verse into fa-la-la's, jigs represent

a movement away from semantic sense toward something different, toward kinetic sensation. At the same time, jigs can be heard as accentuating aural effects that have been present all along in the performance of the play's verse and prose: alliteration, assonance, rhyme, the regularities of the iambic pulse. In either case, jigs enact a movement toward [o:]. They serve as reminders that speech belongs, after all, to a larger cycle of sounds, a cycle within which semantic sense occupies only a certain segment.

In their choreographical capabilities the Fool in "The Wooing of Nan," Jockey in "Fool's Fortune," and the titular trickster in "Singing Simpkin" are altogether typical of the *saltationis personae* of stage jigs: it is the fool who turns out to be the best dancer of all. As such, the role was a natural for the acting companies' clowns. In their own time, Richard Tarlton and William Kemp were as well known for the jigs they danced when the play was over as for the jests they pulled while the play was in progress. The jig was only the culmination of alphabetical anarchy that the clown had been causing all along. All the parts that Wiles proposes Shakespeare devised for Kemp— Launce in *The Two Gentlemen of Verona*, Grumio in *The Taming of the Shrew*, Costard in *Love's Labors Lost*, Bottom in *A Midsummer Night's Dream*, Peter in *Romeo and Juliet*, Launcelot Gobbo in *The Merchant of Venice*, Falstaff in *1 and 2 Henry IV*, and Dogberry in *Much Ado About Nothing*—cast the clown as an inveterate punster, a mauler of other men's words, a deconstructor of language *avant la lettre*. The riddle Peter puts to the musicians in *Romeo and Juliet* 4.4 is altogether typical. Singing the song he wants the musicians to play—and giving the audience a preview of the singing and dancing he will do at the play's end—Peter stops at the line "Then musique with her siluer sound." Why *silver* sound, he asks. The musicians can't tell. "Ile speake for you," Peter replies. "I saye Siluer sound, because such Fellowes as you haue sildome Golde for sounding"—and off he bounds with the bag of gold he is supposed to have paid them (Q1597: 4.4.154, 164–166). Sound triumphs over sense, the nimble body over verbal rigor. The clown's primary "language," Wiles argues, is physical, not verbal, and it reaches its apogee in jigs (Wiles 1994: 56).

The transformation that clowns like Tarlton worked onstage, turning the playwright's written text into body, sound, and laughter, was attempted in reverse when printers published jest books like *Tarltons Newes out of Purgatorie* (1590), issued two years or so after the famous comedian's death, or broadside ballads like "The Crowe sits vpon the wall, / Please one and please all" (1592), signed "R. T." Shortly after leaving the Lord Chamberlain's Men in 1599, Kemp himself cashed in on the textualization business. *Kempes nine daies wonder . . . written by himselfe to satisfie his friends* (1600) puts the whole process into reverse: body, sound, and laughter are turned into a written text. The title page to *Nine Days' Wonder,* centered on a woodcut of Kemp dancing the morris, scarf flying, while a musician taps his tabor and blows his pipe, presents one of the best known visual images from early modern English culture. But who among us has *read* the book? How, indeed, can one "read" a dance, much less a dance that extended from London to Norwich over a period of four weeks? The subtitle describes the text as "Containing the pleasure, paines, and kinde entertainment of William Kemp" between London and Norwich, but pleasure, pains, and kindness are ultimately sense experiences, things that texts can*not* "contain." A text "wherein is somewhat set down worth note" needs the emphasis on "somewhat." Kemp's book can be read as a bid for first-person presence against a number of deconstructive forces: ballad-makers, hearsay, the crowds of people who kept impeding his progress along the way, and the nonverbal essence of "wonder" itself.

Ballad-makers constituted the greatest threat to Kemp's integrity as a performer. To Mistress Anne Fitton, the pamphlet's dedicatee, he complains, "A sort of mad fellows, seeing me merrily dispos'd in a Morris, haue so bepainted mee in print since my gambols began from London to Norwich, that (hauing but an ill face before) I shall appeare to the world without a face if your fayre hand wipe not away their foule colours." Their "pittifull papers" have been "pasted on euery poast" (1884: 3, 34). Kemp's language here is graphically visual; his revenge is to be graphically aural. *Kemp's Nine Days' Wonder* is constantly pressing the limits of textualization, even as it tries to exploit textualization for Kemp's own ends. Kemp's narrating "I" remains remarkably colloquial from start to finish, replicating, Wiles claims, some the verbal maneuvers that made Kemp famous onstage (Wiles 1994: 29–30). Through it all Kemp reveals a keen ear for sound. His companion's pipe, he reports, "might well be heard in a still morning or euening a myle" (1884: 1). Conversations with people along the way get reported verbatim. Two inserted ballads, written by a friend, perform the antics of a garrulous host and a fat Maid Marian who danced with Kemp from Sudbury to Melford. The Sudbury lass "had a good eare," Kemp says, and "daunst truely"—truth in dancing being measured by the ear as truth in building is measured by the eye (1884: 18). On at least one occasion the narrative turns into song. Out of Chelmsford, Kemp runs into February mud:

> my Taberer struck up, and lightly I tript forward; but I had the
> heauiest way that euer mad Morrice-dancer trod; yet,
>> With hey and ho, through thicke and thin,
>> The hobby horse quite forgotten,

> I follow'd, as I did begin,
>> Although the way were rotten.
>>> (1884: 15)

In such moments the printed narrative threatens to run off into song and ambient sound—and from song and ambient sound into nonsense. At Stratford Langton, Kemp's well-wishers know about his love for bear-baiting and try to stage one for him, "but so unreasonable were the multitudes of people, that I could only heare the Beare roare and the dogges howle; therefore forward I went with my hey-de-gaies to Ilford" (1884: 9). "Hey-de-gaies" constitute Kemp's true subject as a dancer of jigs.

UNSPEAKABLE PLEASURES

"Loue, loue, nothing but loue, still love still more," sings Pandarus to Helen and Paris in *Troilus and Cressida*.

> These louers cry, oh ho they dye.
> Yet that which seemes the wound to kill,
> Doth turne oh ho, to ha ha he,
> So dying loues liues still,
> O ho a while, but ha ha ha
> O ho grones out for ha ha ha—hey ho.
>> (Q1609: 3.1.111, 117–122)

The history of [o:] is ultimately about the erotics of [o:]. For St. George fighting the Turkish Knight, no less than for Helen of Troy disarming a series of chivalric lovers, to die in [o:] is to rise again in [ha] [ha] [ha]. Laughter, uncontrollable spasms of the diaphragm, is the bodily experience gests of all kinds rush to excite. *Anima*ted bodies are, literally, bodies full of breath. Letting out the air—or, rather, allowing the air to force its way out willy nilly—gives the whole body a good shaking out. It squeezes the gut, convulses the lungs, loosens the arms and legs, vibrates the throat, airs out the mouth, stretches wide the lips, shoves back the head, partly closes the eyelids, and charges the outer air with rhythmic bursts from deep inside. In its most vehement form, laughter is something heard and felt, not something seen: constriction of the muscles beneath the eyes is an involuntary effect of smiling, and smiling is an involuntary effect of laughing. Try it and see—or rather, try it and *not* see.

In the course of the sixteenth and seventeenth centuries, as Keith Thomas observes, bodily control became an ever more visible index of social hierarchy. Etiquette books like Della Casa's *Galateo* are quite explicit on the subject. Along with spitting, snorting, and breaking wind, loud and open laughter became a sign of inferior social status.

"The eccentric Sir Thomas Browne," as Keith Thomas notes, "could still commend laughter for its 'sweet contraction of the muscles of the face, and . . . pleasant agitation of the vocal organs.' But the orthodox view was that the dignity of the great was seriously impaired by such 'undecent cachinnations,' such 'ridiculous wrinkles in the face.'" By the eighteenth century, Addison and Steele confidently contrast fools giving vent to "the most honest, natural open laugh in the world" with men of wit allowing themselves "but a faint constrained kind of half-laugh" (Thomas 1977: 77–81). Laughter is a matter not only of physiology but of politics. Because it involves the entire body, laughter is a synaesthetic experience: it happens when multiple senses are stimulated, when multiple orifices are pleasured. Dancing, as Phillip Stubbes knew all too well, is just such an experience. Praising God through dancing may be mentioned in the Old Testament, Stubbes concedes, but what he sees about him in the festivities of early modern England hardly qualifies as a form of worship. Every sense is allured: seeing ("leapinges, skippings, & other unchast gestures"), hearing ("pyping, fluting, dromming, and such like inticements to wantonnesse & sin"), touching ("what clipping, what culling, what kissing and bussing"), tasting ("what smouching & slabbering one of another"), implicitly even smelling ("what filthie groping and vncleane handling") (1877–1879: xi, 155).

Stubbes's worst fears are borne out by witnesses to the street dancing at Wells in 1607. The May lord, George Greenstreet, and his lady, Mistress Thomasine White, danced at the head of a retinue that numbered as many as a hundred. Accompanied by two fiddlers, a taborer, and perhaps two drummers, the crowd danced through the streets, according to one witness, "with muche hooping & halloweing." When they reached the door of Christopher Croker, clothier, the May lady's husband "did then will the reste of the Companye to hoope sayeing hoope, roagues hallowe roagues, hee him self then vehemenlye hoopinge hallowinge and leapinge in his dauncinge." The party's destination, Croker reports, was the sign of the George, "where (as this deponent hath creadiblye heard) they did euerye of them drinck a pynte or quart of wyne" (Stokes 1996, 1: 347). Among the synaesthetic delights of gests, drinking occupies a privileged place: the whooping, hollering drinker takes in pleasure through the mouth while letting it out through the same orifice. We have been taught to conceive the grotesque body in visual terms: *Enter Ursula, dripping.* Gests invite us to conceive it also in aural terms: *Enter Moll Cutpurse, roaring.*

The danced [o:] traces out a circle of suspicion. If the basic form of dance is a circle, what lies within? The physical and temporal space defined by moving bodies and percussive sounds is a liminal space, in which the usual rules of language, like the usual rules of social conduct, are relaxed or inverted. It is a space inhabited by clowns, fools, and transvestites who disrupt the logic of language as they disrupt the protocols of social difference. In Luhmann's terms, gests open up particularly wide, particularly permeable "zones of consensuality" between body and media, between body and society, between body and psyche. The most threatening of these border crossings, to the likes of Stubbes, may be the last. Serres has drawn attention to the way dance as a centrifugal force obliterates personal identity and turns the dancer into a cipher in a larger whole:

> The more I dance, the more I am naked, absent, a calculation and a number. Dance is to the body proper what exercise of thought is to the subject known as I. The more I dance, the less I am me. . . . The dancer, like the thinker, is an arrow pointing elsewhere. He shows something else, he makes it exist, he makes an absent world descend into presence. He must thus himself be absent. The body of the dancer is the body of the possible, blank, naked, nonexistent. This location is polysemy come down into the limbs. (1995: 39–40)

Around the edges of the danced circle lurk the adulterous, the demonic, the mad, the sodomitical. Anxieties about those specters are registered in stage representations of morris dancing. Timothy Tweedle, the piper for the morris in *Jack Drum's Entertainment* (Paul's Boys, 1600), equates playing with his pipe and playing with his yard. "The wenches, ha," he reminisces to Jack Drum, the taborer for the morris, "when I was a yong man and could tickle the Minikin, and made them crie thankes sweete *Timothy*, I had the best stroke, the sweetest touch." The song that accompanies the dancing leaves no doubt about the connection between coupling and copulating: "Skip it, & trip it, nimbly, nimbly, tickle it, tickle it lustily, / Strike up the Taber, for the wenches favor, tickle it, tickle it, lustily" (Marston 1939, 3: 181–183). In *The Witch of Edmonton* it is the Devil himself who fiddles for the dancers in the form of a dog, Mother Sawyer's familiar. The morris dancers in *The Two Noble Kinsmen*, finding themselves at the last minute short a woman, press the Jailer's mad daughter into their train. In her deranged state she is eager to join in. "[R]aise me a devill now," she says to the Schoolmaster who is organizing the entertainment, "and let him play / *Quipassa*, o'th bels and bones" (Q1634: 3.5.86–87). When the instruments are ready to strike up, the mad daughter leaps to take the lead. Flirtations with sodomy, as we have seen, are rife in the Fool's wooing rite and in "Robin Hood and the Friar." Hence, perhaps, Stubbes's insistence on the "filthiness" of dancing—a connection with defecation that the early modern imagination associates also with the flesh, with the rabble, with ballads, and quite specifically with sodomy (Smith 1996: 103–105).

To Puritan bystanders, the dancing body offered itself as a particularly threatening embodiment of the grotesque body of *carne*val. It lacks the discipline of spirit over flesh; it threatens insubordination. Even Stubbes will countenance dancing under certain circumstances: "being vsed in a mans priuat-chamber, or howse, for his Godly solace and recreation in the feare of GOD; or otherwise abroad, with respect had to the time, place and persons, it is in no respect to be disalowed" (1877–1879: xi). Dancing, that is to say, should happen preferably indoors, in a private domestic space presided over by a male authority figure, among people of one social sort, at carefully circumscribed times. As eruptions of high spirits among lowly folk, gests are dances of another ilk entirely. Impelled by the taborer's tapping and the piper's tweedling, such dances might go on forever—or at least for a thousand years, like the gest noted in a commonplace book begun about 1600, now in the Folger Library: "In hereford was a morrice dance taken upon ye Welsh side y^t made above a thousand years, ye one supplying w^t ye other wanted of a hundred: & one philip squire ye tabourer, & bess quin ye maid-marian made above 100 years apeice" (Folger MS V.a.381: 29). If such a thing were possible, it is because the dance took place on the far side of the border, in the realm of irrepressible orality known as Wales.

What Stubbes and his sort want to do in political terms—what they in fact do in their printed pamphlets—is to take a four-dimensional bodily sensation and turn it into a two-dimensional text. Something heard, felt, *en/joyed* becomes, in their hands, something seen, known, *mastered*. To turn gests into objects of surveillance is, in effect, to turn them into texts. What gests mean is presumed to exist on the outside, as a message that an onlooker can read. But gests take place in four dimensions, not two: they sprawl in the three dimensions of physical space, they pulsate in the fourth dimension of time. They possess *depth:* in space, in time, above all in the interiors of the sounding bodies that *do* them. As such, gests are not finally readable. To assume that they are is to confuse two distinct fields of experience. The perceptual field you occupy in reading this book is visual: you hold the book at a certain distance, you maintain a linear separation between yourself as reader and the subject at hand. The perceptual field occupied by participants in early modern gests is, by contrast, aural: sound surrounds the participants, wells up within them, fills the air around them, making it impossible for them to keep their distance (Ihde 1976: 74–76; Ihde 1993: 73–87). The dynamics of meaning-making in the visual and the aural fields are altogether different. Vision marks the difference between two equally present alternatives, between "this" and "that." Hearing knows only where the hearer happens to be: it can distinguish between "within" and "without," but it knows "without" only by inference. "Now am I, frere, within, and thou, Robin, without," boasts Friar Tuck in the combat-game appended to Copland's *A Mery Geste of Robyn Hood.* When Robin joins the fray, both of the fighters are very much "within."

The increasingly successful attempts to regulate and suppress gests in early modern England entailed some form of inscription: printed proclamations were circulated to constables like Hole, pamphlets like Stubbes's were published and sold, printed scripture was searched, sermons were written out for delivery, notes were taken by pious listeners. The politics of mirth is, in part, the conflict between two different ways of locating meaning: in *sounding* bodies and in *inscribed* bodies. As Stubbes realized, the latter are much more susceptible to political control than the former. If homoerotic byplay thrives in gests and escapes the notoriety of sodomy, it may be because no name is needed for what is going on, because nothing is being written down.* What Puritans fear most is just what gests are, in more ways than one, *about:* "the unspeakable."

> For o for o
> ba ba ba
> ri forlaurel laddy ri forlaurel lay
> fa la la la la la la la la
> fa la la la la fa la la la la la la
> O O ha ha ha
> O O ha ha ha

*On early modern strategies of denying sodomy see Alan Bray (1982: 58–80). For the suggestion that dance and sodomy are allied in being fugitives from print surveillance I am grateful to Michael Blackie.

Ballads Within, Around, Among, Of, Upon, Against, Within

Where do ballads belong?

Most obviously, perhaps, on a sheet of paper, printed on just one side. Calculating from entries in the Stationers' registers and from the number of surviving titles that were never entered, Hyder Rollins projects that upwards of 4,000 broadside ballads were printed before 1600; in the registers they outnumber books and plays from 1557 to 1642 (Rollins 1924: 1; Rollins 1919: 281–282, 296; Livingston 1991: 32). Perhaps also on a post, on a door, on a wall. As the most conspicuous place for putting up notices of all kinds, St. Paul's Cathedral offered two sites for self-publishing balladeers: the church doors and the churchyard, where cleft sticks were set up bearing both handwritten pasquinades and advertisements for printed wares on sale nearby (Rollins 1919: 324–326). Within doors, broadside ballads, replete with woodcuts, might do the work of a painted cloth, a painted panel, or a bit of painted wall, as Jonson's Old Cokes attests when he sees ballads for sale at Bartholomew Fair: "O Sister, doe you remember the ballads ouer the Nursery-chimney at home o' my owne pasting up, there be braue pictures" (Rollins 1919: 336–338; Jonson 1925–1963, 6: 74).

If not on a post, a door, or a wall, then in the street. Natascha Würzbach has used speech-act theory to ground the singing sellers of printed ballads in the streets of London. "Good people all to me draw neer, / and to my Song a while attend," "I Pray good People all draw near, / and mark these lines that here are pen'd," "Give eare, my loving countrey-men, / that still desire newes": in the act of hawking a ballad the seller would perform the text in such a way that the absent, figurative "I" of the story takes shape as the aggressively self-presenting seller, the absent, figurative "you" as the listeners (1990: 39–104).

From the street to the stage was only a matter of steps—about fifty, in fact, if the Globe was 99 feet in diameter. On the stage, ballads might be not only commented upon and quoted, but performed and *meta*performed. Shakespeare's Cleopatra is only the most famous of the characters who voice contempt for "scald Rimers" who will "Ballad vs out a Tune"—in the same breath that she scorns the "quicke Comedians" who "extemporally will stage vs" (F1623: 5.2.211–213).

Such ritual snubs to a commercially rival medium do not, however, prevent characters from seizing on popular ballads in moments that require lyric intensity, as Desdemona does with "Willow, willow" just before her death or Benedick with "The God of loue that sits aboue" in the throes of his love for Beatrice (*Othello* F1623: 4.3.38–55; *Much Ado* Q1600: 5.2.25–28). Performance of ballads onstage did not stop with quotation, however. The jigs that concluded performances in the South Bank theaters, even of tragedies, are no more than narrative ballads that happen to have been danced as well as sung. *Meta*performances of ballads onstage—that is to say, performances of performances of ballads—occur most famously in *The Winter's Tale* and *Bartholomew Fair*, where Autolychus and Nightingale ply their wares to rural and to urban customers who are alike in their eager gullibility.

In manuscripts and books ballads might also find a place. Elderton's "The gods of love" survives, in fact, only in miscellany of the early seventeenth century, where it lodges with lute music, twenty-eight other poems, cookery recipes, and home remedies (Osborn 1958: 11). In print it is worth noting that two ballads from oral tradition—"A Lytell Geste of Robyn Hode" and "Adambel, Clym of the cloughe, and Wyllyam of cloudesle" (Child 1957: 117, 116)—as well as John Skelton's "A balade of the scottysshe Kynge" are among the earliest chapbooks published in England (Livingston 1991: 36).

To some contemporaries the only place for ballads was on a dung hill. Sir William Cornwallis in his book of essays modeled after Montaigne tells about standing as a bemused onlooker before just the kind of street scene Würzbach describes. Delighting in the poet's "strained stuffe" and the crowd's exuberance, he notes that both parties are well paid, "they with a filthy noise, hee with a base reward." The occasion for Cornwallis's essay "Of the observation, and use of things" is an hour's conversation with an unlettered husbandman, a man of estimable wit, "if it were refined, and separated from the durt that hangs about it." The filthiness of the ballad-seller's singing and the dirt on the body of the husbandman are very much in Cornwallis's mind when he talks about his appetite for cheap print—and the use to which he puts the paper when he is done:

All kinde of bookes are profitable, except printed Bawdery; they abuse youth: but Pamphlets, and lying Stories, and News, and two penny Poets I would knowe them, but beware of beeing familiar with them: my custome is to read these, and presently to make use of them, for they lie in my privy, and when I come thither, and have occasion to imploy it, I read them, halfe a side at once is my ordinary, which when I have read, I use in that kind, that waste paper is most subject too, but to a cleanlier profit. (1600: I6–I7ᵛ)

Cornwallis's habits help to explain, perhaps, why so few copies of broadside ballads survive, particularly from the sixteenth century. Of the 4,000 ballads likely published before 1600, only about 260 sheets survive, including fragments and duplicates. The filth and dirt that even so genial an observer as Cornwallis attaches to ballads in 1600 belongs not only to the pleasured lower extremities of his body but to the pleasured lower extremities of the social order. A mode of communication that had attracted poets like Lydgate toward the end of the fifteenth century had, by 1600, become identified as proletarian entertainments (Burke 1978: 58–64, 270–281; Marcus 1995: 106–139; Dorson 1968: 1–43).

In the first instance at least, ballads belong not on a broadside, not on a post, not in the street, not on the stage, not in a manuscript, not in a book, not on a dung hill—but in lungs, larynx, and mouth, in arms and hands, legs and feet. In its origins, a ballad has as much to do with moving one's body as it does with making sounds. Via Middle English *balade*, a "dancing-song," the term *ballad* ultimately derives from late Latin *ballare*, "to dance." Sixteenth- and seventeenth-century pronunciation of *ballad* as *ballet* (compare *salad* as *sallet*) aurally makes a connection that visually and somatically was made in jigs and in broadside ballads that invite listeners to jump in and dance. To begin with lungs, larynx, mouth, arms, and feet, we must begin not with the ballad but with the balladeer. One of the great scenes in Deloney's *Jack of Newbury* (registered 1596–97) puts early modern ballads in place. Perhaps the most extravagant fantasy in the book is the power of Jack's wealth as a capitalist clothier to attract the notice of King Henry VIII and his court, who come for a visit. After dining his royal guests as lavishly as at court, Jack gives his visitors a factory tour. First they look in on a room fitted out with a hundred looms, each worked by two men, who "pleasantly" sing a song in praise of weavers, beginning,

> When *Hercules* did vse to spin,
>   and *Pallas* wrought vpon the Loome,
> Our Trade to flourish did begin,
>   while Conscience went not selling Broome.
>   Then loue and friendship did agree,
>   To keepe the band of amitie.
>   (1619: F1ᵛ)

When King Henry and his company come to the spinsters and carders, they discover a roomful of women, "who for the most part were very faire and comly creatures, and were all attired alike from top to toe." After occasioning some lewd jokes from the clown Will Sommers and offering the king "due reuerence," the maidens "in dulcet

Musical quotation 7.1

It   was   a   Knight   in   *Scot - land*   borne,

Fol - low   my   loue,   leape o'er   the   strand   Was   ta - ken   pris - 'ner   and

left   for - lorne,   euen   by   the   good Earle   of   *North - um - ber - land.*

(Tune adapted from Bronson 1976: 32, text from Deloney 1619: F3)

manner chaunted out this Song, two of them singing the Ditty, and all the rest bearing the burden." The song turns out to be just the sort of narrative ballad Deloney himself was famous for contributing to the broadside trade. "Instant history" (for example, Queen Elizabeth's review of the troops at Tilbury) was Deloney's specialty, but the ballad he supplies on this occasion is the earliest recorded instance of a ballad of border braveries that continued in oral tradition as Child ballad number 9, "The Fair Flower of Northumberland"* (musical quotation 7.1).

First the two lead singers supply the ditty: "It was a Knight in *Scotland* borne. . . ." Then all the women join in the burden: "follow my loue, leape ouer the strand. . . ." Again comes the ditty: "Was taken prisoner and left forlorne. . . ." Then again the answering burden: "euen by the good Earle of *Northumberland.*" And so on for thirty-four more stanzas, with "follow my loue, leap ouer the strand" as the invariable first part of the burden and variations on "Northumberland" as the second: "euen by the good Earle of *Northumberland,*" "and she the faire flower of *Northumberland,*" "to wend from her father to faire *Scotland,*" "and I the faire flower of *Northumberland.*" Body rhythms in this case are provided not by dance steps but by hand work. The story that the singers tell is, in Deloney's view at least, as appropriate to women as the praise of the clothing trade is to men. "*The Maidens Song,*" as Deloney titles it, recounts the passion of the

---

*The earliest surviving tune for "The Fair Flower of Northumberland" was not recorded until the 1880s, but it fits exactly with the rhythms and emotional contours of Deloney's text.

Earl of Northumberland's daughter, who chances to pass by the prison where her father has immured a captured knight from across the forbidden border that haunts so many ballads from oral tradition. The Scottish knight spies her, pleads his love, promises to marry her. "The faire flower of *Northumberland*" succumbs, steals her father's ring, uses it to secure the knight's release, and rides off with him over the border. At the gates of Edinburgh, he confesses that he already has a wife and five children. "Now chuse (quoth he) thou wanton flower, / . . . / Whether thou wilt be my paramour." Taking her horse away, he orders her to walk.

> O false and faithlesse Knight quoth she
>   follow my loue, come ouer the strand;
> And canst thou deale so bad with me,
>   and I the faire flowre of *Northumberland?*

She is abandoned. At length, two gallant knights of fair England come riding by, hear her confess her offense against her father, and take her home. Be warned, fair maids: "Scots were neuer true, nor neuer will be" (Deloney 1619: F3–G1ᵛ).

Fantasy though it be, Deloney's story casts into high relief several important details. First of all, the singers, like most of the informants from whom ballads were recorded in the nineteenth and early twentieth centuries, are women. The ballad they sing has been appropriated by Deloney from oral tradition, but despite its appearance in print it continued to be passed along in oral tradition down to the nineteenth century and beyond. Our only access to that oral tradition, in the form of Deloney's novel, comes via print. Furthermore, Deloney implies five distinct contexts for situating the individual singer: (1) in her own fair and comely person, (2) in the workroom as an acoustic field, (3) in the social circumstances that bring her together with other people as a woman and as a hand laborer, (4) in her political subjection under the surveillance of king, factory owner, and author, and (5) in the psychological space she occupies as a result of all the other factors.

In Deloney's scenario, a ballad is more than a genre, a certain verse form and certain set of subjects; it is more than a medium, an oral art that operates both inside writing and without. It is, rather, a complete system of communication, involving certain people in certain kinds of situations communicating certain kinds of experiences in certain kinds of ways. Partly as a result of the commerce of print, partly as a result of the rift between "the great tradition" and "the little tradition," ballads became ever more conspicuously autonomous in the course of the sixteenth century. In Niklas Luhmann's terms, early modern ballads form an "autopoietic" system of communication, self-referential and self-reproducing (1990: 1–20). As such, it touches other systems—the human body, the built environment, the economic system, the political system—in what Luhmann calls "zones of consensuality." Each of these other systems constitutes a "horizon" for ballads. (In Chapter 1 appears a diagram of how systems are related to one another in Luhmann's model.) The capacities of women's bodies for singing, the physical organization of Jack of Newbury's factory, the shared social experience of "spinsters and carders," the apparatus of state control over what people may say and do—all of these factors impinge on ballad-singing. But ballad-singing maintains a logic of its own. Within these horizons, ballads constitute, in Luhmann's terms, a highly "resonant" medium: that is to say, they interact in highly volatile ways with the physical body, with soundscapes, with speech communities, with political authority, with the singer's sense of self. Although in print we may never be able to specify just where ballads belong, we can approach that site via these intersecting horizons.

THE PHYSIOLOGICAL HORIZON

Sir William Cornwallis, seated in his privy, was hardly alone in associating ballads with the material body, even if other sites on the body occupied most other people's imaginations. With William Elderton, author of "The gods above" and at least sixteen other extant ballads, it was the nose (Rollins 1920: 213–214, 218–220). His was red. Elderton's lost ballad "Eldertons Jestes with his mery Toyes" (registered 1561–62) seems to have provoked an attack from a London hosier named Leach. William Fulwood joined the fray with "A Supplication to Eldertonne, for Leaches vnlewdnes: Desiring him to pardone, his manifest vnrudeness" (1562: BB 78), ostensibly comforting Elderton against so unworthy an attacker but in fact insulting Elderton—especially his nose:

> And him me thinkes you should not blame,
>> that can wel shape a hose:
> For he may likewise cut and frame,
>> a case for your riche nose.
>
>> (Fulwood 1562)*

The Martin Marprelate tracts *Pappe with an hatchet* (1589), usually attributed to John Lyly, invokes Elderton as the archetypal ballad-writer, one whose "ballets come out of the lungs of the licour" (Rol-

---

*Broadside ballads are cited in the text by the catalogue number (BB#) assigned to them in Carole Rose Livingston, *British Broadside Ballads of the Sixteenth Century* (1991).

lins 1920: 231, 234). It is in an alehouse that Richard Brathwait's characteristic ballad-monger concocts his wares, "extracted from the muddie spirit of Bottle-Ale and froth" (1631: B4).

Censure from the likes of Gabriel Harvey was more intellectual than moral, but specifically religious attacks on balladry throughout the sixteenth and seventeenth centuries persistently brand it as lecherous. Elderton's "The god of love that sits above," for example, provoked not only an attack but a pious parody. The attack comes from Thomas Brice in a ballad "Against filthy writing / and such like delighting" (BB 74, 1561–62). Brice's persona is no less assertive ("What meane the rimes that run thus large in euery shop to sell? / With wanton sound, and filthie sense, me thinke it grees not well") than savvy about ballad-lovers' tolerance for religious instruction ("come bak, where wil ye go?") (Brice 1561–62). Brice wonders whether "the god of love that sits above" is Christ or Cupid; William Birch provides his own answer in "The complaint of a sinner, vexed with paine, / Desyring the ioye, that euer shall remayne. After W. E. moralized" (BB 81, 1563), which begins, "The God of loue, that sits aboue, / Doth know vs, Doth know vs" (Birch 1563). "Wanton sound" and "filthie sense" are more than clergymen's fantasies in ballads like "The Carman's Whistle" and "Watkin's Ale," which turn the male member into something seen, heard—and danced. In "A Ditty delightfull of mother watkins ale / A Warning wel wayed, though counted a tale" (BB 237, c. 1590) a lad happens to overhear a maiden's lament that she doesn't want to die a virgin. He promises to give her some "Watkin's ale." Once she has tasted it, she can't get enough. From, of all places, *The Fitzwilliam Virginal Book* comes the tune that sets human feet to the dance of metrical feet in the ballad's penultimate stanza (musical quotation 7.2). When the balladeer sings about a *"cuntry* dance," he has more than geography on his mind.

The physicality of Cornwallis on his jakes, Elderton in his cups, and the carman at his whistle also characterizes Deloney's maidens at their spinning wheels. The rhythm of their singing, in steady four-beat lines, is synchronized with the rhythms of their hands as they comb out the tangles in the wool and set the strands to the turning wheels. Turns in their work—from the top of a tuft to the bottom, from one tuft to another, from tuft to wheel, from wheel to thread—are heard as turns in their song, as two singers, line by line by line, hand over "the ditty" and all the others hand back "the burden." In this exchange it is the ditty in lines 1 and 3 that advances the story; "the burden" in lines 2 and 4 meditates upon it, almost like the chorus in a Greek tragedy. More typically in orally transmitted ballads, the burden would sound three beats to the ditty's four, but Deloney has preserved the narrative rhythm of advance/pause/advance/

Musical quotation 7.2

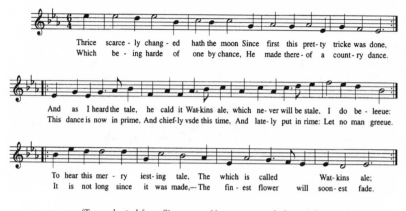

Thrice scarce - ly chang - ed   hath the moon Since first this pret- ty tricke was done,
Which   be - ing harde of   one by chance, He made there- of a   count- ry dance.

And   as I heard the tale, he cald it Wat-kins ale, which ne- ver will be stale. I do be - leeue:
This dance is now in prime, And chief-ly vsde this time, And late- ly put in rime: Let no man greeue.

To hear this mer - ry   iest- ing tale, The which is   called   Wat- kins   ale;
It is not long since it was made,— The   fin - est flower will soon- est fade.

(Tune adapted from Simpson 1966: no. 494, words from *A Ditty Delightful* 1590)

pause that distinguishes oral performances from written documents.
Moving the story forward by choosing from a stock of tried-and-true
motifs and phrases, a singer needs time to choose what will come
next (Jones 1961: 97–112). (The fact that *two* singers are able to do
this simultaneously may be a mark of Deloney's literacy in imagining
the scene.)

In "The Flower of Northumberland," as in most oral ballads, the
burden in lines 2 and 4 is more than filler. The exactly repeated burden in line 2 (*"follow my loue, leap ouer the strand," "follow my loue, come
ouer the strand"*) returns obsessively to the episode in which the
knight makes the maid follow him across a stream that is *"rough and
wonderfull deep."* The varied repetitions of the final line—"Northumberland" and "Scotland," *"good Earle"* and *"faire flower"*—evoke
a mental and ethical geography for the events of the ballad, maintained through all the variations. Altogether, the assonances of the
burden—"love"/"come"/"Humber," "follow"/"flower," "follow"/
"o'er," "follow"/"north," "Scot"/"o'er," "strand"/"land"—create a
satisfyingly regular sequence of muscle movements in the mouth—
[f], [a], [u], [r], [e], [f], [a], [r], [u], [a]—that is heard as an open sound
field bounded by [r]. The fact that these incantations are self-
produced helps each singer to internalize the events of the story. Victor Zuckerkandl's explanation of meaning in vocal music, quoted by
Mark W. Booth in *The Experience of Song*, seems particularly apt for
ballads: "The singer who uses words wants more than just to be with
the group: he also wants to be with things, those things to which the
words of the poem refer. . . . [Tones] remove the barrier between person and thing, and clear the way for what might be called the singer's

inner participation in that of which he sings" (1981: 19). Chanted again and again, the phrase "*And I the faire flower of* Northumberland" turns the subject position into each singer's own. The fictional "I" of the story becomes the singing "I" of the performance. That happens, not through a cerebral act of imagination, but through the pressure of air in the lungs, certain constrictions of the throat, certain movements of the tongue, the opening and closing of the lips.

Wherever Deloney got the ballad, from oral singers or out of his own head, "The Fair Flower of Northumberland" bears many of the structural marks of body-based remembering that David Buchan has traced in border ballads collected from oral tradition (1972: 87–144). As written out by Deloney, the ballad follows the traces of a spatially central design, in which the narrative moves by stages toward a central event—the knight's escape from prison in stanza 17—and then out again. Along the way, episodes in the first half of the journey have their counterparts in the second half: the knight's first plea for pity six stanzas from the beginning is answered by his abandonment of the maid six stanzas from the end. The maid's protestations about her honor seven stanzas from the beginning are echoed in her plea that he "Dishonour not a Ladies name" seven stanzas from the end. The knight's assurance "I haue no wife nor children I" ten stanzas from the start is reversed ten stanzas from the end when he confesses, "For I haue wife and children fiue."

In Child's compilation of texts, the next written notice of "The Fair Flower of Northumberland" after Deloney's version is a manuscript of about 1826, taken down from recitation of a Scottish informant, Miss E. Beattie (Child 1957, 1: 115, 5: 398). The way from *Jack of Newbury* to Miss E. Beattie is literally uncharted territory. Nineteenth-century versions of the ballad, none of them nearly as long as Deloney's text, variously transform the maid into a provost's daughter or a bailiff's and domesticate the story by telling of her homecoming and by giving her one censorious and one forgiving parent. How did these nineteenth-century singers know the ballad? Probably not from Deloney's text. Was there an oral tradition from which Deloney transcribed his version—an oral tradition that continued independent of print down to the nineteenth century? Or was Deloney's printed ballad the inspiration for a subsequently oral tradition? Both possibilities are known from other ballads. The Stationers' register contains six entries before 1600 for now-lost broadsides of ballads that seem to have come from oral tradition: "The Jolly Pinder of Wakefield" (Child no. 124, registered 1557–58), "Dives and Lazarus" (Child no. 56, 1557–58, 1570–71), "Tam Lin" (Child no. 39, 1558), "The Lochmaben Harper" (Child no. 192, 1564–65, 1565–66), "King Edward IV and a Tanner of Tamworth" (Child no. 273, 1564, 1586, 1600, 1615, 1624), and "The Lord of Lorne and the False Steward" (Child no. 271, 1580, 1624). A version of "The Tanner of Tamworth" survives as a chapbook (1596, 1613), and both "The Tanner of Tamworth" and "The Lord of Lorne" survive in late seventeenth-century broadsides. On the other hand, at least one orally transmitted ballad, "The Daemon Lover" (Child 1957: 243) is first recorded as a late seventeenth-century broadside, "A Warning for Married Women, being an example of Mrs. Jane Reynolds (a West-country woman) born near Plymouth, who, having plighted her troth to a Seaman, was afterwards married to a Carpenter, and at last carried away by a Spirit, the manner how shall presently be recited" (Pepys 1987, 4: 101; Crawford 1890: 1114; Holloway 1971: 377 and 378).

The printed broadsides hawked around the countryside by a ballad-monger might grow so common "as every poore Milk maid can chant and chirpe it under her Cow; which she useth as an harmelesse charme to make her let downe her milke" (Brathwait 1631: B4ᵛ). Wye Saltonstall says the same of a peddler at a country fair: "If his Ballet bee of love, the countrey wenches buy it, to get by heart at home, and after sing it over the milkepayles" (1951: no. 21). What happens between page and pail is impossible to predict. Carole Livingston has convincingly demonstrated that one of the ballads collected in Thomas Percy's folio manuscript, "Bishoppe and Browne" (1756, 2: 265), is an oral reconstruction of William Elderton's broadside "A new Ballad, declaring the Treason conspired aginst the young King of Scots, and how one Andrew Browne an Englishman, which was the Kings Chamberlaine, preuented the same" (BB 200, 1581) (Livingston 1991: 880–885). The movement of ballads into and out of print in the sixteenth and seventeenth centuries stands as a signal illustration of Michel de Certeau's point that orality and literacy, far from being polar opposites, exist only in terms of each other—and in terms that are constantly changing. One of the two, furthermore, must function as the dominant factor, as the standard by which the other is judged to be different (de Certeau 1984: 133). For sixteenth- and early seventeenth-century broadside ballads, the dominant factor is clearly orality.

Traces that the singing, remembering human body has left on printed ballads are not hard to find. The earliest surviving printed broadside (BB 1) is an untitled sheet of verses "spoken" in first person by the seven virtues, possibly intended to be cut apart as scrolls to be put above or beneath figures representing the seven Virtues on the wall of a room, as many of Lydgate's verses are known to have been (Bradshaw 1889: 100; Schreiber 1928: 51–54). "*Titulus* verse" is Alan Nelson's name for what amounts to a distinct genre (1982: 189–210). Furthermore, printed ballads and other broadside texts conventionally identify the author via "quod" or "quoth" at the end of the text, followed by the author's name or initials. In effect, the foregoing text becomes something the author "spoke," "said," "declared." Finally, the printed text contains all sorts of cues to the ballad as a somatic experience: tune designations (especially the likes of "To a pleasant new tune") require the purchaser to get the tune from hearing it, woodcuts provide bodily habitations for first-person voices, scrolls above speakers' heads furnish shorthand mnemonics of lengthy speeches.

Even as they stand in for the human body, printed ballads declare themselves to be artifacts in the literal sense of the word: something made by the writer's hand but not involving other members of the writer's body. Such touches are present even in the earliest broadside ballads. Sir Thomas Smyth's contribution to a flurry of ballads on the fall of Thomas Cromwell in 1540 sings the language of carols and rounds but features the pen as a weapon. To William Gray, one of Cromwell's ballading attackers, Smyth offers this offense:

¶If with the poynte of my penne, I do you so spurre and prycke
That therby you be greved and greatly styred to yre
Yet doubte I not syt sure / all though you wynche and kycke
Fast closed in my dewty / to save me from the myre.

                  (Smyth 1540b)

In printed broadsides it is not uncommon to find the author "saying" and "writing," both in the same line. Set down in writing but pointed toward performance, the ballad often becomes a "bill" or a "letter." The execution of John Felton, who affixed to the Bishop of London's palace a copy of the papal bull excommunicating Queen Elizabeth, prompted Steven Peele to write a pair of ballads, both of them cast as letters. In the first, "A letter to Rome, to declare to ye Pope, Iohn Felton his freend is hangd in a rope" (BB 146, 1570), the persona presents himself as a messenger, beating on the gates of the Vatican. The tune is "Row well ye Mariners" (musical quotation 7.3). The reply— "The pope in his fury doth answer returne, To a letter ye which to Rome is late come ... " (BB 147, 1570)—likewise speaks two languages at once: the language of a letter and the language of dramatic presence. "When I had pervsd your byl," the pope says, "It greued mee those lines to vew / Were written in your name" (Peele 1570b), even though the first ballad allows a singer to imagine himself bellowing the words of the "byl" in the pope's own ears.

The ambiguous status of a broadside ballad—written and yet sung, seen and yet heard—is registered most acutely, perhaps, in "A new balade entituled as foloweth, / To such as write in Metres, I write / Of small matters an exhortation, / By reading of which, men

Musical quotation 7.3

Who keepes Saint An - gell gates? Where li - eth our hol - ly fa - ther say?   Sir
I muze that no man waytes, Nor comes to meete me on the way.   Bowe

Pope I say? yf you be nere   Come forth, be - sturre you then a pace,
downe to me your list - ning eare: For I haue newes to show your grace,

Stay not, come on, That I from hence were short - ly gon:
Harke well, heare me, What ti - dings I haue brought to thee.

(Tune adapted from Simpson 1966: no. 401, words from Peele 1570a)

may delite / In such as be worthy commendation" by one "R. B." (BB 142), published the same year as Steven Peele's musical missives to the pope. (No tune is specified.) R. B. is to broadside ballads what Sir Philip Sidney is to poesie. He provides criteria for judging excellence, an enumeration of kinds, models for emulation, a politics of writing, and (overtopping Sidney in this respect) concern with the material medium. R. B.'s touchstone of judgment is nothing less than Horace's *Ars Poetica*—evidence enough that the "low culture" status of broadside verse had not quite been established by 1570. Horace, like persons of discernment today, was pestered by bad poets:

¶Read in his bookes, and then vnderstand,
They vexed his eares, they troubled his eyes
With Metres in number, compared to yᵉ sand:
And lacked not such, as wolde to the skyes
So prayse their workes (such was their guyse). . . .

After implicitly comparing ballads to spices—both hawked by peddlers, both sold to "Master" and "Clowne" alike, both "good to cast downe / When ye haue doen"—R. B. goes on to catalog three distinct genres: "balades of loue," "newes," and "open sclander" from Catholic exiles in Louvain. As models, ballad-writers should emulate Chaucer, Lydgate, Wager, Barclay, and Bale. Last of all, he admonishes printers to be careful in manufacturing broadsides as commodities for readers' consumption: "The Printer must vse good paper and inke. / Or els the reader may sometime shrinke / When faulte by inke or paper is seene." And finally an Althusserian gesture: "And thus euery day before we drinke / Let vs pray God to saue our Queene. Amen." Or is it a Dionysian gesture?

THE PHYSICAL HORIZON

"The Fair Flower of Northumberland" may begin within each singer's body, but Deloney's singing maidens project the ballad into the space round about them. They fill the workroom with their sound. As an enclosed space, likely built of sound-deflecting stone or reverberant plaster-over-lath, the workroom would be only lightly damped, making for a high degree of resonance between the sounding voices and the acoustic surround (Handel 1989: 47). The whirring of the spinning wheels and the scrape of combs through wool fibers—both low-amplitude, multifrequency sounds—would be apprehended as unlocalized "white noise." To the ear, the room would *be* the women's song. As consciously crafted sound, singing pushes at the boundary between the human body as an autopoietic system and the built environment as an autopoietic system. Through singing, the body projects itself into space and claims that space as its own.

Even self-consciously literate appropriations of ballads preserve this quality of self-assertion. After the Virtues' speeches and five more or less fragmentary sheets, the next surviving printed ballads are eight volleys in a verse combat carried out among William Gray, Thomas Smyth, R. Smyth, and "G, C" in 1540 (BB 6–13, Livingston 1991: 821–828). The arrest and impending execution of Thomas Cromwell, chief minister to Henry VIII, provided the occasion for the ballads, but hurling insults at each other occupies most of the poets' attention—an example of "agonistically toned" oral discourse if there ever was one (Ong 1982: 31–77). Figuratively at least, the participants throw their bodies about as they do so. The first salvo survives only in a reprint by Thomas Percy in *Reliques of Ancient English Poetry:* "A newe ballade made of Thomas Crumwell called TROLLE ON AWAY" (Percy 1756, 2: 58). With the exception of the final gesture toward reconciliation (BB 13), described as a "Paumflet" by its author "G, C," all the succeeding assaults are likewise cast as "ballads," and most of them ask the reader to "troll." With respect to the human body, to troll is to roll, ramble, or walk about; with respect to the ancient Yuletide carol (or similar), to troll is to sing the parts of a round. "Synge trolle on awaye, synge trolle on away / Heve and how rombelowe trolle on away" goes the refrain to Sir Thomas Smyth's "A balade agaynst malycyous Sclanderers" (BB 6), a reply to "TROLLE ON AWAY." In the absence of specified tunes, it is unclear whether these ballads were sung as a whole, chanted as a whole, or sung in the refrains and chanted in the verses (Livingston 1991: 824). What *is* certain is Smyth's swagger:

¶Trolle into the way / trolle in and retrolle
Small charyte and lesse wytte is in thy nolle
Thus for to rayle upon a christen soule
Wherfore me thynke the worthy blame
Trolle into the way agayne for shame.

In the exchanges among Gray, the Smyths, and "G, C" we can observe a fusion of at least three distinct cultural practices. With its heaving and hoing, its singing and trolling, Smyth's "A balade agaynst malycyous Sclanderers" is a piece of performance art, a public mockery like the *ad hominem* attacks in Greek Old Comedy. In its written state it is also a proclamation, a satiric inscription in a public place, a pasquinade. It is, finally, a commodity that must have earned quite a few pennies for John Gough, Richard Grafton, Richard Banks, and other printers and booksellers involved in the whole business. The political implications of such goings-on were serious enough to land Gray, Thomas Smyth, Gough, and Grafton in the Fleet and, if Livingston is correct, to give the Tudor monarchy its first cue that broadside ballads were a medium to be carefully censored (Livingston 1991: 826–827).

Propelling the singer's body into space is only a figurative gesture in the "TROLLE ON AWAY" ballads, but a handful of ballads in manuscript and print invite the listener to leap into the circle and join the dance. As it happens, the occasion for most of these ballads is a wedding. "A newe Ballade intytuled, Good Fellowes must go learne to daunce" (BB 110, 1569) makes the invitation in jubilant monosyllables that never miss the beat of four and three. The tune, unspecified on the printed sheet, appears along with one stanza of the text in Bodleian MS Mus. e.1–5 (from which it is transcribed in musical quotation 7.4). The French "brawl," to hear Moth describe it in *Love's Labour's Lost*, was an especially flirtatious dance. Having sung a song for Don Armado *inamorato*, Moth is commanded to call a servant to take a letter to Jacquennetta. "Maister," he asks, "will you win your loue with a french braule?"

[ARMADO]
How meanest thou? brawling in French.
BOY
No my complet Maister, but to Iigge off a tune at the tongues
ende, canarie to it with your feete, humour it with turning vp your
eylids, sigh a note and sing a note. . . . (Q1598: 3.1.5–10)

Less jocular witnesses suggest a couples' dance, alternately in partners and in lines, that sometimes involved the trading of kisses all

Musical quotation 7.4

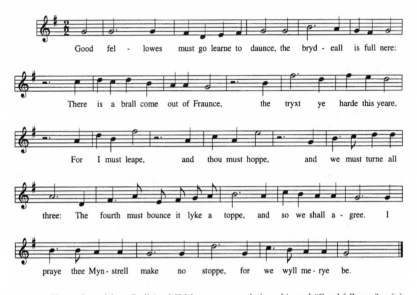

(Tune adapted from Bodleian MS Mus. e.1–5, words from [Anon.] "Good fellowes" 1569)

around (Baskervill 1929: 347–349). What the singer of "Good fellowes" is imagining is not only a brawl but a morris. Let us deck ourselves with a band of bells, he proclaims, and "A shurte after the Moryce guyse, / To flounce it in the wynde a." A staff-bearing whiffler will clear the way for bringing in May. The singer's final gesture is to summon one and all to join in: "Drawe to dauncinge, neyghboures all, / Good fellowshyppe is best a. . . ." A whole series of ballads on "Jocky and Jennie" also orchestrate fantasies of bodies trolling about with the ballad's sound. One of the cacophonous best, by John Wallis, appears in a manuscript likely assembled by the minstrel Richard Sheale in the early 1560s. After a narrative ballad about Jock and Jenny's courtship, banns, and wedding comes an invitation to the dance: "Now play us a horn pype, Jocky can say; / Then todle lowdle the pyper dyd playe." The dance itself, partly pentameter couplets and partly in ballad stanza, spins along as a swirl of shouts:

> "Torn rownde, Robyne! kepe trace, Wylkyne!
>   Mak churchye pege behynde;"
> "Set fut to fut a pas," quod Pylkyne;
>   "Abowt with howghe let us wynde."

> "To, Tybe, war, Tom well," sayd Cate;
>   "Kepe in, Sandar, holde owte, Syme.

Nowe, Gaff, hear gome abowt me mat;
　　Nyccoll, well dansyde and tryme."
　　　　　　　(Sheale 1860: 123–124)

The very apotheosis of balladeering is the jig: a ballad that thrusts
bodies, along with sound, into ambient space. It does so in leaps and
whirls (Baskervill 1929: 357–358). When Beatrice describes a typical
lover's wooing as "hot and hasty like a Scotch ijgge [sic] (and ful as
fantasticall)" (Q1600: 1.3.67–68), she could be looking ahead to the jig
that will end the performance of Much Ado, since the plot of most
surviving jigs is clever seduction, usually of someone else's wife. A
jig preserved among Edward Alleyn's papers at Dulwich College sets
one of the most famous jig characters, Rowland, on his usual quest,
a dance of sexual conquest. Nan's favors will go to whoever dances
best. This time, however, Rowland and three other wooers are foiled
by who else but the Fool of traditional gests. In "The Wooing of Nan"
singing, dancing, and sexual performance amount to the same thing.
Sings Nan to the disappointed Gentleman among her suitors, "& if
you be Ielous god give you god night / I feare you are a gelding you
caper so light." "Exnt" Nan, Fool, et al., leaving the jilted Gentleman
to sing the moral:

I thought she had but Iested & ment but to fable
but now I doe see she hath playd wth his bable
I wishe all my frends by me to take heed
that a foole com not neere you when you mene to speed.
　　　　　　　(Baskervill 1929: 432–436)

Ballads lay claim to space through lungs and larynx, through arms,
hands, legs, feet—and genitals.

THE SOCIAL HORIZON

"Newes newes newes newes / ye never herd so many newes,"
goes the burden.

A[s I sat] vpon a strawe
Cudlying of my cowe
Ther came to me Iake dawe
newes newes

our dame mylked the mares talle
The cate was lykyng the potte
our mayd cme out wyt aflayle

and layd her vnder fotte
newes newes

In ther came our next neyghbur
frome whens I can not tell
but there begane a hard stouer
aw youe any musterd to sell
newes newes

a cowe had stolyn a clafe away
and put her in a sake
forsoth I sel no puddynges to day
maysters what do youe lake
newes newes

Robyne ys gone to hu[n]tyngton
To bye our gose afayle
lyke spip my yongest son
was huntyng of a snalle
newes newes

our mayd Iohn was her to morowe
I wote not where she ferwend
our cate lyet syke and takyte grete sorow[....]
(Seng 1978: 6–7)

Dairyman, mistress, "next neighbor," peddler, Robin, son Spip, Joan
the maid, even the cat, first licking the pot, then lying sick and sor-
rowful: the circle convened in this ballad from Henry Savile's manuscript (British Library MS Cotton Vespasian A-25, 1570s) typifies the
social horizon that encircles all ballads as they are said, sung, and
danced, as they are gotten by heart, as they are passed along to others. Ballads may begin *within*, they may reverberate *around*, but they
have their social being *among*. Ballads help to confirm a speech community's identity. In the case of Deloney's spinsters and carders, that
act of confirmation is carried out collectively, with two singers calling
out the story and all the rest physically adding their consent. "The
Maidens Song," as Deloney entitles the text, affirms their *maiden*ness,
just as "The Weauers Song" affirms the men's membership in a vener-
able and honorable trade. "Then loue and friendship did agree, / To
keep the band amitie," they sing to round out each verse. "Good Fellowes must go learne to daunce" ends by drawing "neyghboures all"
into the circle of fellowship. A *good* fellow is one who will not refuse.
Even if the appeal is not so direct, ballads give voice to an implicit
"us."

For there to be an "us," there has to be a "them." In a way, then,
all ballads are *border* ballads. Often the semiotically necessary Other

is palpably present: a demon, the Sheriff of Nottingham, a Scottish knight, a seditious Catholic. "Open sclander" from Catholic exiles is one of the major genres of printed broadsides that R. B. specifies in "To such as write in Metres" (1570)—a distinction justified by Rollins's calculation that during the years 1569–1571 ballads on Catholic plots and on the Catholic-inspired Rising in the North constitute the majority of ballads entered in the Stationers' register (Rollins 1920: 208–211). Steven Peele's pugnacious "A letter to Rome, to declare to ye Pope, Iohn Felton his freend is hangd in a rope" (1570) provides a signal example. In taunting the pope with news of Felton's gruesome execution Peele speaks, or sings, for all true Englishmen. A subtler, more ambiguous sense of "them" is conveyed by the implicit geography of sixteenth- and early seventeenth-century ballads in English. "The Fair Flower of Northumberland" is far from being the only orally transmitted ballad that delights in recounting "the matter of the North." To the North belong not only the martial deeds of Otterburn (Child no. 161) and Chevy Chase (Child no. 162) but the freewheeling physicality of "Jocky and Jenny":

> In all the nouthe land, my Jocky,
>   As it pleanly doth apear,
> Was not syk anothar weddyne
>   This fyve and forty year.
>   (Sheale 1860: 124)

In "Good fellowes must go learne to Daunce" that place of libidinal license is closer at hand:

> O where shall all this dauncing bee,
>   In Kent or at Cotsolde a?
> Oure Lorde doth knowe, then axe not mee,—
>   And so my tale is tolde a.

Such dislocations are easy to understand for Deloney's merchant-class readers in London. Kentish villagers don't *need* a ballad bidding them to dance a morris. To what degree is the poet of "Good fellowes" putting words into Kentish dancers' mouths? To what degree is he staging a country spectacle for urban consumption, like the bride-ale staged before Elizabeth at Kenilworth in 1575? To what degree is he staging, for urban consumption, a country spectacle into which Kentish dancers might want to project themselves once they have heard—or even read—the new ballad? Such questions indicate the complications of trying to fix the social horizon when the medium of communication is print. One thing seems certain: even in orally transmitted ballads the way to adventure most often lies north by northeast, more rarely south by southwest. With respect to singers and listeners as a community wherever they may be—in London, in Kent, in the Cotswolds—the events of a ballad are *there* in fiction but *here* in sound and body.

Dislocations of the sort to be witnessed in "Good fellowes must go learne to Daunce" are made possible by print. Ballads with no place but lungs and larynx are less *portable* than ballads with a place on the printed page. From early in the sixteenth century, broadside ballads demonstrate their power to colonize oral culture. "TROLLE ON AWAY" and its progeny may bear marks of orality, but none of them is an oral ballad. Rather, Thomas Smyth and his cronies have appropriated a folk form in just the way Henry VIII and his courtiers did with the "greenwood" songs, poems, and plays that captured the court's fancy in the teens and early twenties (Stevens 1961: 177–187; Hutton 1994: 66–67). Two later typographical flytings are similar appropriations of country mores. Sixteen broadside ballads exchanged among Thomas Churchyard, Thomas Camel, and assorted others in 1552 started out as ridicule of Churchyard's ballad "Dauy Dycars Dreame," an innocuous enough confrontation of Piers Ploughman with mid-sixteenth-century mores, and ended up as a souvenir quarto of the whole affair in *The Contention bettwyxte Churchyeard and Camell, upon David Dycers Dreame . . .* (1560). It was for delectable insults like this that purchasers put down good money for broadside or book:

> Is this the order, that Camels do use?
> Bicause you are a beast, I must you excuse.
> A Camell, a Capon a Curre sure by kynde,
> I may you well call, synce so I you fynde.
> (Churchyard 1560: A3–A3ᵛ)

Even more obviously than in the "TROLLE ON AWAY" ballads, Churchyard and his shake-pens are appropriating—and mocking—oral practices. Davy Dicer can have such revelatory dreams in the first place because he is an unlettered husbandman. When Geoffrey Chappel joins Churchyard and Camel in the fray (if "Chappel" is not in fact a pseudonym for *Church*yard himself), a fiction is built up that country bumpkins are delivering to Camell and to Chappel "bills" on which the insults are written. First, Chappel sends to Camel "A supplicacion" in the person of "Harry Whobal." Then "Steven Steple" (another fixture of the *Church*yard?) joins the fray as a messenger from Chappel to Camell in "Steuen Steple to mast Camell." As the flyting goes on, the dialect gets thicker and thicker. "Steuen Steple" speaks a nonce language made up of Cotswold *ich*'s, West Country *z*'s for *s*'s and *v*'s for *f*'s, and Dutch *d*'s for *th*'s that has little to do with his supposedly native Kent. At Geoffrey Chappel's biding, Steven Steple returns the "bill" containing Camel's latest ballad:

> And her cha brought yor byl ayen, corrupt it iz ich go,
> Vor vende godes vorbodman I zedge, to let it go vorth zo:
> But well ich zee yor braine is dicke, your wits be curstly vext,
> Prey God ye be not zyde yor zyelf, er be to morow next:
> Deruore go couch and sleap a now, and dan com to yor parte,
> And dyte a wyser dyng dan dat, or all is not wort a vart.
>
> (Churchyard 1560: E2ᵛ)

In the galumphing fourteeners of Churchyard and his cronies, the medium seems indeed to be the message. The later skirmish touched off by Churchyard's "A Farewell cauld, Churcheyeards, rounde / From the Courte to the Cuntry grownd" (BB 99, 1566) likewise takes place within the imaginative precincts inhabited by simple country folk, though not in their dialect.

In the process of transforming everyday practices into commodities for consumption, broadsides must forever proclaim their newness. "A new ballad entitled . . . ," "To a pleasant new tune": titles like these insinuate the ballad's claims on a purchaser's attention. Along with newness comes, to Mospa's ears at least, a guarantee of truth. Enter Autolychus, with a false beard and a pack of true ballads:

> CLOWNE What hast heere? Ballads?
>
> MOPSA Pray now buy some: I loue a ballet in print, a life, for then we are sure they are true.
>
> AUTOLICUS Here's one, to a very dolefull tune, how a Vsurers wife was brought to bed of twenty money baggs at a burthen, and how she long'd to eate Adders heads, and Toads carbonado'd.
>
> MOPSA Is it true, thinke you?
>
> AUTOLICUS Very true, and but a moneth old. (WT 4.4.257–265 in F1623: 311)

A moneylender's wife who gives birth to twenty sacks of cash: here's news to put "The Fair Flower of Northumberland" to shame. In one way or another, printed broadsides manage to be insistently topical. In R. B.'s catalog of kinds—"balades of loue," "newes," and "open sclander"—only the first would seem to escape topicality, but even love-lyrics proclaim their newness. The most famous of all Elizabethan broadside ballads first appears in print as "A *new* Courtly Sonet, of the Lady Green sleeues. To the *new* tune of Greensleeves," in the collection entitled *A Handefull of pleasant delites, Containing sundrie new Sonets and delectable Histories, in diuers kinds of Meeter* (Robinson 1584: B2–B3ᵛ, emphasis added). Four years earlier Richard Jones, the publisher of *A Handefull*, had been licensed to print "A *newe* northern

Dittye of yᵉ Ladye Greene Sleves." But "new" thrice over was not new enough. In short order, the Stationers' register received entries for "ye Ladie Greene Sleves answere to Donkyn hir frende," "Greene Sleves moralised," "Greene Sleves and Countenaunce in Countenaunce is Greene Sleves," "A merry newe Northern songe of Greensleves begynninge the boniest lasse in all the land," "A Reprehension againste Greene Sleves by William Elderton," and "Greene Sleves is worne awaie, Yellowe Sleeves Comme to decaie, Blacke Sleeves I holde in despite, But White Sleeves is my delighte" (Simpson 1966: 269). Traditional ballads are never "new" in quite the same way. Each singer's performance, drawn from a repertoire of phrases and motifs, may be different, but "The Fair Flower of Northumberland" from Deloney's spinsters in 1596 and from Miss E. Beattie in 1826 is recognizably the same story. "Newes" is a phenomenon of print. "Newes newes newes newes / ye never herd so many newes": it is, perhaps, the phenomenon of broadsides that gives the dairyman's ballad its locally topical wit.

THE POLITICAL HORIZON

As Deloney sets the scene, the weavers in their praise of weaving and the spinsters and carders in their song of womanly daring are literally *over*seen and *over*heard by economic authority in the person of Jack and state authority in the person of Henry VIII. Figuratively that is the case with all ballads, both oral and printed. The difficulty is first in knowing just where to place the political horizon, and then in judging its constrictive power. As voice projects the singer into the acoustic space around him, as the singer takes her place in a speech community, so the ballad ranges outward to grasp authority figures and draw them by force into the singer's song. *To ballad* is to make a political gesture. Intransitively, one ballads by making up a song; transitively, one ballads by making someone or something the ballad's object. In sixteenth-century terms, Steven Peele "ballads" John Felton, he "ballads" the pope, he "ballads" Catholics. Implicit in both senses of the verb *to ballad* is the idea of taking an event or a person and performing it: giving it a voice, giving it a body, appropriating it by *becoming* it. A ballad is said to be "of" thus-and-so, "upon" thus-and-so, "against" thus-and-so. The politics of ballads consists in the relationship between the balladeer and the thing being "balleted."

In the ballad of "Agincourte Battell," included in Thomas Percy's folio, distinctions between "against" and "of" are especially sharp. The ballad is against the French, in particular the person of Charles VI; it is "of" Henry V. Even the tune makes a political statement. A printed copy of the ballad in the second edition of *A Crowne-Garland of Golden Roses* (1659) specifies the tune as "Flying Fame"—the very tune to which the Ballad of Chevy Chase was sung. Charles is briefly impersonated, but in altogether odious terms. Hearing the message of defiance sent by "our King" via "our ambassador," the not-our French king frames a nasty reply (musical quotation 7.5). Tennis balls are duly sent, duly received, duly revenged. Closely following the plot line of Shakespeare's *Henry V* (and certainly postdating the play's first performances), "Agincourte Battell" avoids all of the political ambiguities that post-World War II critics have attempted to find in Shakespeare's play. The ballad is clearly "of" a brave and outspoken Henry. When he speaks in his own person, as he does on the eve of the battle, the singer is proud to become him.

> "regard not of their multitude,
> > tho they are more then wee,
> for eche of vs well able is
> > to beate downe ffrenchmen 3. . . ."

News that the French have taken "all our Iewells & treasure" and "many of our boyes haue slaine" quite pointedly comes *before* Henry orders the English soldiers to kill their French prisoners (Hales and Furnivall 1867–1868: 166–173).

Anti-Catholic ballads provide another occasion for declaring and patrolling borders. It needed no encouragement from above for professional ballad-writers to give true-hearted Protestants a voice for railing against John Felton, against the northern lords, against the pope, against whoever it was who cast a "Papisticall Bill" in the streets of Northampton in 1570. The wittiest among them, hands down, is "A Lamentation from Rome, how the Pope doth bewayle, That the Rebelles in England can not preuayle" (BB 132, 1570) by Thomas Preston. The author of *Cambises* pursues a raging vein—in the person of a fly who happens to be lodged in the pope's nose when

Musical quotation 7.5

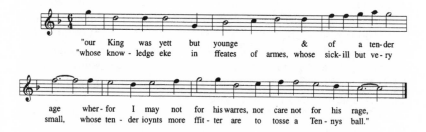

(Tune adapted from Simpson 1966: no. 63, words from Hales and Furnivall 1867–1868: 166–173)

news comes about the Catholic uprising in the north of England. "All
you that newes would here, / Geue eare to me poore Fabyn Flye,"
the ditty begins. First report has it that the rebels have won. The pope
rejoices so stoutly "From that his nose he blew me out." But Fabyn
creeps back in while the Pope is asleep. When news comes in the
middle of the night that the rebellion has been put down, the pope,
the cardinals, and assorted friars are distraught. They say mass, the
pope swoons, the cardinals try to help him, the pope rages and
throws stools against the wall. Fabyn is terrified. The tune is "Row
well, ye mariners," the same tune Steven Peele uses in his pair of
ballads against the pope (musical quotation 7.6). With his wings Fabyn beats a hasty retreat, just before the pope knocks down the center
post, pulling down the house, leaving no harbor for even a mouse.
For "balleting" the pope three different strategies are mapped out by
Steven Peele's "A letter to Rome," the same author's "The pope in his
fury doth answer returne," and Preston's "A Lamentation from
Rome." Peele's "letter" invites the singer to joke, jibe, and jab in the
proud person of an English subject. The pope's "answer" allows the
singer to *become* the pope and thus to enjoy the effects of his or her
own jibing. Preston's way with the pope combines the two strategies:
Fabyn gives the singer access to the pope's very body, allowing the
singer to be inside the pope and outside the pope all at the same time.

In the capacity of ballads to let just anyone become a figure of
authority lies their political danger. To "ballet" King Henry V at Agin-
court was, for the 1590s, safe enough. To "ballet" the Earl of Essex
was something else again. Margaret Allde, the publisher of "A lamentable Dittie composed upon the death of Robert Devereux late Earle

Musical quotation 7.6

He courst me so a - bout, In the house   I   coulde finde no roome,   Then
Loth I   was to   go   out, And   shrind my selfe   vn - der a Brome.   With

by   and by downe he   was set,   He   rubd   his el - bowe on   the Wall,
an - ger he was one   a swet,   So   fell   a - rayl - ing on   Saint Paule

Fye   fye   bloud hatte, He scratchde him selfe   till   he dyd smart,
poll nose   rube eye,   Grash the teth drawe mouth a - wrye.

(Tune adapted from Simpson 1966: no. 401, words from Preston 1570)

of Essex, who was beheaded in the Tower of London, upon Ashwednesday in the morning. 1601," seems to have waited fifteen
months—until the accession of James I—before she printed the ballad up and put it on sale. "God saue the King" is the text's last line.
Her caution was justified, since the ballad offers a sympathetic line
on the earl's career, in more ways than one. The very tune, "Welladay,"
invited singers and listeners to remember the earl's father, Walter, and
his death by poisoning, as recounted in the ballad "Essex's last good
night" (Chappell 1855–1859, 1: 174–177; Collmann 1912: 106–110).
Walter, first Earl of Essex, died in Ireland, the very arena of heroic
endeavor that brought his son into royal disfavor. Through a series of
risings and fallings, the tune modulates from lamentation (in the section marked A) to defiance (in B) to a qualified lamentation (in C)
(musical quotation 7.7).

Meditative pauses in the narrative, first on the phrase "Welladay,
welladay," then on the phrase "evermore still," are realized by the
singing voice as mellismas, as vocal movements free from the march
of syllables. These phrases, especially the first, become the emotional
heart of the ballad. As the story moves along, "welladay, welladay"
becomes "gallantly gallantly" (for Essex's performances in tilts before
the queen and in exploits in Ireland, France, and Spain), "gratiously

Musical quotation 7.7

(Tune adapted from Simpson 1966: no. 496, words from [Anon.], "A lamentable Ditty" 1603)

gratiously" (for the queen's clemency to all the conspirators but Essex), "mournefully mournefully" (for the words of the lieutenant of the Tower to the condemned lord, for Essex's private prayers, and for Essex's pronouncement of forgiveness to his enemies), and finally "cruelly cruelly" (for the fall of the executioner's axe). Although it sometimes advances the narrative, "evermore still" likewise becomes a series of emotionally charged phrases: "well is it known," "like him before," "more was the pittie," "as it is sayd," "to die tomorrow," "of this my death," "priuate to pray," "for this your death," "that had him wrong'd," "For all the blowes." What all these variations solicit is empathy. In these moments particularly the singer *becomes* the event. Indeed, the ballad invites him or her to *become* the Earl of Essex. Third-person narrative begins the ballad, but the twenty-five stanzas are structured like a ballad remembered from oral tradition in their move toward the center, to the moment when Essex sings in his own person, in the singer's person:

> I haue a sinner been
>   welladay welladay
> Yet neuer wrong'd my Queene
>   in all my life,
> My God I did offend,
> Which grieues me at my end,
> May all the rest amend,
>   I doe forgiue them.

It is precisely here, toward the middle of the ballad, in a phrase at the very top of the melodic crest, that Essex expresses his love for the common people of England:

> To the state I ne're ment ill
>   welladay welladay
> Neither wisht the commons ill
>   in all my life,
> But loued all with my heart,
> And alwaies tooke their part,
> Whereas there was desert,
>   In any place.

Although the narrative returns briefly to third person, it finishes in first person plural, in a gesture of solidarity with the executed hero:

> His soule it is at rest,
> in heauen among the blest,
> Where God send us to rest,
>   When it shall please him.
>   ("A lamentable Ditty," 1603)

"A lamentable Dittie composed vpon the death of Robert Lord Deuer-eux" helps to explain why Essex was a popular hero. In the act of "balleting" Essex, the ballad "subjectifies" Essex, making it possible for hero and singer to speak as one person. Frederic Gerschow, visiting from Germany just a year after the earl's execution, testifies to the popularity of a song on the event—perhaps, indeed, this very ballad. Having been shown the site of the earl's death in the Tower, Gerschow remarks,

> How beloved and admired this Earl was throughout the kingdom, may be judged from the circumstance that his song, in which he takes leave of the Queen and the whole country, and in which also he shows the reason of his unlucky fate, is sung and played on musical instruments all over the country, even in our presence at the royal court, though his memory is condemned as that of a man having committed high treason. (1892: 15)

The contrast with balletings upon Queen Elizabeth could hardly be greater. Two ballads contemporary with Elizabeth's review of the troops at Tilbury are remarkable for denying the queen—and the singer—a subject position. In T. I.'s "A Ioyful Song of the Royall receiuing of the Queenes most excellent Maiestie into her highnesse Campe at Tilburie in Essex: on Thursday and Fryday the eight and ninth of August, 1588" (BB 222, 1588) there is only the slightest flirtation with such a position in the line "Then might she see the hats to flye." The queen's famous speech is reported in only the most general terms—and in third person. Thomas Deloney's "The Queenes visiting of the Campe at Tilsburie with her entertainment there" (BB 221, 1588) does allow the singer to become the queen, but compresses her famous speech into just one stanza. The tune is "Wilson's Wild" (musical quotation 7.8). Immediately the singer resumes the subject position that dominates the rest of the ballad, that of Elizabeth's loyal subjects:

> This done the souldiers all at once,
>   a mightie shout or crye did give:
> Which forced from the Assure skyes,
>   an Eccoo loud from thence to driue.
>   (Deloney 1912: 478)

The ballad finishes with a bystander's view of the queen proceeding to the Lord Chief General's tent for a feast, then boarding her barge and being rowed away. Instructed by Deloney, the singer respectfully keeps his or her distance. Elizabeth herself remains a third-person Other. The same arrangements hold true for Richard Harrington's Accession Day ballad "A famous dittie of the Ioyful receauing of the

Musical quotation 7.8

(Tune adapted from Simpson 1966: no. 526, words from Deloney 1912: 478)

Queens moste excellent maiestie, by the worthy Citizens of London the xij day of Noumber, 1584. at her graces comming to Saint Iames" (BB 210, 1584). "Come ouer the born bessy / come ouer the born bessy / Swete bessy come ouer to me": "A Songe betwene the Quenes maiestie and Englande" (BB 92, 1564), from early in Elizabeth's reign, seems to be a charming exception to the rule. A ballad-singer can become Henry V, become the pope, become the Earl of Essex, but he or she may not become the queen.

At quite the opposite extreme is the chance to become a common criminal: a cutpurse, a highwayman, a murderer. Confessions sung in the person of condemned prisoners—indeed, of already executed prisoners—were prominent in the ballad-seller's stock in trade. Such songs found their inspiration partly in the spoken confessions required by sentencing and partly in the songs felons sang on the way to the gallows. Orazio Busino, chaplain to the Venetian ambassador in 1617–18 and in general a trustworthy witness, seems to be speaking from personal observation when he describes the rituals of public execution:

They take them five and twenty at a time, every month, besides sudden and extraordinary executions in the course of the week on a large
cart like a high scaffold. They go along quite jollily, holding their sprigs
of rosemary and singing songs, accompanied by their friends and a
multitude of people. On reaching the gallows one of the party acts as
spokesman, saying fifty words or so. Then the music, which they had
learned at their leisure in the prisons, being repeated, the executioner
hastens the business, and beginning at one end, fastens each man's hal-
ter to the gibbet. (1995: 148–149)

Ballads like "Luke Huttons lamentation: which he wrote the day before his death, being condemned to be hanged at Yorke this last assises for his robberies and trespasses committed" (BB 246, 1598) may,
then, be making some claims to authenticity. (The specified tune,
"Wandering wauering," has apparently been lost.)

> Adue my louing frends each one,
>   ah woe is me woe is me for my great folly
> Thinke on my words when I am gone,
>   be warned young wantons, &c.
> When on the ladder you shal me view,
>   thinke I am neerer heauen then you.
>   ([Anon.] 1598)

What is offered in ballads like these seems to be, in fact, a *double*
subject position: the singer gets to be the criminal, but she also gets
to be the criminal's judge. Opportunity to be both the victim and the
executioner, particularly in scenes of erotic violence, is one of the
things ballads share with stage plays (Smith 1996: 421–443).

The politics of *upon* and *against* can be explained in part by shifts
in the social status of ballads and balleting in the course of the sixteenth century. In the general divergence of "the great tradition" from
"the little tradition" ballads occupy a crucial place. At the beginning
of the century, literate ballads bore the pedigree of Chaucer and Lydgate and were cultivated by the likes of the writers in R. B.'s list:
Alexander Barclay, John Bale, William Wager. By the end of the century literate ballads had become, as Natascha Würzbach demonstrates, a self-consciously proletarian medium (1990: 13–27). Sir William Cornwallis, hanging back at the edge of the crowd, typifies the
ambivalent relationship of cultivated men and women to ballads on
the street and ballads in print. Their ambivalence may have had as
much to do with the commodification of ballads as with the smellin-
ess of the crowd. Well before 1600, the titles emblazoned on broadside
ballads register no embarrassment whatsoever about their commercial status—a far cry indeed from books that often apologize for their
very existence, particularly when written by someone like Sir William

Cornwallis. His collected *Essayes* carry a dedicatory epistle from Henry Olney, who asks pardon for putting his friend's works into print "although I know that worthy Knight, the Author of these Essayes, hateth nothing more then comming in publick" (Cornwallis 1600: A2). From the standpoint of the authorities, ballads were dangerous not only for *what* they might say but for *how* they might say it. To ballet a subject was to commandeer the subject. Within the political horizon at least, Carole Livingston's thesis seems just: the main thing distinguishing printed ballads from oral ballads is the constant threat of censorship (1991: 902–910). The Commonwealth would have none of them. After 1642, players on the stage fell silent. After 1649, so did balleters in the street (Rollins 1919: 321).

THE PSYCHOLOGICAL HORIZON

Positioned at the center of four intersecting horizons—the physiological, the acoustic, the social, the political—the balleting subject finds her self-identity in the fifth. Her sounding voice reverberates inside her body, it projects itself into the space around her, it rings out with the voices and the bodies of her peers, it strikes the baffles of political authority—and sometimes penetrates them. At the fifth horizon the ballad returns to the interiority from which, as sound, it first issued, but it returns with resonances from all the other horizons it has touched. Early modern singers and listeners were themselves aware that it was the first-personhood of ballads that made them so interesting. When Cokes asks Nightingale to point out any cutpurses he sees lurking in Bartholomew Fair, Nightingale whips a ballad out of his packet: "Sir, this is a spell against 'hem, spicke and span new; and 'tis made *as 'twere in mine owne person,* and I sing it in mine owne defence. But 'twill cost a penny alone, if you buy it" (3.5.42–45, emphasis added, in Jonson 1925–1963, 6: 74). Buy several, he implies, and I'll lower the price of each. Potent or not against cutpurses— Philip Stubbes in *The Anatomy of Abuses* cites performances of "Caveats against Cut-Purses" as just the time when the peddler's confederates would be hard at work fleecing the crowd (Rollins 1919: 320)— ballads "made as 'twere in mine own person" were especially seductive when the subject was love. The decades-long popularity of William Elderton's "The gods of love" derives from its capacity to give sound and rhythm to the desires of thousands of men and women (musical quotation 7.9).

Just as "A lamentable Dittie composed vpon the death of Robert Lord Deuereux" finds its most intimate, most personal moments in meditative pauses on "welladay welladay," so "The gods of love" insinuates itself as the singer's very own in the phrase "And knows me,

Musical quotation 7.9

The God   of love   that   sits a- bove And knows me, and knows me
Grant my re - quest that   at the least She show me, she show me

How   sor-row - ful   I   do serve;   That ev- ery brawl   may turn   to bliss,   To
some   pi - ty when I   de-serve.

joy   with all   that joy - ful is   Do this my dear   and bind   me   For -
                                           And as you here do find me,   So

e - ver   and e - ver your own,   For   till I hear this u - ni - ty I   lan -
let   your   love   be shown:

guish   in ex - trem - i - ty.

(Tune adapted from Simpson 1966: no. 163, words from Osborn 1958)

and knows me," echoed shortly in "She show me, she show me." In
those two phrases are invested the song's erotic longing: the singer's
narcissistic pleasure cries out for sexual admiration. The tune in each
case accentuates the effect of strong personal feeling by a change in
the rhythmic pattern and a coming to closure. In later stanzas, the
phrase "And knows me, and knows me" becomes a series of equally
charged meditations: "Uprightly, uprightly," "To find ye, to find ye,"
"Full truly, full truly," "To speed me, to speed me." The second
phrase, "She show me, she show me," undergoes similar metamorphoses: "So lightly, so lightly," "To mind ye, to mind ye," "As duly, as
duly," "You need me, you need me." In the last stanza in particular,
full of figurative words given physical force, the combination of self-
indulgence and physical desire is, for the singer at least, irresistible:

> With courtesy now, so bend, so bow,
>   To speed me, to speed me,
>   As answereth my desire;
>
> . . . .
> Unworthy though to come so nigh
>   That passing show that feeds mine eye,

> Yet shall I die without it,
>   If pity be not in you. . . .
>   (Osborn 1958: 11)

Small wonder that Benedick should think of the song in the very height of his frustrated love for Beatrice.

It is in ballads upon love, perhaps, that the singer most conspicuously becomes the thing sung about. But narrative ballads offer the same possibilities for self-projection in the telling and self-absorption in the singing. First-person dialogue, not third-person narration, brings the events of orally transmitted ballads into aural and visual presence. In certain of these ballads—"Edward" (Child no. 13) and "Lord Randall" (Child no. 14) are probably the best-known examples today—there is no third-person narration at all. The singer who traverses the city streets and travels the countryside with Brathwait's characteristic ballad-monger changes his voice to suit changing subjects—and perhaps even changing customers. A veritable *"Chantel-eere"* (someone who sings and ogles at the same time?), he can even impersonate Puritans, as he

> sings with varietie of ayres (having as you may suppose, an instrume[n]tall *Polyphon* in the cranie of his nose). Now he counterfeits a naturall *Base*, then a perpetuall *Treble*, and ends with a *Counter-tenure*. You shall heare him feigne an artfull straine through the Nose, purposely to insinuate into the attention of the purer *brother-hood:* But all in vaine; They blush at the *abomination* of this knave, and demurely passing by him, call him the *lost childe*. (1631: B5)

As performance pieces, jigs in particular take shape as a series of dramatic exchanges. Even when the story-singer sets the scene, connects episodes separated in narrative time, and rounds out the story with a moral, as Deloney's maidens do in "The Fair Flower of Northumberland," dramatic speeches remain the heart of the ballad. The reason is not hard to discover: it is precisely in those moments of dramatic presence that the singer becomes the subject—or rather the *subjects*. Narrative ballads offer a *range* of opportunities for becoming people other than oneself by assuming their voices. In "The Death of Queen Jane" (Child no. 170), for example, the singer becomes both Jane Seymour and Henry VIII. Queen Jane's death after birthing Prince Edward in 1537 inspired a ballad that did not receive literate attention until the late eighteenth century (Child 1957, 3: 372–376) but that presumably goes back to the event itself. Certainly the ballad enjoyed wide circulation in the nineteenth and twentieth centuries, both in Britain and in North America. Mrs. Kate Thomas's version, taken down by Cecil Sharp and Maud Karpeles in Lee County, Kentucky, sometime between 1916 and 1918, gives voice to a queen

Musical quotation 7.10

(Sharp and Karpeles 1968: 45)

and a king who had been dead for four hundred years (musical quotation 7.10). Mrs. Thomas sang these long-dead voices, but she did
so on her own terms. When Sharp told Mrs. Thomas who the people
were about whom she sang, she replied, "There now, I always said
that song must be true because it is so beautiful" (Sharp and Karpeles
1968: 45, 104). Even an informant in England, even an informant
closer in time to the historical events of the ballad, might have made
the same reply. If Mark W. Booth is correct, identification of the singer
with the song is the very thing that assures a song's popularity and
survival (1981: 14–17). What Mrs. Thomas offers in her version of
"The Death of Queen Jane" is a personalizing of two historical figures. For Queen Elizabeth, at least during her reign, that was not a
possibility for ballad-singers.

One transformative mark of print on ballads is greater dependency on third-person narrative (Livingston 1991: 895). Thus the text
of "The wofull death of Queene Iane, Wife to King Henry the eight:
and how King Edward was cut out of his mothers belly," printed in A
Crowne-Garland of Goulden Roses (1612), probably from a lost broadside, tells the story almost entirely in third person. Neither Henry nor
Jane is given a single dramatic speech. The one person who is given
voice, "a lady" who reports Jane's dire condition to the king, points
toward the ballad's implicit subject position: the people of England.
"Oh mourne, mourne mourn faire Ladies, / Iane your Queene the
flower of England dies," goes the refrain at the end of each stanza. In
the final stanza this refrain is given topical immediacy. It is Prince
Edward's successor Elizabeth whose death is being mourned: "Oh
mourne, mourne, mourne faire Ladies / Elizabeth the flower of Englands dead" (R. Johnson 1612: C2ᵛ–C4). There is a voice that recommends "The wofull death of Queene Iane" to singers, listeners, and
readers in the first decade of the seventeenth century, but it happens

not to be the voice of a mother dying in childbirth or the voice of an unyielding husband.

Third-person narrative could be regarded as a corruption of oral immediacy, but printed broadsides offer a singer or a listener a wider choice of subject positions, both *among* ballads and *within* ballads.

> Ballads! my masters, ballads! Will ye ha'any ballads o' the newest and truest matter in all London? I have of them for all people, and of all arguments too. Here be your story-ballads, your love-ballads, and your ballads of good-life; fit for your gallant, your nice maiden, your grave senior, and all sorts of men beside. Ballads! my masters, rare ballads! Take a fine ballad, Sir, with a picture to't. (Tite 1845: 44–45)

The ballad-hawker's cry in this Restoration pastiche of *Bartholomew Fair* gives some sense of the possibilities, not only for the "Sir" who will pay but for the "nice maiden" who will enjoy her consort's purchase. Print, and the topicality fostered by print, made it possible for a ballad-singer to become not only the faithless lover, the constant lover, the knight, and the trickster of oral tradition, but Luke Hutton the highwayman (BB 246, 1598), King Solomon (Heber 18339, c. 1600), Richard Tarlton caught in a flood (BB 154, 1570), a fly in the pope's nose (BB 132, 1570), a soldier at Tilbury (BB 222, 1588), the Earl of Essex (1603), a morris dancer in the Cotswolds or Kent (BB 110, 1569), a military deserter (Shirburn 47, 1600), the town of Beckles, Suffolk, speaking for itself after being burnt to the ground (BB 217, 1586), a naked love-crazed madman on a rampage among his neighbors (Euing 201, 1637), even God Himself (BB 240, 1592). Something of the same freedom is offered *within* a single ballad. The singer becomes first the narrator, then this character, then that character, then perhaps a third. Rather than foreclosing the possibilities for dramatic impersonation, printed ballads open them up. The result, as Frederick O. Waage has argued, is the capacity of ballads to speak to the contradictions in early modern culture, allowing the singer to sympathize both with the condemned criminal who speaks in first person and with the system that condemns him (1977: 731–742).

To a much greater extent than texts in "the great tradition," popular ballads invited singers and listeners to occupy a female subject position. In addition to "The Fair Flower of Northumberland" and "The Death of Queen Jane," Deloney's singing spinsters and carders might have included in their repertoire any number of ballads from oral tradition that give voice and presence to women: "Tam Lin" (Child no. 39, licensed for printing 1558), "The Lord of Lorne and the False Steward" (Child no. 271, licensed 1580), "Fair Margaret and Sweet William" (Child no. 74, quoted in *The Knight of the Burning Pestle* 1611), "The Knight and the Shepherd's Daughter" (Child no.

110, first printed c. 1660, possibly quoted in *The Knight of the Burning Pestle*), "Little Musgrave and Lady Barnard" (Child no. 81, printed 1630), "Georgie" (Child no. 209, possibly identifiable with "George Stoole," printed c. 1630, and "George of Oxford," printed 1683), "The Daemon Lover" (Child no. 243, printed c. 1650), "The Twa Sisters" (Child no. 10, printed 1656), "Lord Thomas and Fair Annet" (Child no. 73, printed c. 1670), "The Bailiff's Daughter of Islington" (Child no. 105, printed 1670–1696), and "Bonny Barbara Allan" (Child no. 84, printed 1685–1692). From ballads circulating, in part at least, in writing they might have sung "The nutt browne mayd" (Percy MS, ed. Hales and Furnivall 1867–1868, 3: 174–186, datable to 1502), Ellen Thorn's song (BL MS Cot. Ves. A-25, ed. Seng, no. 34, datable to the 1570s), "The complaint of a woman Louer" and "The lamentation of a woman being wrongfully defamed" (printed in *A Handful of Pleasant Delights*, 1584), "Marye Aumbree" (Percy MS 1756, 1: 515–519, cited in *Epicoene* 1609), "Ladyes: ffall" (Percy 1756, 2: 246–252, datable to 1595), and "Queene Dido" (1756, 3: 499–506, printed 1620). Lovers—constant, faithless, murderous, dying—loom large in these narratives, but in length, breadth, and intensity they sing on equal terms with men. In "Little Musgrave and Lady Barnard" the women rivals dominate the dialogue, and in Ellen Thorn's song men have no singing part at all. In "Georgie" and "Queene Dido" male singers exist primarily to give the female protagonists their cues. In strength of presence, the most remarkable of these female protagonists is Mary Ambree, whose prowess as a soldier was proverbial enough to earn contemptuous references from Jonson in *Epicoene* (4.2.123), *A Tale of a Tub* (1.4.22), and *The Fortunate Isles* (393).

To insist on gender exclusivity is, in a way, to miss the point: what ballads offer the singer and the listener is the possibility of becoming many subjects, by internalizing the sounds and rhythms of those subjects' voices. A male singer of "The Lord of Lorne and the False Steward" must sing in the person of the duchess who discovers the young lord's identity and marries him, as well as in the persons of the young lord, his father, and the steward who tries to convince the duchess and her father that the young lord is a common drudge. Likewise, a female singer of "Tam Lin" must impersonate the fairy knight who changes into the shapes of beasts as well as the princess who wins him by holding on tightly through all his metamorphoses.

In their allurement to fantasies of identity, ballads in performance are like plays in performance. Commonwealth authorities recognized as much when they closed the playhouses in 1642 and began to persecute ballad-singers at about the same time. After 1649 it was illegal to sing ballads in the streets. In general, links between plays and ballads are not hard to identify. Both media had become, by the 1590s, capitalist commodifications of cultural performances that had once been community-produced without a profit motive (Weimann 1996: 113–119; Bristol 1996: 31–41). Several people made successful careers out of both. William Elderton, for example, not only produced some of the most quoted broadside ballads of the century but directed boy actors in plays before the queen. Churchyard in his flytings with Camel *et al.* cast himself as a vice, a plain-speaker who calls a spade a spade:

> Nay, any, some one must speake,
>   although the vice it bee:
> Or els the play were done ye wot,
>   then Lordinges pardon mee.
>   (Churchyard 1566)

Richard Tarlton's fame as a performer of jigs was enough to induce four different printers to attribute ballads and chapbooks to his talents, the two surviving examples (BB 154, 1570, and BB 239, 1591–1592) being remarkably unjig-like. But why not? Jigs were simply ballads put on the stage, a way of rounding out a good comedy or turning a tragedy on its head. The custom made a strong impression on, for one, Thomas Platter, taking in a performance of *Julius Caesar* at the Globe. "Their wont" in England was to break down a play into a danced ballad: "Portia" becomes the likes of "Nan," "Caesar" the likes of "Rowland," "Brutus" the likes of "The Fool."

Plays could become ballads without such radical changes in the plot. In addition to "Agincourte Battel," strollers in the street and browsers in bookstalls could buy any number of ballads "upon" the subjects of famous plays: "A ballad of the life and deathe of Doctor Ffaustus the great Cunngerer" (registered 1589, MS copy c. 1616 as Shirburn 15, surviving imprints 1624, 1658–64, 1686–88, c. 1693, c. 1695), "A newe ballad of Romeo and Juliett" (reg. 1596), "A new song, shewing the crueltie of Gernutus a jew, who would have a pound of flesh" (c. 1620), "The Spanish tragedy, containing the lamentable murders of Horatio and Bellimperia" (c. 1620), and "A lamentable song of the Death of King Leir and his Three Daughters" (collected in Richard Johnson's *The Golden Garland of Princely Pleasures* [1620]). In some cases the ballad and the play may derive from the same printed source (for example, "Faustus" and possibly "Gernutus"), but dates of surviving imprints suggest that in most cases ballads were designed to cash in on the stage successes of plays. "Residuals" is the term in modern show business. The fact that John Danter in 1594 registered "a booke intituled *a Noble Roman Historye of TYTUS ANDRONICUS*"—a play that turns out, in print, to be Shakespeare's—and, in the very next entry, his copyright for "the ballad thereof," suggests that playscripts and ballads may, for some purchasers at least, have served the same mnemonic function of keeping the play alive by keeping it in their mouths and ears (Arber 1875, 2: 644).

"The Lamentable and Tragical History of *Titus Andronicus*" (surviving imprints c. 1625, 1658–59, 1675, 1690, and 1700) provides, in fact, a good example. The tune, "Fortune my foe," is one of the most widely circulated among broadside ballads of the late sixteenth and early seventeenth centuries. As the setting for laments such as those of numerous broadside ballads upon murders, natural disasters, and deathbed confessions (Simpson 1966: 225–231), it was likewise a tune especially well suited to first-person "complaints" such as Titus's (musical quotation 7.11). Like most broadside ballads, "The Lamentable and Tragical History of *Titus Andronicus*" is designed to retail "news," and to do it in the most arresting, expeditious way possible, which typically entails more direct narrative than dramatic dialogue. In this case, however, the narrator is the protagonist: all the events of the story—even his own death—are sung by him. The only other character who "speaks" is the tongueless Lavinia, whose inscribed words are distinguished in this imprint by Roman typeface:

> For with a staff, without the help of hand,
> She writ these words upon a plat of Sand:
> The lustful Sons of the proud Empress,
> Are doers of this hateful wickedness.
>   ("Titus Andronicus," 1658–59)

In singing the ballad alone, in performing it for others, even in performing it *with* others, the singer perforce becomes the titular hero: all

Musical quotation 7.11

(Tune adapted from Simpson 1966: no. 144, words from "The Lamentable and Tragical History of *Titus Andronicus*" 1658–1659)

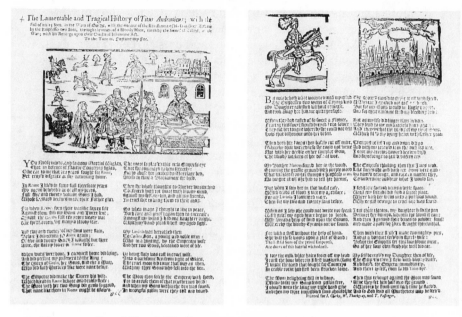

Figure 7.1. Broadside of "The Lamentable and Tragical History of *Titus Andronicus*" (c. 1658–1659). Reproduced by permission of the Pepys Library, Magdalene College, Cambridge, and the Folger Shakespeare Library, Washington, D.C.

the words he or she sings, even Lavinia's message, are Titus's words as well as the singer's.

That act of aural internalization is cued by the woodcuts in the 1658–59 edition and later. The purpose-made woodcut directly under the title in the 1658–59 imprint shows signs of wear in its losses and age in its wormholes, suggesting that it dates from now-lost earlier editions, perhaps even Danton's (fig. 7.1; the earliest surviving imprint, c. 1625, has no woodcuts). Just to the right of center a generic warrior, lance at the ready, gives Titus a body—and the singer a visible cue to the dominant subject position. To the knight's right, a generic city, crowned with a hero's laurel-wreath, fills in for Rome, the city for which Titus has won so many battles. In the story-specific woodcut on the left some of the ballad's major events are depicted within a single frame. At the top right is a view of "the stately tower of Rome" to which Titus returns from his military conquests at the ballad's beginning. Ancient Rome, of course, the woodcut is not. It depicts, rather, the gabled houses and church towers of a city like early seventeenth-century London, complete with a large building in the center that looks very much like renderings of the South Bank amphitheaters in contemporary view-maps. Note the drooping flag (the taut flags are weathervanes) and the horizontal lines to the left of the structure, shadowing roundness. To the left in the upper band are two of the atrocities visited upon Titus: in the center, Lavinia using a staff to write her message on the sand, and on the left, the capture of Titus's sons while hunting. Featured more prominently at the bottom are the revenges Titus carries out: the half-burial of the Moor, the torture of the empress's sons by Lavinia and Titus, and the banquet of king and empress on "two mighty Pies" containing the sons' powdered flesh. Each of these images offers a range of shifting subject positions, moments of indirect discourse, within the dominant subject position assumed by Titus. As different as it may be from Shakespeare's play in details of plot and sequence, "The Lamentable and Tragical History of *Titus Andronicus*" shares with the play the same fundamental appeal: both play and ballad give voice—vibrating, ringing, reverberating voice—to the experience of ungratefulness and cruelty. Despite their lack of verbal sophistication, ballads may tell us a great deal about what it was like to hear a play in the public theater—but with this difference: in possessing the ballad as a physical object, and in getting it by heart, one can perform the play in one's own voice.

Where do ballads belong? Within, around, among, of, upon, against, within.

Within the Wooden O

In a culture that favored conspicuous display—in the façades that buildings turned to the street no less than in the garments in which people presented themselves for public viewing—London's South Bank theaters stood out by their difference. On the outside they presented bare walls. City views by Norden, Visscher, and Hollar may vary in certain details about the amphitheaters, but all of them show plain, flat exteriors relieved only by small windows. It was within the narrow depths of London's best houses, Fynes Moryson notes, that the city's real splendors were to be found (1617: KKK4). The South Bank theaters present a similar case: they were built not to display but to contain. Inside, not outside, provided their very reason for being. What they contained, most obviously, was spectacle: many-sided galleries, surrounding the thrust stage as a focal point, gave much better sight-lines than a square structure would for viewing not only the play but other members of the audience. Extrapolating from the Fortune contract, no one in the Fortune or the 1599 Globe was more than fifty feet from an actor standing downstage, at the focal center of the space. What the theaters contained, less obviously, was sound. That same actor, standing at the center of the visual space, stood also at the center of an aural space. "Sit in a full Theater," says the delineator of "An excellent Actor" in Sir Thomas Overbury's expanded collection of characters, "and you will thinke you see so many lines drawen from the circumference of so many eares, whiles the *Actor* is the *Center*" (1616: M2). The South Bank amphitheaters were, in fact, instruments for producing, shaping, and propagating sound.

Evidence that theaters were thought about as sound-devices is not hard to come by. For special occasions it was common for large households—schools, colleges, the inns of court, the court of the realm—to erect temporary theaters inside an existing hall. Alan Nelson has reconstructed the elaborate timbered structure that was erected within the hall of Queens' College, Cambridge, for putting on college plays each season beginning in 1546 and continuing into the 1640s. Not only a stage but galleries for spectators were part of the structure, made out of marked timbers that were dismantled and stored away at the end of each season (1994: 16–37). In effect, the theater was not so much a building in itself as a large, free-standing object that could be erected inside a preexisting building. Its multiple planes and all-wood construction would have provided richer resonance than the masonry room itself. Orazio Busino's description of the pre-Jones banqueting house at Whitehall likewise suggests a box-within-a-box. The external brick walls contained an interior structure of wood and plaster—complete with colonnades and a coffered ceiling covered with *putti*—that offered not only visual interest but the resonators and baffles required for good sound distribution in a large space (1995: 137).

Theoretical justification for such structures, if any were needed, could be found in Vitruvius, who designed the ideal theaters in *De Architectura* first and foremost around sound. Bronze vases were placed at regular intervals along the rising tiers not just for ornament but to catch sound waves of particular frequencies and amplify them. These vases worked like water glasses, filled to various depths, in a glass organ: when touched, each one produced a different pitch (5.3–5 in 1931, 1: 262–283). A Vitruvian theater could be played by the actors as if it were a musical instrument. According to Daniel Barbaro's influential commentary on Vitruvius (it was Barbaro who turned Vitruvius' *scaenae frons* into a proscenium arch with illusionistic scenery beyond), architecture presents a convergence of all the arts—including rhetoric. When it comes to theaters, an architect needs to be both a natural philosopher and a musician: "paying attention to motions of the voice, observations about numbers, and the practicalities of sound (which I take to be the principles of mathematics and the rules of music), he should shape theaters accordingly, so that the space resonates all the more." The shape of that space, Barbaro insists, should approximate the shape of sound itself: a sphere (Vitruvius 1567: 1–2, 172, my translation). Vitruvius's ancient precepts and the exigencies of early modern practice are reconciled in Sebastiano Serlio's *Architettura* (1545), which turns Vitruvius' designs for permanent stone-built structures into a wooden contraption that can be set up inside a great hall and taken down again (Smith 1988: 84–85).

Theaters as instruments for the production and reception of sound ask to be thought about in different ways than theaters as frames for the mounting and viewing of spectacle. What were the acoustic properties of the instruments themselves? What were they made of? What kinds of sounds could they produce? What constituted the repertory of sounds on which playwrights and actors could draw? What qualities of the human voice figured in this repertory? To answer such questions let us inspect the instrument itself before we attempt to inventory the range of sounds, first artificial, then human, that could be played on—and within—the largest, airiest, loudest, subtlest sound-making device fabricated by the culture of early modern England.

WOOD, PLASTER, THATCH, MORTAR, AIR

When the Lord Chamberlain's Men were forced to vacate The Theater in Shoreditch in April 1597, they took with them their play-books, their props, and their costumes. Two years later, after playing in rented quarters at The Curtain, they went back and got their reso-nator. Laying legally dubious claim to the building their father had put up on leased land, Cuthbert and Richard Burbage dispatched a builder, Peter Streete, to dismantle The Theater's wooden frame, transport it across the river, and re-erect it on land they had just leased on the South Bank (Gurr 1996: 292–293). In that act the Bur-bages were not just moving a building: they were transporting part of the company's professional equipment, like viol-players bringing their instruments with them to a concert. The 1599 Globe was an instrument to be played upon, and the key element in that instrument was wood. The oak timbers that framed the structure were a foot square and up to thirty feet long. The 1599 Globe was not just an instrument but a *vintage* instrument: by the early seventeenth century fire regulations and the rising price of timber meant that most new theaters, in the City at least, were built of brick (Gurr 1992: 141).

As a device for propagating sound, the 1599 Globe was extraordi-narily efficient. In its tubular shape it approximated the shape of the human vocal tract. In a theater, as in the human body, production of sound requires three things: (1) an energy source, (2) something that vibrates in response to that energy, and (3) something that propagates those vibrations into ambient space. In the case of the human body, the energy source is the lungs, the vibrator is the larynx, and the propagator is the throat, mouth, and sinuses (see fig. 1.1). If the struc-ture of the Globe is imagined as the vocal tract, the energy source was either lungs (for vocal sounds and wind instruments) or arms and hands (for plucked and bowed instruments, drums, and sound effects). The vibrator was the stage. The propagator was the architec-tural surround. In the production of "theatrical" sound, the building itself functions as the larynx, mouth, and sinuses do in the produc-tion of purely vocal sound: they give the sound its harmonic profile and influence its volume. While one might assume that those quali-ties of timbre and volume were lost forever when the structure was pulled down sometime after 1642, evidence does exist for recon-structing the Globe as a distinctive acoustic space. Three factors de-serve consideration: (1) the materials out of which the theater was built, (2) the size of the listening space, and (3) its shape.

The primary materials out of which the Globe was constructed—wooden beams, plaster over lath, and wooden boards over joists—all return to the ambient air a high percentage of the sound waves that strike them. Within the frequency range of adult male voices (with a mean of 120 cycles per second) plaster over lath absorbs only about 14 percent of these waves of energy, giving it an "absorption coefficient" of 0.14. That is to say, plaster over lath reflects back 86 percent of the soundwaves that strike it. Within the frequency range of adolescent male voices (with a mean of about 240 cycles per second) plaster over lath absorbs even less, about 10 percent, giving it a coefficient of 0.10 or a reflectivity of 90 percent (Egan 1988: 52; Fry 1977: 44; Curry 1940: 48–62). The 10 to 14 percent of sound that enters the plaster is due to the air space between and behind the laths: plaster over brick, by contrast, turns back fully 99 percent of the sound waves that strike it. In comparison with plaster, wood is more absorbent of sound, but with sufficient air space behind, beneath, or within, wood can act as a resonator, just as it does in a guitar body. Wood boards over joists in a structure like the stage of the Globe reflect about the same percentage of sound waves as plaster over lath: 85 percent within the frequency range of male voices, 90 percent in the range of adolescent voices (B. J. Smith 1971: 48).

The result of these reflections from wood and plaster within the wooden O is a plenitude of what acoustical science calls "standing waves"—stationary patterns of vibration formed by many reflected sound waves, coming from many different surfaces, all superimposed on one another (Handel 1989: 33). Auditors experience these steady waves of energy as full, present sound, uniform throughout the listening space. The effect is enhanced even by the energy the wood and plaster do happen to absorb. As a medium for transmitting sound waves, wood is highly "damped": in comparison to, say, metal, wood more rapidly loses the vibrations that strike it. At the same time, wood is more rapidly excited in the first place than metal is, and it has the distinctive characteristic of reaching a high amplitude across a wide range of frequencies. Metal reaches maximum amplitude at only a narrow range of frequencies and takes longer to get there (Handel 1989: 551; Fry 1979: 24–27). Vibrations in wood may be short in duration, but wood catches the harmonic complexities of ambient sound. In effect, the stage of the Globe acted as a gigantic sounding board: made of reverberative material, it translated vibrations in the air above into standing waves in the air underneath, producing a harmonically rich amplification of the voices of actors positioned on top. In this respect it worked like the wooden choir stalls in England's churches. Some of these structures positioned ceramic vessels under the floor boards—medieval versions of Vitruvius's vases—and provided microphones of a sort in circular openings along the sides close to the stone floor.

Within the acoustic environment of the Globe there were only three highly sound-absorbant materials: the arras, the surface of the yard, and human bodies. In the lower-frequency range of male voices, heavy woven fabrics absorb about the same amount of sound as plaster over lath and boards over joists: just 14 percent. But for higher-frequency adolescent voices, they can absorb as much as 35 percent. In square footage, of course, the arras occupied only a small portion of the available space. Much more significant were the surface of the yard and the density of spectator-auditors. If the yard of the Globe resembled at all the yard of the excavated Rose—15 inches deep in hazlenut shells, ash, and clinker—it could have soaked up as much as 60 percent of the sound waves striking its surface. Clothed human bodies are also highly absorptive of sound, stopping up to 80 percent of the sound waves that strike them. As with the arras, so with clothed bodies: the higher the frequency, the greater the absorbency (Egan 1988: 52–53). Human bodies, then, presented the greatest obstacle to the efficient propagation of sound. Fortunately, the size and shape of the Globe mitigated that damping effect.

We tend to think of early modern theaters as having been small and crowded, but the 1599 Globe apparently offered a volumetric listening space per auditor that actually surpasses that of modern theaters. Assuming dimensions projected from partial excavacations of the site in 1989, the Globe was a twenty-sided polygon 99 feet in diameter (Orrell 1990: 95–118; 1997: 50–65; Blatherwick 1997: 66–80) (fig. 8.2). Assuming that this polygon was raised to the same height as the galleries in the Fortune contract, we have a structure 32 feet high that approximates the shape of a cylinder. To arrive at an estimate of the volume of that cylinder we can multiply the square of the radius (49.5 feet × 49) by the height (32 feet) by pi (3.14), to arrive at a volume of 243,714 cubic feet. Deducting the space occupied by the tiring house and the "cellarage" under the stage, we arrive at a listening space of about 231,028 cubic feet.* By modern standards, this

---

*Estimating the dimensions of the tiring house and the area beneath the stage is tricky, of course, since physical remains are lacking and documentary evidence is subject to different interpretations. Assuming, however, from The Fortune contract that the stage extended halfway into the yard and had a width of 43 feet and a depth of 27 feet 6 inches, and assuming further that it was 5 feet high, we can calculate the volumetric space beneath the stage as 5,913 cubic feet (43 × 27.5 × 5). If the tiring house, like the one at The Fortune, was built into the framework of the galleries at a depth of 11 feet 3 inches (allowing 3 inches for the thickness of the walls) and ended at a gallery or balcony 14 feet above the stage, then it took up another 6,773 cubic feet (43 × 11.25 × 14). Deducting these two figures from the overall volume (243,714 − 12,686), we arrive at a listening space of 231,028 cubic feet.

is a very large space indeed. The Olivier Theatre in London, for example, contains just 158,922 cubic feet, and the Barbican Theatre even less: just 48,854 cubic feet (Mulryne and Shewring 1995: 120–123).* What may be most remarkable about the 1599 Globe, however, is not the sheer volumetric size of the place but the volume of listening space per patron. Modern standards of acoustic engineering suggest an optimal space of about 98.9 cubic feet per person for speech, with 173 cubic feet as a maximum (Smith 1971: 44). If the capacity of the Globe was 3,000 people, then the listening space per person works out to 77.01 cubic feet—somewhat less than in most modern theaters. Compare the 137 cubic feet per person for the 1,160 listeners in the Olivier (Mulryne and Shewring 1995: 120). When, however, the Globe operated at less than full capacity—and Henslowe's diary suggests that most of the time it probably did—the listening space per auditor would have approximated or even exceeded modern spaces. Transferred to the Globe, the Olivier's 1,150 auditors would find themselves surrounded by 50 percent more listening space, 201 cubic feet per person to the Olivier's 137.

Among the factors that gave the 1599 Globe its distinctive sound, the most crucial was the structure's shape. The standing waves that create harmonically rich, in-filling sound are produced by reflections off many surfaces. In general, the more surfaces there are, the fuller the acoustic effect. As a twenty-sided polygon, the Globe provided plenty of reflective surfaces. The interplay of sound waves within the polygon can be appreciated by looking at the theater from two angles: from above (see fig. 8.2) and from the side (fig. 8.1). In each case an actor is imagined as standing at the rear of the stage, just in front of the tiring house. Figure 8.2 shows in gray the area where the actor's broadcast speech would be optimal, an area 70 degrees to his left and 70 degrees to his right. Although speech sounds at low frequencies (less than 500 cycles per second, the region of most vowels) are diminished very little to a speaker's sides or even to his rear, higher-frequency sounds (more than 4,000 cycles per second, the region of most consonants) tend to fade out in these areas (Egan 1988: 83). Dotted lines in Figure 8.2 show how concave surfaces like those of the Globe would have served to focus sound in the center of the yard, filling the space with standing waves. (With sound waves, as

---

*Comparing the volumetric listening spaces of these theaters with the Globe is, in more ways than one, not straightforward, since both the Olivier and the Barbican have an enormous fly-tower above the stage that is not reckoned in measures of the listening space. However, most of the sound waves that travel up into the fly-space are not reflected into the auditorium, and much of the speaking in these theaters takes place on the platform thrust out from under the fly-tower.

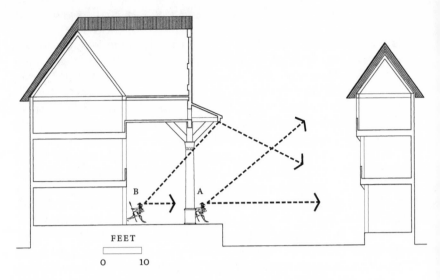

Figure 8.1. Side view of the Globe Theater (1599).

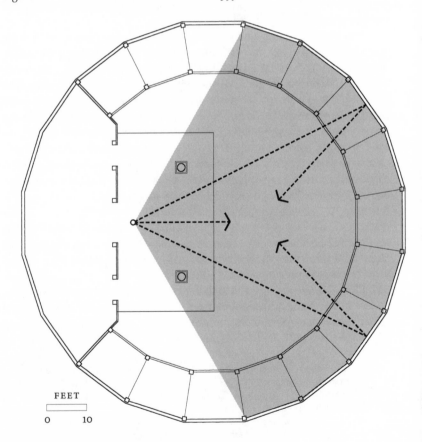

Figure 8.2. Top view of the Globe Theater (1599).

with light waves, the angle of incidence equals the angle of reflection.)

This pattern of sound concentration would have been complicated, however, by two structural factors: the absence of a roof over the yard and the canopy over the stage. In a cylindrical structure like the Globe, open at the top with nothing for soundwaves to strike against and closed at the bottom with highly absorbent material in the form of human bodies, sound waves would have been reflected mainly from side to side, not from top to bottom. The result would have been a "broad" as opposed to a "round" sound (Handel 1989: 41). Such a quality would be ideally suited to the epic sweep of history plays. Depth of the canopy over the stage proved to be one of the most controversial features in the 1990s reconstruction of the Globe on the South Bank, since the Fortune contract is silent on the subject, the visual evidence of contemporary views of London is contradictory, and most of the archeological evidence is still unexcavated. For the purposes of analysis, let us assume that the canopy covered the stage, if for nothing else than to protect expensive costumes from the rain. With the canopy, there would be two primary paths for sound waves, depending on where an actor happened to be standing (see fig. 8.1). If he were standing at the front of the stage, under the edge of the canopy (position A), the sound waves would have gone out directly to listeners standing in the yard and and sitting in the galleries. If, on the other hand, he were standing at the rear of the stage (position B), sound waves would have gone directly out into the yard, but they would also have been reflected off the underside of the canopy. Using as a reference point a patron seated at the furthest point from the speaker, in the rear of the bottom gallery, some 65 feet away, we can calculate that the difference between the direct path of sound and the indirect path would have been approximately 20 feet, producing a time delay between the two signals of less than 0.02 seconds. Modern acoustical engineering ranks such a delay as being just within the range of conditions "excellent for speech and music" (Egan 1988: 96).*

Contrast between the largely horizontal sound that an actor projects when standing downstage and the vertical as well as horizontal sound he projects further upstage calls into question some received ideas about early modern theaters that are based on visual analysis alone. Experience in the reconstructed Globe in London has demonstrated that the position of greatest dramatic power is not all the way downstage, where some theater historians imagine soliloquies to have been spoken, but several feet back, somewhere in between the

---

*I am grateful to Andrew Gurr for these calculations.

two pillars holding up the canopy. An actor may occupy the position of greatest *visual* presence at the geometric center of the playhouse, but he commands the greatest *acoustical* power near the geometric center of the space beneath the canopy. The canopy of the reconstructed Globe also demonstrates the excellent acoustics enjoyed by occupants of the Lords' Room in the upper recesses of the *scenae frons*. In terms of both vision and hearing, the Lords' Room offered an optimal situation: one could not only see and be seen but hear and be heard: the canopy would have projected the lords' voices as well as the actors'. Thanks to the absence of a roof over the yard, auditors in the yard and in the galleries would have found themselves in a perceptibly different relationship to the auditory events going on all around them. In a cylindrical space listeners can locate sounds horizontally far more accurately than they can in a space enclosed on six sides. Applause sounds on the left and the right, not all around; loud laughter comes from over *there*, a rude comment from over *there*. Performers in the reconstructed Globe in London have commented on the way audience response can start in one part of the theater and then spread laterally to the rest.* The experience of broad sound comes not only from the actors onstage but from one's fellow auditors.

In all three respects—building materials, size, and shape—indoor theaters like the Blackfriars presented an altogether different acoustic environment. Although many crucial details about the Blackfriars theater, which the King's Men occupied from 1609 to 1642, are not known, Burbage's 1596 deed of purchase and subsequent lawsuits provide a very good list of the materials out of which it was built: stone walls, paved flooring, galleries (presumably built of wood), a stage (also presumably built of wood), seats (likely wooden benches), window glass, "wooden windows" (probably shutters), and plenty of wainscotting that had formerly divided part of the space into seven separate rooms (H. Berry 1987: 46–73; Wickham 1972, 2.2: 123–138; Hosley 1969: 74–88; I. Smith 1964: 471–475, 302). Stone walls are even more highly reflective of sound than wood and plaster, returning 98 to 99 percent of the energy waves that strike them. Paved flooring is almost as reflective, returning 97 percent of sound waves. Even the small panes of glass in the theater's windows would have bounced back 80 to 90 percent of the ambient sound. Judged by its outer shell, the Blackfriars theater would have been a very "live" space. The theater's distinctive acoustic properties, however, were tempered by

---

*I come by these observations through interviews with actors at the Globe in October 1997 and through personal experience as an auditor at various places in the reconstructed theater.

wood. As the later legal documents make clear, galleries (*porticibus*) and seats (*sedilibus*) were expensive parts of the property and were deemed to convey with the title (H. Berry 1987: 68–71). As with the Globe, the primary damping medium was the audience.

Precise dimensions of the Blackfriars listening space are provided by the bill of sale and later legal documents: 66 feet north to south and 46 feet east to west. At 3,036 square feet, the floor area of the Blackfriars theater would have been slightly larger than the yard of the Fortune and slightly smaller than the still extant great halls at Hampton Court and the Middle Temple (I. Smith 1964: 102). Unresolved is the crucial question of how high the Blackfriars listening space may have been. That all depends on just where in the three-level property James Burbage fitted out his new theater in 1596 and how high the ceiling was. If, as E. K. Chambers and most other theater historians have concluded, Burbage's theater was located in the upper chamber where Parliament had once met, it could have included anywhere from 78,004 cubic feet of listening space to 96,565, depending on the ceiling height (fig. 8.3).* When people are added to the picture, the Blackfriars theater, like the Globe, shapes up as a space in which individual auditors enjoyed a listening space exceeding modern standards. If, as Irwin Smith proposes, the Blackfriars accommodated 512 people, we end up with per-person listening spaces of 152 cubic feet under a 32-foot-high flat ceiling, 188 cubic feet under a 38-foot flat ceiling, and 233 cubic feet under a 53-foot vaulted ceiling. (Compare the Globe's 77.8 cubic feet per person at full capacity or 155.7 at half capacity.) By the standards of modern acoustical engineering, the figures for the Blackfriars fall somewhere in between the optimal listening spaces recommended for speech

---

*Irwin Smith argues from the proportions of the halls at Hampton Court and the Middle Temple that the side walls must have been 38 to 40 feet high beneath a vaulted ceiling extending up another 15 feet (I. Smith 1964: 103). Hosley assumes the side walls to have been 32 feet high and to have ended at a dropped ceiling dating from the days when the hall had been partitioned into seven chambers (Hosley 1969: 79, 83, 87). At 66 × 46 × 32, Hosley's reconstructed space would have had a listening area of 97,152 cubic feet. Assuming that the tiring house took up part of this space, extending the full width of the hall (46 feet), at the same depth as The Fortune tiring house (11 feet 3 inches, allowing for the thickness of the walls), up to the height of the second gallery (25 feet), we can deduct 12,938 cubic feet, plus another 6,210 cubic feet for the size of Hosley's conjectural stage (46 feet wide × 30 feet deep × 4.5 feet high). The listening space would then have been 78,004 cubic feet. Other proposed heights would, of course, yield other estimates of the listening space. Smith's 38-foot walls with a 53-foot vault would (minus tiring house and stage) produce 96,565 cubic feet—about the same volumetric size as the great halls at Hampton Court and the Middle Temple. Adopting the 38-foot height of these extant structures but assuming that a dropped ceiling stayed in place, the volume would be 96,220 cubic feet.

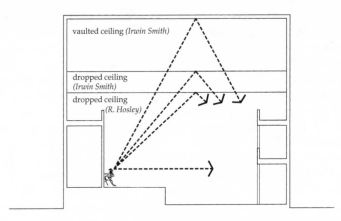

Figure 8.3. Side view of the Blackfriars Theater (1596).

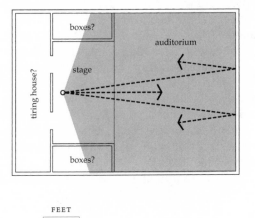

Figure 8.4. Top view of the Blackfriars Theater (1596).

(98.9 cubic feet per person) and for music (251 cubic feet per person) (I. Smith 1964: 42; Egan 1988: 96).

In its shape the Blackfriars theater fostered a very different kind of sound than the Globe. However stage, galleries, and open seating may have been configured, the Blackfriars was a rectilinear space. As such, it *dispersed* sound waves throughout the room rather than *focusing* them in the center (fig. 8.4). Standing at the rear of the stage, a speaker commanded a 140-degree broadcast area (shaded gray) that covered much more of the available listening space than it would have for an actor in the same position in the Globe. What is more, the sound the speaker sent out into the hall would not immediately have been returned to the center, as it would have at the Globe, but would have struck the back wall, bounced to the sides, and only then returned to the center. This dispersal effect would have been enhanced by the multiple planes of the galleries. However deep they may have been, whether or not they ran the full perimeter of the room, the galleries provided a series of differently angled, resonant wood surfaces that contributed to the dispersal of sound in its full range of frequencies.

The reverberant quality of this sound can be calculated by positioning an actor at the rear of the stage and plotting the path of sound to a person seated at the opposite end of the room (see fig. 8.3). The direct path for sound waves would be approximately 50 feet. A dropped ceiling 32 feet above the floor would give reflected sound an angled path of about 70 feet; a dropped ceiling at 38 feet, an angled path of about 80 feet; a vaulted ceiling at 53 feet, an angled path of about 100 feet. For listeners, these differences would be significant. The lowest of the three ceiling heights, 32 feet, would produce a difference between direct and indirect sound of 20 feet—about 0.02 seconds—the same as in the Globe. A difference of 20 feet or less is rated by modern acoustical engineering as ideal for both speech and music. Higher ceilings would have yielded less propitious results. A ceiling height of 38 feet would have produced a difference of 30 feet, still rated as good for speech but only fair for music. A vaulted ceiling at 53 feet would have resulted in a difference of 50 feet between direct and indirect sound, rated as marginal to unsatisfactory (Egan 1988: 96). The scattering effect of hammer-beams, intercepting some of the sound waves before they reached the roof and sending them back towards the floor, might have mitigated the absolute disparity in distance. With or without a vaulted ceiling, the rectilinear surfaces of the Blackfriars theater would have produced a "round" sound quite different from the "broad" sound of the Globe— just the reverse of the effect suggested by the physical shapes of the two structures.

TRUMPETS, DRUMS, HAUTBOYS, CORNETS, RECORDERS, VIOLS

When noise-sensitive Morose vows to stay clear of the theater, he is quite specific about the sounds he does *not* want to hear: "fights at sea, drum, trumpet, and target." Playhouses rank high on Morose's list of the noisiest places in London (Jonson 1925–1963, 5: 169–170, 230). Morose's suspicions about what he might hear are confirmed by the petition raised by neighbors in the Blackfriars in 1596, when

James Burbage bought the Upper Frater and started fitting out a new acting space for the Chamberlain's Men. In addition to the traffic, the neighbors complain, "the same playhouse is so neere the Church that the noyse of the drummes and trumpetts will greatly disturbe and hinder both the ministers and parishioners" (Gurr 1996: 283). Their fears on this point, as it turned out, were misplaced: drums and trumpets do not figure prominently in the plays that were actually performed in the new indoor theater. The outdoor amphitheaters, established for twenty years, were something else again. To popular imagination, brass and percussion seem to have been what these playing places were all about. Customarily it was three trumpet blasts, filling all 231,028 cubic feet of the acoustic space, that signaled the start of performances at the Globe. Thomas Dekker seems wittily mindful of the difference between reading a play and hearing a play when he starts off the printed text of *Satiromastix*, acted at the Globe in 1601, with a kind of prologue *"Ad Lectorem"* ("To the Reader") in which he casts the ensuing list of printing mistakes as a "Comedy of Errors." What the reader sees on the page becomes an equivalent for what he or she would have heard in the theater: "In steed of the Trumpets sounding thrice, before the Play begin: it shall not be amisse (for him that will read) first to beholde this short Comedy of Errors, and where the greatest enter, to give them in stead of a hisse, a gentle correction" (Dekker 1953–1961, 1: 306). Dekker's Epilogue to the same script also invokes the power of trumpets to "set men together by the eares." The members of the audience who especially needed it, or so Dekker's Epilogue implies, were the standees whose proximity to the play is given a distinctly sexual turn: "Gentlemen, Gallants, and you my little Swaggerers that *fight lowe:* my tough hearts of Oake that *stand too't* so valliantly, and are *still within a yard* of your Capten: Now the Trumpets (that set men together by the eares) have left their Tantara-rag-boy, let's part friends." From the "Swaggerers" who are standing below him the speaker then transposes his speech upward to "the Gentle-folkes (that walke i'th Galleries)" (1953–1961, 1: 385, emphasis added). If Dekker can be trusted, plays in London's public theaters began with the auditory focusing of trumpet calls. The plays that ensued were full, not just of human voices, but of sound effects.

Instruments for providing some of those effects are detailed in Henslowe's inventory of the Admiral's Men's goods, drawn up in 1598. As the company's costumes and props make up a palette for *visual* design, so their musical instruments and other sound-producing devices make up a "palette" for *aural* design. Included on Henslowe's list are four groupings of musical instruments: (1) "a trebel viall, a basse viall, a bandore, a sytteren," (2) "j sack-bute," (3) "iij tymbrells," and ₄) "iij trumpettes and a drum." Henslowe's diary for 1598–99 includes sizable payments (up to 40 shillings each, equal to the takings from 480 standees) for a sackbut, a bass viol, and "a drome when to go into the contry," as well as other unspecified "enstrumentes" (1907: 114–118; 1961: 101, 102, 122, 130). What some of the other instruments may have been are suggested by a speech in Dekker's *Old Fortunatus*, acted by the Admiral's Men the year after Henslowe had made his inventory. Shadow comes on while Andelocia is being charmed asleep by a lullaby. "*Musicke still*," reads the stage direction: "*Enter Shaddow*." In describing what he hears, Shadow in effect reiterates Henslowe's first entry and adds one other instrument: "Musicke? O delicate warble [recorder or flute] . . . O delicious strings [viols]: these heauenly wyre-drawers [cittern and bandore] " (Dekker 1953–1961, 1: 138–141; Chan 1980: 31). The cittern and the bandore were both guitar-like instruments, the bandore providing the bass to the cittern's treble (Munrow 1976: 80–83). With the addition of a recorder or flute, the instruments grouped in Henslowe's first entry make up an ensemble that sixteenth- and early seventeenth-century musicians knew as a "broken" consort. (It was "broken" because it was made up not of just one "family" of instruments, like a consort of recorders or viols, but of representatives from several different "families.") Morley's *First Booke of Consort Lessons*, published the same year *Old Fortunatus* was performed, calls for just this ensemble of flute, treble viol, bass viol, cittern, and bandore, with the addition of a treble lute. Philip Rosseter's *Lessons for Consort* (1609) is scored for the same set of instruments. In the case of the Admiral's Men, the standard broken consort might have been supplemented by two other instruments in Henslowe's inventory: the trombone-like sackbut and one or more of the tambourine-like timbrels (Long 1961–1971, 1: 28–29, 34). In such ensembles it was the bowed and blown instruments that carried the tune, the plucked and tapped instruments that provided rhythm. A lutenist, if one was handy, might have offered virtuouso variations on the melody (Chan 1980: 33).

To a different category of sound entirely belong the three trumpets on Henslowe's list, along with the drum. Lacking valves, early modern trumpets were restricted to the equivalent of bugle calls. In the theater their main use was for flourishes, fanfares, and military signals (Long 1961–1971, 1: 25). Other items in Henslowe's inventory are percussion instruments. An entry for "ij stepells, & j chyme of belles, & j beacon" has been interpreted by Michael Hattaway as sets of various kinds of bells: clock bells ("steeples"), hand bells ("a chime"), and a bell for ringing alarums ("a beacon") (1982: 32). David Munrow describes a chime as something more like a set of miniature cymbals: hung in a wood frame, the hemisphere-shaped chimes were struck with hammers, not rung by hand (1976: 34–35). Certain items inventoried by Henslowe among the company's props should be thought about not only as visual icons but as sound-making devices: "j longe sorde," "viij lances," "j copper targate, & xvij foyles," "iiij wooden targates, j greve armer," "j buckler," "j shelde, with iij lyones," and "j gylte speare" would have contributed their distinctive crashes, clinks, and thuds to sounds within the wooden O.

Guns do not figure in Henslowe's inventory, but stage directions occasionally call for the firing of an unspecified form of "ordnance" or "piece" or, sometimes more precisely, of "chambers," small pieces of unmounted ordnance customarily used for firing salutes (OED, "chamber" 10). It was the stage direction "*Drum and Trumpet, Chambers dischargd*" in Act One, scene four, of *Henry VIII* that set the Globe on fire in 1613 (F1623: 1.4.50). Fireworks, like firearms, fail to make Henslowe's list, but exploding squibs were a standard aural event whenever devils arrived on the scene from hell (Leggatt 1992: 67–70). Another stupendous sound effect, usually the aural sign of supernatural happenings, was thunder. Ben Jonson, using the prologue to *Everyman in His Humor* (1616 text) to justify his disdain for such gimcrackery, divulges how thunder was made, by a bullet rolled about, presumably along a wooden timber (3: 303). From ethereal recorders to finely grained viols to blasting trumpets to booming artillery, the outdoor theaters of early modern London were full of sounds besides those made by human voices.

An unspecific stage direction for "*Musicke*" in scripts for the public playhouses is more likely to indicate an individual instrument or a pair of instruments than a full broken consort. The extent of the musical resources of the professional acting companies between 1590 and 1610 is not altogether clear. John Long and Mary Chan have each proposed that before 1590 actors themselves doubled as musicians but that after 1590 they depended more on professional musicians (Long 1961–1971, 1: 30–31; Chan 1980: 32–33). To judge from surviving scripts, ensemble music was required only occasionly in outdoor performances. When something more than trumpets, hautboys, and percussion was called for, it is possible that "waits" or "noises" were hired for the occasion. Professional musicians operated under separate licenses from the Revels Office (Hattaway 1982: 62–63). The establishment of boys' companies in indoor theaters after 1600 changed the aural scene considerably. For one thing, sounds were scaled back in volume. Differences between the two venues can be appreciated through the moment in *The Knight of the Burning Pestle* (acted at the Blackfriars in 1607) when the Grocer calls for shawms. The obliging boy actors tells him that, alas, the company has only recorders (Gurr 1992: 176). Stage directions in Marston's *Sophonisba*, written for the same venue, substitute domesticated cornets for the battlefield trumpets described in the script (Gurr 1992: 176). Even gunfire was toned down. *Love's Pilgrimage*, played at the Blackfriars in 1635, contains an order for a cannon to be shot off, accompanied by the book-holder's direction *"Joh. Bacon ready to shoot off a Pistol"* (Gurr 1992: 177).

The most significant difference between the indoor and the outdoor houses, however, involved music. The boy actors were trained as singers, and music was a major part of the entertainment they offered (Chan 1980: 14–15; Sternfeld 1963: 14–20). Certainly it was music as much as the play that charmed Frederic Gerschow, the Duke of Stettin-Pomerania's secretary, when he and his companions went to the Blackfriars Theater in September 1602 and took in a performance by the boys' company then in residence. "For a whole hour before," Gerschow reports, "a delightful performance of musicam instrumentalem is given on organs, lutes, pandores, mandolines, violins, and flutes" (1892: 29). Stage directions to the 1606 printing of Marston's *Sophonisba* "as it hath beene sundry times Acted at the *Black Friers*" confirms Gerschow's report that "organs" were part of the theater's equipment. The organ's capacity to cover the full range of pitches and volumes of all the other instruments, as well as to blend with the human singing voice, is suggested by the varied combinations specified in *Sophonisba*: "cornets, Organ, and voices" perform a wedding song in act one, "Cornets and Organs playing loud full Musicke" mark the act's end, "Organ mixt with Recorders" does the same for act two, "Cornets and Organs playing full musick" accompany a sacrifice scene in act three, "Organs Violls and Voices" perform between acts three and four, and "Orgaine and Recorders play to a single voice" as funeral music at the tragedy's end (2:1, 12, 18, 32, 36, 43, 63). When the King's Men took over the theater in 1609, the musical consort stayed on and continued to provide pre-play concerts and, on some occasions at least, music between the acts, as well.

In Andrew Gurr's view, the musical consort "brought the largest single alteration to the King's Men's practices when they took over the Blackfriars playhouse." The addition of music proved so popular that the company retrofitted the Globe to include a curtained music room in the balcony above the stage (Gurr 1996: 367–368). In acoustic terms, the effect of these musical preludes was to fill the aural field with sounds across a wide range of pitches. Michael Praetorius catches the effect in his description of broken consorts in *Syntagma Musicum* (1619):

> The English give the name Consort to what is very appropriate to a grouping of instruments (*consortio*), when several persons with various instruments, such as a Clavicymbal or a large Spinet, a large Lyra, a

Double Harp, Lutes, Theorboes, Bandores, Penorcons, Citterns, Bass Viol, a little Treble Fiddle, a Transverse Flute or a Recorder, sometimes also a soft Trombone or a Racket, all together in a Company or Society play with very quiet, soft and sweet accord and harmonize with one another in pleasing symphony. (Galpin 1965: 202)

Contrast with the trumpet blasts that had heralded the start of performances in the outdoor amphitheaters could hardly be sharper. One effect of consort music was to situate the audience within a wider, more fully articulated field of sound than in the outdoor amphitheaters. By the time the play began, audiences had acclimated their hearing accordingly. Human voices emerged from a matrix of bass viol, bandore, treble viol, cittern, and recorder.

LUNGS, LARYNX, MOUTH

Morose to the contrary, what is scored to be played upon in theatrical performances is not primarily drums or trumpets, hautboys or cornets, but human voices. The difficulty, for psychoacoustics, is that the sounds of drums, trumpets, hautboys, and cornets are much easier to specify than the sounds of human voices. Even if Edward Alleyn, Richard Burbage, Will Kemp, and Nathan Field were available to submit their voices to a spectrograph, we still would find it hard to explain their stage success. What we understand by "voice" is, after all, not a *thing* but an *effect*. The thing-ness of voice consists of (1) the body tissues of lungs, larynx, and mouth, (2) moving molecules of air, and (3) the cartilage, flesh, bones, and nerves of the ear. The effect of voice, for speakers and listeners alike, is something more than the sum of these material parts. Quintilian acknowledges as much in the treatise that codified Roman rhetoric for Renaissance schoolmasters: "just as the face, although it consists of a limited number of features, yet possesses infinite variety of expression, so it is with the voice: for though it possesses but few varieties to which we can give a name, yet every human being possesses a distinctive voice of his own, which is as easily distinguished by the ear as are facial characteristics by the eye" (11.3.14–15, with some modifications to the English translation). Quintilian goes on, however, to distinguish two features in the "physiognomy" of voice: *quantitas* and *qualitas*. "Quantity" is the easier of the two to describe, "since as a rule it is either strong or weak, although there are certain kinds of voice which fall between these extremes, and there are a number of gradations from the highest notes to the lowest and from the lowest to the highest." Quantity, that is to say, can be gauged in two ways: (1) by the volume of sound the speaker produces and (2) by the range of pitches he uses. Both things can be measured and specified. "Quality" is another matter altogether, "for the voice may be clear or husky, full or thin, smooth or harsh, narrow or diffuse, rigid or flexible, sharp or blunt, while lung-power may be greater or lesser" (11.3.14–15, in 1921–1922, 4: 251, with some modifications to the English translation).

In this particular passage Quintilian takes breath control to be a question of quality, but elsewhere in the *Institutio Oratoria* he includes it, along with volume and pitch, as a quantitative concern. Depth and frequency of breathing can, after all, be measured, and Quintilian is able to provide precise guidelines, complete with phrase-by-phrase examples (11.3.33–39, 43–57). Quintilian's concern in these passages is with *spatium*, with intervals, timing, rhythm. Managing *spatium* is as much a matter of discipline as volume and pitch are. Breath control is fundamental to all three. Thus, Aristotle specifies rhythm (*rhythmos*) as a third factor to be considered in rhetoric along with volume (*megethos*) and harmony (*harmonia*) (*Rhetoric* 3.1.4 in Aristotle 1941: 1435). Under the rubric of *"Action, or Pronunciation"* these three boundary markers of the vocal field are epitomized in *A Briefe of the Art of Rhetorique. Containing in substance all that ARISTOTLE hath written in his Three Bookes of that subject, Except onely what is not applicable to the English Tongue:* "Tragaedians* were the first that invented such *Action*, and that but of late; and it consisteth in governing well the *Magnitude, Tone*, and *Measure* of the *Voice*; a thing lesse subject to *Art*, then is either *Proofe*, or *Elocution*" (Aristotle 1637: 152).

It is, perhaps, the noncerebral nature of volume, pitch, and rhythm—their brute physicality—that explains why early modern rhetorical treatises typically give much less attention to vocal delivery than to invention and argument (Ong 1968: 39–69). However briefly they may treat the practicalities of speaking what the orator has so meticulously been trained to invent, early modern rhetorical manuals all direct attention to the same three aspects of sound. "Magnitude," "tone," and "measure" are the delineators of voice that were carried over from Aristotle, Cicero, and Quintilian into early modern rhetorical manuals, thence into early modern schoolrooms, and quite possibly thence onto the stages of early modern theaters. At the very least, volume, pitch, and rhythm give us three quantitative reference points for plotting the repertory of voice sounds that scripts for the public stage imply. In treating an actor's voice as a sound-producing instrument possessed of a certain range of volume, a certain range of pitches, and a certain range of rhythms we are following the example of early modern rhetoricians. The *qualitas* of Burbage's voice may be beyond recovery; the *quantitas* is not. Let us consider these three factors one by one.

Magnitude of vocal sound is fundamentally a measure of *space*. In any sound, the air molecules that are being displaced will move in waves of greater or lesser amplitude, depending on how much force has been exerted against the vibrating surface. Those waves strike listeners' ears with correspondingly greater or lesser force. Sounds perceived to be loud take up more space than sounds perceived to be soft. It should come as no surprise that plays designed for the Globe and other large outdoor amphitheaters betray an acute awareness of volume in capturing, holding, and guiding an audience's aural attention. Thomas Dekker, for one, presumes that the audience will be noisy, a roaring crowd that an expert actor can charm into silence. "Giue me *That Man*," says the Prologue to *If This be not a Good Play, The Diuell is in It*, who when a bad play starts emptying the house "Can call the *Banishd* Auditor home, And tye / His Eare (with golden chaines) to his Melody" (Pro.26–36 in Dekker 1953–1961, 3: 121–122). The Prologue to *The Whore of Babylon* goes so far as to pronounce a charm to establish calm within the quadrilateral spaces of the Fortune Theater: "The Charmes of silence through this Square be throwne, / That an vn-vsed Attention (like a Iewell) / May hang at euery eare" (Pro.1–3, 2: 499). To establish aural command, Dekker's usual strategy is to send out a Prologue and have *him* take possession of the acoustic field. If that tactic is successful, any volume of sound can come next. The Prologus to *The Roaring Girl*, for example, warns the audience that "our Scoene, / Cannot speak high," since the subject is mean, "A Roaring Girle (whose notes till now neuer were)." Having piqued the audience's interest ("I see attention sets wide ope her gates / Of hearing, and with couetous listning waites, / To know what Girle, this Roaring Girle should be"), the Prologus yields the stage to Mary Fitz-Allard her/himself, who starts off the play in anything but a roar, in a private scene with the servant Neatfoot (Pro.7–14, 3: 12).

With respect to volume, the Roman rhetoricians had recommended carefully plotted modulations in the course of a single speech. The *Rhetorica ad Herennium*, for example, counsels a "calm tone" [*sedata vox*] in the beginning and "a sustained flow" [*continens vox*], apparently at a relatively high volume, toward the end: "and does not this, too, most vigorously stir the hearer at the Conclusion of the entire discourse?" (3.12.22 in [Cicero] 1968: 195). Entire scripts seem to follow this advice. As plays tend to begin with high-intensity sound, so they tend to end. Most of Shakespeare's scripts, for example, end in public scenes presided over by an authority figure whose political power is presumably measured by his aural power as well as his physical presence. Out of the 39 surviving texts, only a handful do not end with a speech from such a figure, if not with a flourish, a drum roll, or a declamatory epilogue. Notable exceptions are *Henry VI Part One* (which ends with a confidential exchange between Gloucester and Suffolk), *Love's Labor's Lost* (the dialogue between Winter and Spring), *The Merchant of Venice* (quips among the four lovers, with silent Antonio standing by), *Twelfth Night* (Feste's song), and *The History of King Lear* (Albany's two couplets, followed by an *"Exeunt"* without the drum-rolls of the *Tragedy's "dead march"*).

The volume level in early modern performances was a function not only of narrative line but of subject matter, acoustical space, date of performance, and the age of the actors. In general, history plays and tragedies call for more noise than comedies do. As instruments for the actors to play upon, outdoor amphitheaters accommodated these large-volume sounds more comfortably than indoor theaters. One remembers the cornets substituted for trumpets in *Sophonisba* at the Blackfriars in 1606 or the pistol for a cannon in *Love's Pilgrimage* in the same space thirty years later. As designs in sound, Shakespeare's scripts reflect these general trends. At the same time, they exploit the human voice's full range in magnitude, from the whispering nobles in *Henry VIII* (3.2.S.D. before 204) at 30 decibels to Richard III's shout "A horse! A horse! My kingdom for a horse!" at 75 dB. Since decibels are logarithms—each 10 decibels measuring an increase in intensity by a factor of 10—the latter sound is more than 10,000 times greater than the first. Within those parameters range the volume of vocal sounds in Shakespeare's scripts.

The highly reverberant acoustics of the reconstructed Globe suggest that a speaker's output need not have been anything like 75 dB to fill the wooden O. The volume level of normal conversation is about 60 dB at three feet away from the speaker. Within a single bit of speech, variations of up to 26 decibels are possible between the most intense sounds and the least (Handel 1989: 7–72; Fry 1977: 40–60). To some degree, those variations in volume are cued by the phonemes that happen to make up the speech. Unvoiced consonants, for example, come out relatively low in volume—[th] is the weakest—while some vowels come out stronger than others (fig. 8.5). These relative intensities remain more or less constant across changes in the overall force of the speaker's breath, i.e. across changes in the overall volume of the speech. The strongest phoneme of all is [o:]. "O for a Muse of Fire": when the Prologue to *Henry V* attempts to silence the audience gathered within the wooden O, he begins with the most intense phoneme the human voice can make in English speech, followed by the tenth, sixth, and ninth most intense in [u:] and [ai]. Because listeners are bracketing the whole speech as a phenomenon, they will not necessarily perceive [o:] to be louder than other phonemes—listeners *need* [o:] to be louder than [u:] and [ai] in order to hear [o:] as [o:]—but the physical fact of the sound's relative intensity remains. By contrast, a concentration of consonants—particularly [m], [l], [n], and [ŋ]—positively require that the actor playing Ophelia speak relatively softly when he says, "My Lord, I haue re*membr*ances of yours / That I haue *longed long* to rede*l*uver" (*H5* F1623: Pro.1; *Ham* Q1604: 3.1.99–100, emphasis added). In individual lines, as in opening scenes, volume control is written into scripts for the stage.

The projected volume of Ophelia's speech would also have been shaped by the vocal apparatus of the boy actor who pronounced the line. When Orsino tells Viola/"Cesario" "thy small pipe / Is as the maidens organ, shrill, and sound" (*TN* F1623: 1.4.32–33) he is measuring not only the physical size of the windpipe but the volume of sound it makes. With respect to smallness as well as shrillness, the interchangeability of boys' and women's voices is attested by the

oː	29	m	17
o	28	tʃ	16
aː	26	n	15
ʌ	26	dʒ	13
əː	25	ʒ	13
a	24	z	12
u	24	s	12
e	23	t	11
i	22	g	11
uː	22	k	11
iː	22	v	10
w	21	ð	10
r	20	b	8
j	20	d	8
l	20	p	7
ʃ	19	f	7
ŋ	18	θ	—

Figure 8.5. Comparative intensities (in decibels) of English phonemes. From Dennis Fry, *The Physics of Speech* (1979). Reproduced by permission of Cambridge University Press.

Wooer in *The Two Noble Kinsmen*, who tells of hearing the Jailer's Daughter (played by a boy) singing:

> I heard a voyce, a shrill one, and attentive
> I gave my eare, when I might well perceive
> T'was one that sung, and by the smallnesse of it
> A boy or woman.

(Q1634: 4.1.56–59)

The change of voice that males undergo at the age of puberty was understood by early modern physiology to be the result of an increase in heat, which in turn produces a larger body—and a larger voice. "*Why are boyes apt to change their voice about fourteene yeeres of age?*" goes one of the questions in *The Problemes of Aristotle*.

> Bicause that then nature doth cause a great and sudden change of age. Experience prooueth this to be true: for at that time we may see that womens paps do grow great, to hold and gather milke, and also those places which are about the hips, in which the yoong fruit should remaine. Likewise mens breasts and shoulders which then beare great and heauie burthens. Also their stones in which the seed may increase and abide: and his priuie member, to let out the seede with ease. Further al the whole bodie is made bigger and dilated, as the alteration and change of euery part doth testifie.

The windpipe participates in this general enlargement of the body, producing a "larger" (i.e., louder) sound. The harshness and hoarseness so characteristic of adolescent speech is imagined to be the result of uneven expansion in the windpipe (Aristotle 1597: L8^v–M1). That fourteen was the age of male puberty comes, not from the source for this anatomical information in Aristotle's *Generation of Animals* 4.8 (where the ages are unspecified), but presumably from the English translator's own observation and experience. Another of the question-and-answer exchanges in the English edition of *The Problemes* confirms fourteen to be the age of puberty for males and fixes twelve as the age for females, again in the absence of any such indications in the Greek text (Aristotle 1597: C1–C1^v).

If volume is a measure of *space*, pitch is a measure of *time*. In physical terms, the perceived pitch of a sound is a function of its frequency, of how long it takes the displaced air molecules to return to the point of stasis from which the sound wave began. Modern acoustics measures the number of oscillations per second in Hertz (Hz), so that middle C on the piano sounds out at 261 cycles per second, or 261 Hz. As one of Quintilian's objective measures of voice, pitch is in part an aesthetic consideration. Cicero in his treatise *Orator* distinguishes three "tones" or "registers" (*soni*) in the very nature of the voice: "shrill" (*actus*), "moderate" (*inflexus*), and "low" (*gravis*). Out of these three is produced "an accomplished and pleasing variety," which Cicero describes as a species of singing (17.57 in Cicero 1939: 346–349). Beneath artistic choices lay the fundamental facts of human anatomy—and the differences in anatomy between adult males and prepubescent boys. The fact that it was originally boy actors who pronounced now-famous lines like "I am Duchesse of *Malfy* still" gave the line special characteristics in pitch and timbre. The complicated cultural coordinates that allowed boy actors to be substituted for women have been set in place by Stephen Orgel and others (Orgel 1996: 31–82; Levine 1994: 1–25; Rackin 1987: 29–41). Physics and physiology contributed to the illusion. To start with, boys and women possess vocal cords of comparable length. Early modern physiology explained the similarities in sound as a function of the coldness and moistness boys' bodies shared with women's bodies—the same factors that produced, in each case, a "smaller" sound than adult male voices (Aristotle 1597: $C1-C1^v$, $I8^v$, $L8^v-M1$). Cicero observes that "in every voice there is a mean pitch": as it happens, the *mode* of pitch for fourteen-year-old boys and adult females has been demonstrated in modern experiments to be exactly the same. That is to say, the pitch most frequently sounded when fourteen-year-old boys are asked to read aloud from a text is the same as the most frequently sounded pitch by adult females: 261.6 Hz, approximately middle C on the piano. Only a small difference separates the *mean* pitch, the average pitch sounded, in each case: 241.5 Hz for boys (just below the B below middle C) and 220 Hz for women (the A below middle C). The *range* of pitches, as well, is roughly the same, although at the extremes women's voices reach somewhat higher frequencies and boys' voices somewhat lower. It is mainly a more extended lower range that distinguishes fourteen-year-old boys' voices from ten-year-old boys' voices. The modes and the means of pitch are in each case only slightly different (Fry 1977: 26; Zemlin 1964: 150; Curry 1940: 48–62).

In terms of what an audience would hear, the most significant difference between boys' and women's voices involves *harmonics*. What the vibrating piano string sets in motion at middle C is not simply a wave of 261 cycles per second, but a complex wave made up of 261 Hz plus integral multiples of 261 Hz at 522 Hz (261 × 2), 783 Hz (261 × 3), 1,044 Hz (261 × 4), 1,305 Hz (261 × 5), etc. These more rapid cycles, moving through the air all at the same time, constitute the "harmonics" of the sound; the lowest frequency, the one that sets the others in motion, provides the "fundamental" above which the harmonics vibrate. Waves set in motion by the vocal cords work in just the same way. If I sing the equivalent of middle C, I am propagating into the space around me a complex sound wave made up of 261 Hz, 522 Hz, 783 Hz, 1,044 Hz, 1,305 Hz, etc. Because I am using my voice, however, and not striking a piano string, some of these cycles are going to be more important than others. On the piano, the fundamental frequency is the strongest, with the harmonics gradually decreasing in strength up the scale. With voice, the size and shape of the vocal tract is such that the higher harmonics become relatively stronger than the fundamental. Because their vocal tracts are relatively narrower and shorter than women's are, boys when they speak give more prominence to the fundamental at the expense of the upper harmonics. The result is a less complicated sound wave, "purer" in timbre and sharper to the ear. Hamlet seems to be describing this quality when he greets his old friends the traveling players, comments on how much the boy actor has grown, and exclaims, "pray God your voyce like a peece of vncurrant gold, bee not crackt within the ring" (Q1604: 2.2.30–31). "Ring" offers a pun on the shape of the coin, the shape of the windpipe, the shape of the theater, and the "shape" of the boy's sound. "I am Duchesse of *Malfy* still": what audiences at the Blackfriars and the Globe heard in 1614 would have been sounds in the same pitch range as an adult female voice, but more carrying and penetrating.

The aural contrast between boys' voices and men's voices, both in pitch and in timbre, would have been striking. The most frequently sounded pitch for adult males (the mode) is 130.8 Hz, approximately the C one octave below middle C; the average pitch (the mean) is 120 Hz, one semitone lower; the range reaches down well into the next octave (Fry 1977: 26; Zemlin 1964: 150; Fairbanks 1960: 124) (fig. 8.6). The small overlap between the two voices in the region of G below middle C to C below middle C would be more than offset by differences in harmonics: for the pitch of each phoneme, adult male voices would resonate across the full range of harmonics above the fundamental, while boys' voices would ring out closer to the fundamental. These differences in harmonics would accentuate the natural tendency for lower-frequency sounds to be heard as filling the ambient space, in contrast to higher-frequency sounds, which tend to be heard as more localized in space (Handel 1989: 88). In effect, speech sounds gendered as male would pervade the wooden O, filling it from side to side; speech sounds gendered as female would be heard as isolated effects within this male matrix.

Like volume, pitch in the early modern theater varied according to four factors: playing place, genre, date, and the age of the performers playing the protagonists. To gauge the interplay among these four variables, let us compare scenes from three plays at three different

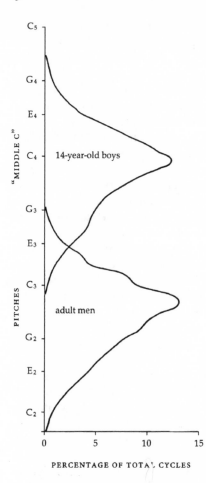

Figure 8.6. Comparative pitch modes: fourteen-year-old boys and adult males.

points in Shakespeare's career. At The Theatre and The Curtain, north of the City, Shakespeare's company mounted plays in which the dominant sounds were male voices and high-energy percussive sound effects. Loud percussive sounds were, indeed, the company's stock in trade in the 1590s, with history plays providing the most extreme examples. The voice parts in *Richard III* are scored overwhelmingly in the male register; the sound effects—flourishes, trumpets, sennets, drums, alarums—are all assaultive. Act Four, scene four, epitomizes this aural design (fig. 8.7). The scene can be analyzed in what modern actors know as "beats," or units of completed action, each signalled by important entrances and exits. Within each of the beats, rectangles enclose the pitch ranges of the several sorts of instruments, human and mechanical, called for in the script. On the upper, treble clef horizontal rectangles indicate the pitch range for boys' voices that speak

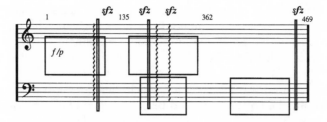

Figure 8.7. Pitches and volume levels, *Richard III,* 4.4.

the lines of the women characters who appear in this scene: Elizabeth, Margaret, and the Duchess of York. On the lower, bass clef horizontal rectangles indicate the pitch range for the adult male voices that speak all the other parts. Slender vertical rectangles extending across both clefs indicate the effective pitch range of trumpets. Finally, jagged vertical lines across both clefs indicate the random frequencies of drumming and other noises. In the first sequence (lines 1–135) Elizabeth, Margaret, and the Duchess of York form a chorus of wailing women. The three actors make their treble lamentations, presumably at a forceful volume (marked *forte,* or loud, in the score), in counterpoint to Margaret's asides, presumably at lesser volume (marked *piano,* or soft, in the score). Their litany is at last interrupted by "*K. Richard marching with Drummes and Trumpetts*" (Q1597: S. D. after 4.4.135), the volume of which is indicated in the score as *sforzando,* or suddenly loud. In the second sequence (ll. 136–362) Richard confronts the women. Although the women are scripted to speak more lines than Richard and his male compeers, Richard repeatedly tries to silence their demands for their murdered husbands and sons:

> A flourish Trumpets, strike Alarum Drummes:
> Let not the Heauens heare these Tell-tale women
> Raile on the Lords Annointed. Strike I say.
> > *Flourish.   Alarums.*
> Either be patient, and intreat me fayre,
> Or with the clamorous report of Warre,
> Thus will I drowne your exclamations.
> > (F1623: 4.4.149–154)

When the Duchess of York persists, Richard threatens more alarums: "Strike vp the Drumme" (4.4.180). Extended speeches between bass-clef Richard and his treble-clef interlocutors give way to one-line exchanges (ll. 274–308), one-line exchanges to a virtual oration by Richard (ll. 328–348) as the king takes firm command of the aural field. The women exit, setting up the final sequence of the scene (ll. 364–469), in which Richard confers with his peers, receives news from four messengers, and prepares to go off to battle. The scene ends with a scripted *"Florish"* of trumpets (S.D. after 4.4.469). The play as a whole ends with a great deal more noise and fanfare—but no more treble voices other than the brief words of Lady Anne's ghost (5.5.113–120).

Brash sound was an effect for which theaters north of the city—the Curtain, the Fortune, the Red Bull—remained famous, long after the indoor theaters of the City were plying subtler designs on listeners' ears. The actors at the Fortune were famous for their "sesquipedales" sound: a deep, resonating bass. However exuberantly Shakespeare's company may have exploited the capacity of amphitheaters for broad, booming sound, they could play the instrument to very different effect in comedy. *Twelfth Night* (1601), like most of Shakespeare's earlier comedies, offers a wider range of pitches than history plays and tragedies, with much more prominence given to higher-frequency sounds. From the beginning of the play to the end, treble-clef sounds move in counterpoint to bass-clef sounds in a manner that comes close to turning the play's musical metaphors into acoustic fact. "If Musicke be the food of Love," *Twelfth Night* provides a rich banquet (F1623: 1.1.1). The heightened range of pitches in the play is a function in part of the large roles assumed by female characters, Viola and Olivia in particular. The quality of spoken sound in the treble register is indicated by Orsino's comment on Viola/"Cesario'"s "small pipe." If Viola and Sebastian are visually twins, then they likely were so aurally as well. Certainly, Olivia hears no difference when she takes Sebastian for "Cesario" in 4.1, 4.3, and 5.1. Sebastian's part almost certainly, therefore, belongs to the treble clef.

Counterpoint to the play's ample treble sounds is provided by the subplot. Aside from Orsino and Antonio, the play's bass-clef sounds are grounded almost entirely in the antics of Sir Toby, Sir Andrew, Fabian, and Malvolio, as played out in 1.3, 1.5, 2.3, 2.5, 3.2, 3.4, and 4.2. Even the actor playing Maria may have contributed to the bass-ness of the subplot. Henslowe's notation of an expenditure "for bornes womones gowne"—William Borne (*alias* William Bird) being an adult member of Admiral's Men at the time of the notation—has persuaded some theater historians that men played some female roles, particularly wily maid-servants and other comic roles (Rutter 1984: 124). In this view, "high mimetic" drag for the heroines would have contrasted with "low mimetic" drag for comic characters, just as in *commedia dell'arte*. Some support for this argument may be found in the 1623 printing of *The Duchess of Malfi*, which states that Robert Pallant, who played the Duchess's maid Cariola, doubled as the Doctor. If so, the effect in scenes of dialogue between the Duchess and Cariola would have been an accentuation of the Duchess's acoustic isolation. If an adult male played Maria in *Twelfth Night*, the effect would likewise have been an accentuation of higher-frequency difference.

In between treble and bass come the musical elements in the play. Music with "a dying fall" opens the play, possibly in the form of a lute solo in the doleful style John Dowland had made popular. (Dowland's *First Booke of Songes or Ayres of Foure Partes with Tableture for the Lute* had appeared in 1597, *The Second Booke* just the year before the play.) Certainly a melancholy note is sounded in each of the songs scripted to be sung by the fool Feste: "O Mistris mine, . . . Youths a stuffe will not endure" in 2.3, "Come away, come away death" in 2.4, snatches of "Hey Robin, iolly Robin, . . . My Ladie is vnkind, *perdie*" in 4.2, and "When that I was and a little tiny boy," which closes the play. A certain droll edge seems to have suited the clown Robert Armin, who had probably joined the company in time to play Touchstone in *As You Like It* as well as Feste in 1601 (Wiles 1987: 144–158). Surviving transcriptions of "When that I was and a little tiny boy" indicate that this song, if not the other three, was performed in a pitch range approximate to a modern tenor.*

When the King's Men took over the Blackfriars in 1609, they moved into an acoustic environment that enhanced the range of pitches sounded in certain earlier plays like *Twelfth Night*. As the first play Shakespeare is likely to have written expressly for the Blackfrairs Theater, *The Tempest* exploits the acoustic potentialities of the new space to the full (Gurr 1996: 367). The script presents an acoustic design with a complexity and subtlety approaching consorted music. Prospero's voice may be the most prominent sound in the mix, but it is surrounded by sounds in a variety of registers, at a variety of volumes. Act Four, scene one, typifies the play's rich acoustic texture (fig. 8.8). The scene can be analyzed in four beats. In the first (ll. 1–33) Prospero exchanges congratulatory speeches with Ferdinand, who has performed the onerous tasks Prospero has set for him. Miranda stands silently by. As Ferdinand and Miranda sit apart, Ariel makes an entry, inaugurating the second beat (ll. 34–163), one of the most acoustically varied sequences in all of Shakespeare's work for the stage. Ariel's speeches, assuming they were spoken by a boy, belong to the treble clef. "*Soft musick*," provided by the Blackfriars consort,

---

*Determining the precise pitch-ranges both for singing voices and musical instruments is difficult, since there were no absolute standards of pitch in early modern performance. Pitch would be established according to the characteristics of the instruments being played, with the result that notated pitch might differ from actual pitch by as much as a fourth or a fifth. For this information I am indebted to Philip Pickett, music adviser to Shakespeare's Globe in London.

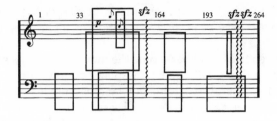

Figure 8.8. Pitches and volume levels, *The Tempest*, 4.1.

accompanies the entry of Iris and the beginning of the wedding masque Prospero conjures for Miranda and Ferdinand (F1623: S. D. before 59). As the score indicates, the full diapason of sounds in the consort's playing, from E_1 more than two octaves below middle C to E_7 more than two octaves above, surrounds and subsumes the pitch ranges of human voices, both as they speak and as they sing. In doing so, the musical consort repeats the effect it had accomplished before the play began: it fills the field of sound across a wide range of pitches, making bass-register male voices part of a much larger whole. Assuming that boys played Iris, Ceres, and Juno, the speaking parts in the masque are all treble, reaching their climax in the song "Honor, riches, marriage, blessing" that the three goddesses sing together. The only lower-frequency sounds in the second beat are the brief speeches Prospero and Ferdinand exchange in between the goddesses' song and the dance of the nymphs and reapers.

The ethereal harmony is interrupted, however, by Prospero's sudden interjection "I had forgot that foule conspiracy of the beast *Calliban*, and his confederates," followed by *"a strange hollow and confused noyse"* that disperses the dancers (F1623: 4.1.139–141 and S.D.). The noise, indicated by jagged lines, ends beat two and starts beat three (ll. 164–193), in which Prospero gives Ariel his instructions for trapping the conspirators with fine apparel. The final beat (ll. 194–264) returns the sound pattern solidly to earth in the speeches of Stephano, Trinculo, and Caliban. *"A noyse of Hunters"* injects cacophony into this lower-frequency matrix, followed by the high-pitched shouts of Prospero and Ariel as they urge on the avenging dog-spirits:

PRO[SPERO]
    Hey *Mountaine*, hey.
AR[IEL]
    Siluer: there it goes, *Siluer.*
PRO[SPERO]
    Fury, Fury: there Tyrant, there: harke, harke.

In the scene's last moments the musical harmony of the masque devolves into what Ariel describes as violent noise: "Harke, they rore" (ll. 254–255, 259, with S.D.). It was just such an acoustic assault—"*A tempestuous noise of Thunder and Lightning*"—that framed the play's polyphony of voices and music in the beginning (S.D. before 1.1.1). For all the acoustic complexity of the play's inner scenes, *The Tempest* ends firmly in the bass register, with the voices of Prospero and other male characters accounting for 90 percent of the sound in Act Five.

The prominence of treble voices in *The Tempest* may be partly an effect of what audiences had come to expect in the Blackfriars in the years just before the King's Men took over the house. Performances by boys' companies at the Blackfriars, in the earlier Paul's playhouse, and elsewhere in the City would have presented an entirely different acoustical profile from performances by predominantly adult companies on the South Bank and north of the walls. As long as the boys' voices had not changed, the pitch range of the sounds they produced would have *all* been in the treble register. The effect would have been altogether delightful in plays like Lyly's *Sapho and Phao* and *Gallathea*, the "musty fopperies of antiquity" (i.e., out-of-date plays from the 1580s) that Paul's Boys and the Blackfriars Boys first put on when they returned to the stage in 1599–1600 after a ten-year absence (Shapiro 1977: 109–110; Gurr 1996: 337–365). But all-treble voices would have offered a curious effect, to say the least, in the "railing plays" for which the companies soon became famous—plays like *A Mad World, My Masters, Michaelmas Term, The Dutch Courtesan, Eastward Ho!, Northward Ho!, Westward Ho!, Epicene, The Isle of Gulls, Satiromastix,* and *The Knight of the Burning Pestle.* Later criticism knows these "railing" plays as "city comedies." While adult companies did perform such plays—Jonson's *The Alchemist*, for example, was written for the King's Men in 1610—it is striking how many "city comedies" were in fact originally performed by boys' companies. It may not be genre alone that accounts for the relatively greater presence of female characters in these plays. For male characters, however, audiences would have heard an aural discrepancy between speakers and speeches that was not unlike the visual discrepancy they saw between boys' supposedly innocent bodies and the often lewd adults they were impersonating. The "rounded" sound of the indoor playing places would have given these treble voices greater presence than in the "horizontal" soundscape of the amphitheaters, but the acoustic effect would still have lacked the "depth" provided by men's voices. All in all, one can imagine a piping, squawking, chattering effect. That, at least, is what Rosencrantz describes when he tells Hamlet about the new boys' troupes in the city, "an Ayrie of children, little

Yases, that crye out on the top of question; and are most tyrannically clap't for't" (F1623: 2.2.340–342). Eyases are young hawks, screechy in voice, aggressive in body. In performances by the boys' companies some aural relief from the higher-frequency mean may have been provided by members of the troupe who had passed puberty. As the companies gradually became professionalized in the course of the second decade of the century, some of the original cast members stayed on, transforming the boys' companies of 1600 into what might better be styled the "youths' companies" of 1615 and later (Gurr 1996: 359–361).

Venue, genre, and date: the three factors that shaped pitch definition in early modern theater interacted in complicated ways. In general, however, we can describe from 1590 to 1615 a move toward greater variety and subtlety in the range of pitches scripted to be heard. Partly that trend was the result of the companies' exploitation of interior spaces as opposed to outdoor spaces, partly the result of new prominence given to consorted music, partly the result of a shift in repertory from history plays toward tragicomedies. In acoustic design as well as in narrative line, *Henry VIII* (1613) stands at the pivot point of this acoustic shift. From history plays of the 1590s come the flourishes of trumpets in 1.4, 2.4, 4.1, 5.3, and 5.4, the cornets that announce the king's entrances in 1.2 and 2.4, the beating of drums in 1.4, and the *"Noyse and Tumult within"* in 5.3—not to mention the *"Chambers discharged"* in 1.4 that set the Globe's thatched roof on fire (F1623: 584, 565). To the indoor theaters of the new century belong the hautboys that accompany the banquet and masque in 1.4, as well as the large roles assigned to Queen Katherine, Anne Boleyn, and an Old Lady.

Through all the changes in acoustic design, however, the adult companies maintained the centeredness of sound in the bass clef. However rich and varied the treble effects may have been, the base line remained the bass line. The equivalent in consort music would be the use of the largest viol as a "ground bass" that supports all the sounds above and defines the shifts in harmony. Among the very few plays that might seem to challenge this pattern *Antony and Cleopatra* (1606) is the most conspicuous. Cleopatra's share of lines in the play is large: she speaks 622 lines to Antony's 766. In the last scene alone she and Charmian account between them for 60 percent of the lines, wresting the mode of pitch into the treble range for the first time in the play. An analysis of 5.2 by beats indicates, however, the way male voices return the play to the lower-frequency norm that has obtained all along (fig. 8.9). Beat one (ll. 1–108) finds Cleopatra, attended by Iris and Charmian, as she receives Proculeius and Dolabella. A flourish of trumpets announces the second beat (ll. 109–203), in which Caesar

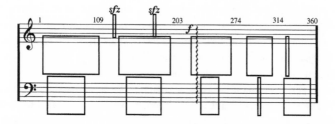

Figure 8.9. Pitches and volume levels, *Antony and Cleopatra*, 5.2.

enters and confronts the queen. In the third beat (ll. 204–274) Cleopatra receives the clown with his deadly basket of figs. In all of these exchanges with male speakers—even with Caesar—Cleopatra maintains her aural command of the field of sound. That authority continues in the fourth beat (ll. 274–314), in which Cleopatra and Charmian are left alone to kill themselves. Only the brief entry of two guards at the end interrupts the higher-frequency register of the play's climactic scene. The proportions are reversed, however, in the final beat (ll. 315–360), in which Caesar reenters to take control: of the stage, of the story, of the field of sound.

Volume and pitch shape the experience of stage plays in complicated ways, but among the three measures of quantity in voice, rhythm is the most basic: it subsumes the other two measures, since changes in volume and pitch occur in regular patterns. The rhythms of speech sound out in a range of phenomena: beats or meter, pace, stress, pauses and attacks, contours of intonation (Handel 1989: 383–459). In physical terms, an actor produces a continuous stream of sounds, through a process speech physiologists call "co-articulation," but he marks certain elements in that continuous stream of sound in certain periodic ways. Members of the audience listen in readiness for those regularities and use them to group the sounds into meaningful patterns. The rhythm of an actor's speech is the aggregate of all these patterns: from split-second iambs to five-second clauses to whole sentences lasting up to a minute. (Modern research indicates a normal speaking rate of 5 syllables per second, or 160 words per minute [Handel 1989: 48].) Many of these rhythmic patterns are acoustic: some phonemes, for example, last longer than others and so call attention to themselves. Other patterns are deliberate, the result of emphasis an actor wishes to give to particular words and phrases. Ultimately, as Stephen Handel argues, all patterns of rhythm derive from the capacities of lung, larynx, and mouth to produce speech sounds:

> Speech involves the complex coordination of many articulatory components, and each component has its own set of dynamic movement con-

straints and possibilities. The lips, tongue, jaw, and glottis cannot open, close, or move instanteously, because of inertia, muscular slack, and limitations of the neuromuscular system. This implies constraints on the possible speech rhythms; some rhythms simply may not be possible to achieve. It is for this reason that ultimately our understanding of speech rhythms must refer to articulatory dynamics. It may be that language capitalizes on the articulatory restraints to generate distinctions between elements. (1989: 420)

The most fundamental restraint of all is breath. How rapidly or slowly an actor speaks, what stresses he gives to particular phonemes, where he pauses and for how long—all of these choices are a function of how often and how deeply a speaker breathes. *Virtus distinguendi*, "excellence in separating," is how Quintilian describes this particular skill, and he devotes detailed analysis to it in the *Institutio Oratoria*. The elements a speaker must learn to distinguish as he speaks are three: *commata, colon*, and *periodus* (11.3.35–39 in 4: 260–263). For Quintilian, these are, in effect, sound bytes of varying lengths, elements that compose the rhythm of speech. Only by a kind of synecdoche has print culture turned commas, colons, and periods into marks on the page. In the terms of Renaissance rhetoric, commas, colons, and periods are not just signs demarcating particular units of speech; they *are* those units of speech. The literal Greek meanings—*comma* is a cutting off, *colon* a member or a limb—situate speech just where it comes from: in the speaker's body, not on the page.

The coporeality of speech-making is the very foundation of John Hart's *An Orthographie, conteyning the due order and reason, howe to write or paint th['] image of mannes voice, most like to the life or nature* (1569). Latin rhetoricians, Hart explains, have translated *colon* as

> *artus membrorum* or *internodium*, which is the space, or the bone, fleshe and skinne betwixt two joyntes, and so (accompting a full sentence, as a complete bodie) these two prickes may well signifie a great part therof,: as of the body, may be taken from the ancle joint to the knée, and from the knée to the huckle or buttock joynt: and knowing thereby that there is more to come, whereas the other first rest of *comma*, doth but in maner devide the small parts (betwixt the joynts) of the hands and féete.

If commas are the bones of the hands and feet, if colons are thighbones, periods figure as the head, "the end of a full and perfite sentence" (1955, 1: 200). The human body provides Hart with one metaphor for speech-making; circularity provides another. The Latin equivalents that Cicero in *Orator* finds for the Greek term *periodos* all suggest roundedness: *ambitus, circuitus, comprehensio, continuatio, circumscriptio* (1939: 3.61.204). "Period," the term in early modern English for the same phenomenon, likewise captures the circular quality of speech-making. Circularity, no less than the joints of the human body, is a "natural" way of visualizing speech. The shape of the field of sound is in fact a circle. Circularity is to the physics of speech-making what the human body is to the physiology. Read on the page, a period is the end of something; heard within the circular field of sound, a period is something in itself, and the "shape" of that something is round. Comma, colon, and period occupy three conceptual categories at once: physiological, rhetorical, and orthographic. They are members of the body, members of speech, and members of a sign system, all at the same time. Spoken within the wooden O, they are also *acoustic* phenomena—dimensions of speech that can be heard.

Such a view of the matter has important consequences for early modern "pointing"—particularly the pointing of texts designed to be declaimed aloud. In fact, not just one system of punctuation obtained in early modern English but two: the older one based on the sound-producing capabilities of the human body existed side by side with—or perhaps beneath—a newer one based on the abstract logic of syntax. "Physiological" and "syntactical" are Walter Ong's terms for these alternative systems of punctuation (Ong 1944: 349–360). "Physiological" punctuation marks the places where a speaker would breathe and raise or lower the pitch of his voice; "syntactical" punctuation marks the separation of sentence elements according to the logic of Latin grammar. Poets, scribes, and grammarians of late antiquity and the Middle Ages had subscribed to the former system; "scientific" students of language in the sixteenth century were arguing for the latter system (Cruttenden 1992: 55–73; Parkes 1993: 50–61). What the rationalizers wanted to do, in effect, was to shift the site of speech from the thorax to the brain—and in that shift to insert one further clause in the Cartesian divorce of mind from body. That ontological step was one that most writers of Shakespeare's generation had yet to take. They wrote in a palimpsest of two different ideas about how writing is related to speech. Take, for example, Ben Jonson's pronouncement in *Timber*: "No glasse renders a mans forme, or likenesse, so true as his speech. Nay, it is likened to a man; and as we consider feature, and composition in a man; so words in Language: in the greatnesse, aptnesse, sound, structure, and harmony of it." The interpolations here of [,], [:], [;], and [.] serve to divide the statement up into logical units: "likenesse," for example, is marked off as an appositive of "man's forme." But the pointings also indicate relatively shorter or longer pauses for breath according to a scheme Jonson articulates in *The English Grammar*. Semicolons require "a meane breathing," commas "somewhat a longer breath," colons and and periods "a more full stay" (1925–1963, 8: 551–552). (Jonson's disagreement with most authorities on the relative time values of colons and semicolons, or perhaps his confusion in *The English Grammar,* is understandable, since [;] had been introduced into printing by Aldus Manutius as recently as 1591 [1925–1963, 11: 209–210].) In this view, the phrase "or likenesse" is less an appositive than a "member"—a "comma"—that asks for a short pause before and after and so acquires a certain aural emphasis. By Jonson's own reckoning, what we have in the whole statement are ten commas, four colons, and two periods. Early modern "pointing" points, then, in two directions at once: toward a linear flow of logic, but even more insistently toward a circle of sound.

The effect of this double orientation is especially pronounced in scripts for the stage. If William Shakespeare's hand is indeed Hand D in *The Booke of Sir Thomas Moore,* it is clear that Shakespeare thought up speeches, wrote them down, and heard them out, all with physiological notation in mind. By modern standards, the 147 lines of Addition II-D to *The Booke of Sir Thomas Moore* is remarkably *under*punctuated. Where modern syntax would distinguish ninety or more separate sentences, the manuscript contains the scribal equivalents of only seven periods (all of them indicated by a [·] well above the base line), one colon, four semicolons (all rendered as [·,]), and 38 commas (Clayton 1969: 22). (By contrast, Stanley Wells and Gary Taylor mark 93 separate sentences in the Oxford text.) Such marks as the writer has chosen to make are all what Richard Mulcaster in his treatise on "the right writing of our English tung" calls "characters signifying . . . but not sounding" (Mulcaster 1582: 148). To "hear" these speeches, we must erase our modern expectations about punctuation, as well as any impulse we might feel to supply silently the punctuation marks that "ought" to be there by modern standards. The result is a set of speeches that *sound* very different from the way they *look* in modernized transcriptions like the one included in the Oxford Shakespeare. The verbal design of *Sir Thomas Moore,* Addition II-D, pits the London mob's one-, two-, and three-line outbursts against More's increasingly longer, ever more eloquent, and finally persuasive speeches. Against the mob's idea of driving away foreigners More argues:

> graunt them remoued and graunt that this yor noyce
> hath Chidd downe all the matie of Ingland
> ymagin that yo^u see the wretched straingers
> their babyes at their backs, w^t their poor lugage   75
> plodding tooth ports and costs for transportacion
> and that yo^u sytt as kings in your desyres
> aucthoryty quyte sylenct by yor braule

and yo^u in ruff of yor opynions clothd
what had yo^u gott ·,· Ile tell yo^u, yo^u had taught   80
how insolenc and strong hand shoold prevayle
how orderd shoold be quelld, and by this patterne
not on of yo^u shoold lyve an aged man
for other ruffians as their fancies wrought
wth sealf same hand sealf reasons and sealf right   85
woold shark on yo^u and men lyke ravenous fishes
woold feed on on another.
                    (Wells and Taylor 1987: 463–467)

What this draft encodes for an actor playing More is not just the
words he should speak but the pace and emphasis with which he
should speak them—or at least the pace and emphasis with which
Shakespeare in the act of writing "heard" such an actor speaking the
words. As punctuated, the entire first nine lines are cast as a single
aural unit that reaches an explosive climax in "What had yo^u gott"
(l. 80). In the tension of slowly exhaled breath that mounts from line
72 to line 80, only a single brief pause is explicitly signalled (after
"their babyes at their backs," l. 75), though other pauses may be implicit in the line-endings that terminate "hath Chidd downe all the
matie of Ingland" (l. 73) and "plodding tooth ports and costs for
transportacion" (l. 76). The idiosyncratic notation of three pricks (·,·)
that follows "What had yo^u gott" signals a major pause—indeed, the
one major pause in the entire speech, which happens as well to be its
logical turning point. The second half of the speech is likewise notated as a single rhetorical turn, punctuated with minor pauses—a
turn that reaches its climactic completion in "woold feed on on another" (l. 87). More's speech has its effect.

Doll   before god thats as trewe as the gospell
lincoln   nay this a sound fellowe I tell yo^u lets mark him

Logically, of course, Lincoln says *two* things in his single line—
(1) "Nay, this' a sound fellow, I tell you" and (2) "Let's mark him"—
and that is just how his line is punctuated in the modernized Oxford
text (l. 98). Physiologically, however, his line is marked as sounds
projected in the course of a single breath.

In the "physiological" scheme of punctuation favored by the
writer of *Sir Thomas Moore*, Addition II-D, commas indicate not just
pauses for breath but the rhetorical events that dictate those pauses—
in particular, shifts in pitch and volume. Take for example More's
peroration at the end of the scene. What if *you* were in the position
of the strangers in England, More asks the London mob. What if *you*
sought refuge in France or Flanders, in Germany, Spain, or Portugal?

Why yo^u must needs be straingers ·, woold yo^u be pleasd   130
to find a nation of such barbarous temper
that breaking out in hiddious violence
woold not afoord yo^u, an abode on earth
whett their detested knyves against yor throtes
spurne yo^u lyke doggs, and lyke as yf that god   135
owed not nor made not yo^u, nor that the elaments
wer not all appropriat to yor Comforts ·
but Charterd unto them, what woold yo^u thinck
to be thus usd, this is the straingers case
and this your montanish inhumanyty   140

The commas in lines 133, 136, and 138 may indicate brief pauses, but all of those pauses are occasioned by rises in pitch and volume on *you* (ll. 133 and 136) and *them* (l. 138). The implicit contrast between *you* and *them* in lines 130 to 136 is made explicit by the full stop that follows "Comforts" in line 137. Before that stop the talk is of *you;* after that stop the talk is of *them*. Syntactical punctuation has no way of indicating this emphasis, short of the italics to which I myself have resorted. For notating dramatic speech, syntactical punctuation is frustratingly rigid and astonishingly inefficient. The fact that none of the earliest printed texts of Shakespeare's scripts is as lightly punctuated as the draft of Addition II-D to *The Booke of Sir Thomas Moore* should make us wary of reading any of these texts as a precise index to stage practice. At the same time, we must realize that the earliest texts, because they were printed and read within an *episteme* that gave primacy to speech, situate character very differently from texts that have been edited according to the standards of syntactical punctuation.

Sound in early modern theater is important not so much for what it *is* as for what it *signifies*. What audiences actually heard in the theater and what they *imagined* they heard may not always have been the same thing. In the printed text of *Coriolanus,* premiered at the Globe in 1608, there appears a stage direction that calls for one set of sounds while the accompanying speech describes another set. Act Five, scene four, is one of the play's several crowd scenes that are amplified by drums and trumpets:

MES[SENGER]
　　Why harke you:
　　Trumpets, Hoboyes, Drums beate, altogether.
　　The Trumpets, Sack-buts, Psalteries, and Fifes,
　　　　Tabors, and Symboles, and the showting Romans
　　　　Make the Sunne dance. Hearke you.　*A shout within.*
　　　　　　　　　　　　　　(F1623: 5.4.49–52)

What the audience in fact hears are trumpets, hautboys, and drums—loud enough in themselves. What the Messenger *tells* them they are hearing is a much wider range of instruments and a volume of sound that, figuratively at least, pushes beyond the theater's walls to the limits of the cosmos. Pierre Iselin has called attention to the way in which music in early modern scripts is *always* framed by language—and usually, Iselin argues, in an ironic way that keeps language firmly in control of musical sounds (1995: 96–113). The moment in *Antony and Cleopatra* illustrates Stephen Handel's point that sound is perceived at three distinct levels (1989: 181–182). In trumpets, hautboys, and drums the audience gathered in the Globe would have heard, first of all, certain *physical* phenomena: a range of distinct frequencies and intensities, particular patterns of attack and decay. At the same time, they would have heard certain *perceptual* phenomena that are not so easy to calibrate: "brightness" in the trumpet, "pointedness" in the hautboys, "dryness" in the drums. The Messenger's speech invites the audience, finally, to hear certain *imaginative* phenomena, to hear the sounds *as objects*. Most obviously those objects are the ones named in the Messenger's speech: trumpets, sackbuts, harps, fifes, drums, cymbals, a mob of people. Beyond that, there is the essence of these individual objects: "trumpet-ness," "drum-ness," "mob-ness." By a process of metonymy, the audience also hears the essence of all these objects taken together: danger, anarchy, chaos. What the sounds mean is the result of all three kinds of phenomena—physical, perceptual, imaginative—impinging on the audience's senses at the same time.

What the audience hears, in the last analysis, is not just physical properties of sound, nor even psychological effects, but the acoustic equivalent of a visual scene—an "aura," perhaps. Evidence from scripts written for the outdoor theaters from 1590 to 1610 invites us to distinguish several distinct "auras" or "aural scenes." Brass instruments define what might be called "the royal scene" or, more broadly, "the power scene." High in pitch, forceful in volume, quick in attack and decay, cornets and trumpets produced sounds that were sharp, hard, and bright—properties that were assumed by the royal personages who made their entrances to such sounds. A different sort of aura, "the hunt scene," was established by wind horns. *"Winde hornes. Enter a Lord from hunting, with his traine"* (F1623: Ind1 S. D. after 13): the broad, plangent bursts required in the Induction to *The Taming of the Shrew* are also scripted to be heard in *Titus Andronicus* 2. 2., *A Midsummer Night's Dream* 4.1, *The Tragedy of King Lear* 1.4, and *A Woman Killed With Kindness* (scene 3). "The combat scene" assaulted the audience's ears in bursts of brass, the rumbling of drums, and the bellowing of gunfire. The explosion of firearms, let us recall, ranks among the very loudest sounds anyone was likely to hear in an age before internal combustion engines. Quick in attack and decay, running the gamut from the trumpet's keening to the drum's riot of multiple pitches to the ordnance's chaos of noise, the sounds of the combat scene served to evoke pitched battles. The same ensemble of sounds might also be used in connection with sword fights. In both the 1604 quarto and the 1623 folio of *Hamlet* gunshots accompany the fencing match between Hamlet and Laertes. "The game scene" takes on aural shape in the tabor's low tap and the pipe's high whistle, as in the morris dance performed in Munday's *John a Kent and John a Cumber*, probably acted by the Admiral's Men in 1589. If pipe and tabor accompanied jigs at the ends of plays, as Kemp's extratheatrical exploits suggest they did, then the game scene of folk festivity provided the sounds ringing in the audience's ears as they left the theater. Although each of these aural scenes has its visual counterpart—presence chamber, woods, battlefield, countryside—each is less a physical place than a kinaesthetic experience. The limits of vision in specifying that experience are indicated by the aura hautboys seem to have created. Technically, hautboys were members of the shawm family, double-reeded instruments whose loud, carrying sound made them a natural for town bands. A persistent distinction in early modern English between shawms in general and hautboys in particular may have turned on pitch or volume or both: "*haut-bois*" means "*high wood.*" Taking the hint, most historians of musical instruments assume that hautboys were shawms in the alto (G_3 to D_5) and soprano (D_4 to A_5) ranges. Their shrill quality, sounding to some witnesses like skirling bagpipes, was proverbial (Munrow 1976: 40–41; Galpin 1965: 123; Long 1961–1971, 1: 20–21). The opening stage direction to *Henry VI, Part II* seems to capitalize on this assaultive quality: "*Flourish of Trumpets: Then Hoboyes*" heralds the entrance of King Henry and his court (F1623: S.D. before 1.1). Why both kinds of instruments? What hautboys could provide that early modern trumpets could not was melody. First the trumpets establish command over the sound field, then hautboys come into play as music for a stately passage over the stage. It must have been the example of shawms in town bands that cast hautboys as aural components of "the processional scene." The ceremonial movement of bodies in space helps to explain the conventional use of hautboys as accompaniments to dumb shows in entertainments at the universities, the inns of court, and the court of the realm, not to mention "The Murder of Gonzago" in the folio text of *Hamlet* (Naylor 1931: 169; F1623: 775). In *Antony and Cleopatra*, acted at the Globe in 1609, the direction "*Musicke of the Hoboyes is vnder the Stage*" underscores the pageant-like scene in which Hercules abandons Antony. Within that highly reverberant space the sound of the instruments must have been very loud indeed—and its totalizing sweep complete. "Musicke i'th'Ayre," proclaims one of the listening soldiers. The other locates it "Vnder the earth" (F1623: 867). It is tempting to describe such a moment as "the cosmic scene," even if Jonson locates that effect in thunder. Power, hunting, combat, game, processional, supernatural: in each of these distinctive fields of sound human voices find a dominant place, but they share the aural scene with artificially produced sounds. When it comes to human voices within the aura of the wooden O, auditors likewise hear an amalgam of physical phenomena (volume and pitch), perceptual phenomena (qualities Quintilian describes as "clear or husky, full or thin, smooth or harsh, narrow or diffuse, rigid or flexible, sharp or blunt"), and imaginative phenomena (the voice of a king, the voice of a maiden disguised as a page, the voice of an air spirit, the voice of an earth spirit). The object the audience hears in a human voice is character.

9   *Circling the Subject*

Whatever name we may give to self-consciousness—"soul," "spirit," "ego"—it feels as if it exists *in here*. Aristotle called this inner something *psyche*—Latin translations render the word as *anima*—and he imagined it to be dispersed through the whole body. Aristotle's post-classical followers have tended to think of it as something unified, as an entity central to a person's sense of his or her physical being. Bartolomeo del Bene's image of the soul as a walled city gives visible shape to this intuitive phenomenon. Psychoanalytical conceptions of "ego" maintain the sense of selfhood as something centered, something that is felt to be inside. *Dico ergo sum:* if Lacan is right, that is the most any of us can say with confidence. We speak ourselves into being. Whether we follow Aristotle and conceive of this existential being as "soul," or Lacan and think of it as "ego," it is the sound of the subject's own voice that centers the subject's "I" in the world. The subject may speak itself into being, but it does so under certain protocols of talking and listening. It finds its identity through the multiple "speech communities" to which it happens to belong, each with its own codes of what is said to the subject and what the subject may say in return. The political relationship of those speech communities to each other helps to determine my sense of self: how strong or how weak, how unified or how fragmented. Protocols for talking and listening are patently social constructions. It should therefore be possible to reconstruct those protocols for a given place and time, like London in 1600. In the process, we may come to understand some of the ways in which early modern subjects achieved selfhood through speech.

Three distinct scenes of speaking are mapped out in Giovanni Della Casa's *Galateo . . . Or rather, A treatise of the man[n]ers and behauiors, it behoueth a man to vse and eschewe, in his familiar conuersation,* as Englished by Robert Peterson in 1576. Two of those scenes are implied in Della Casa's advice that a gentleman should avoid ostentatious words and inverted syntax: "it behoues a ma[n], not onely to shun[n]e this versifying maner of speache, in his familiar and common discourse, or talke: but likewise eschewe ye pompe, brauery, & affectation, that may be suffred and allowed to inriche an *Oration,* spoken in a publike place. Otherwise, me[n] that doe heare it, will but spyte it, and laughe him to scorne for it." In between oratory on the one hand and familiar and common discourse on the other Della Casa positions sermons in church: "Albeit perchaunce, a *Sermon* may shewe a greater cunning and arte, then common talke. But, *Euerie thing must haue his time and place.*" Della Casa fixes these distinctions in terms of occasion, geographical location, and body movement. Everyone can walk, he points out, but not everyone can dance: "For, he that walkes by the way must not daunce, but goe. For, euery man hath not the skill to daunce, and yet euery ma[n] ca[n] skill to goe. But, *Dauncing is meete for feastes and weddings: it is not to vse in the streetes.* You must then take good heede you speake not with a maiestie" (1576: 88). "A public place" on a ceremonial occasion, city streets in the course of everyday life, a church in time of worship: each constitutes a distinct scenario, with its own cast of characters and its own repertory of speech acts. To these three we can add, in the case of early modern London, a fourth scenario: public theaters during performances of plays. Who is licensed to speak in each of these four situations? Who is required to listen? What sort of interchange is possible between speaker and listener? Who has the last word? Each of these scenes positions individual speakers differently vis-à-vis other speakers. Each offers the subject a distinct means of performing self-hood through what she hears others saying and what she may say herself. In Luhmann's terms, each presents a different configuration of body, society, psyche, and sound as a medium, with different "zones of consensuality" among the four intersecting systems. Let us begin with the most controlled of these scenes and proceed to the most open.

ORATORY

The simplest scene of speaking is one in which the talk is all on one side: *I* talk, *you* listen. However innately appealing that scenario may be to most people—at least when they are on the speaking side of the divide—there are several reasons why oratory provided a model of speech particularly well suited to the requirements of early modern culture. It provided, first of all, the handiest theory of discourse available. From classical antiquity early modern writers and educators inherited an extensive, fully developed system of knowledge about language in Aristotle's *Rhetoric*, Cicero's *De Oratore*, the pseudo-Ciceronian *Rhetorica ad Herennium*, and Quintilian's *Institutio Oratoria*. Pedagogical practices based on these texts assured, furthermore, that early modern youths, when prompted to think about their own speech habits, did so in terms of argument and persuasion. Knowledge about most things was, in fact, communicated in the form of speeches—acoustically in the case of university lectures, figuratively in the case of printed dialogues. Not until the seventeenth century did philosophy begin to abandon dialogue for the authoritative single voice. Erasmus wrote dialogues; Descartes, discourses. Even in Erasmus, however, the equal speakers assembled in Cicero's dialogues have already yielded place to *Magister* and *Discipulus* (Wilson 1985: 24–30; Snyder 1989: 48–49). The Master has virtually taken over the dialogue; the Student's functions are to ask questions and to agree. Such an arrangement fitted perfectly the generally hierarchical structure of early modern English culture, in which everyone was someone else's inferior—even the monarch with respect to God.

Oratory represents, in its purest form, the proposition that *all* human speech is a form of persuasion. The first of John Rainolds's lectures on Aristotle's *Rhetoric* offers a particularly resonant instance. The dialogue in Aristotle's Greek text becomes a monologue as Rainolds tells his Oxford students that it was Aristotle's intention "to teach a theory and a way of proceeding by which we may gain the faculty of perceiving whatever is appropriate for persuasion *in any situation*, and of speaking freely and extensively about it" (1986: 95, emphasis added). In all circumstances, it is not just *what* a speaker says that counts—the reasonableness of his argument—but *how* he says it, the physical force he exerts in the act of speaking. In *De Oratore*, Cicero's spokesman Antonius uses the key word *animus*, breath as well as spirit, to describe how rhetoric works on listeners:

> Now nothing in oratory, Catulus, is more important than to win for the orator the favour of his hearer, and to have the latter so affected as to be swayed by something resembling an impulse of the spirit [*impetu quodam animi*] or emotion [*perturbatione*], rather than by judgement or deliberation. For men decide far more problems by hate, or love, or lust, or rage, or sorrow, or joy, or hope, or fear, or illusion, or some other inward emotion [*aliqua permotione mentis*], than by reality, or authority, or any legal standard, or judicial precedent or statute. (2.42.178 in Cicero 1967, 1: 324–325, with emendations to the English translation)

Words like *per-motio* (literally "through-movement") register a sense of oratory as bodily exertion—a perfect example of the "gestalt of *force*" Mark Johnson considers characteristic of human speech in general.

Among the various "force structures" Johnson distinguishes, oratory deploys blockage, diversion, and attraction, but it functions primarily as a form of compulsion (1987: 41–64). (See chapter 2, above.) In psychoanalytic terms, oratory affirms the power of "I" to speak without any need for dialogue. In political terms, it subjects the listener to the speaker's will. In doing so, oratory does overtly what all forms of talk aspire to accomplish—or so it seemed to the writers of early modern conduct books. Nicolas Faret is unusually forthright about the politics of speech in *The Honest Man; or, The Art to please in Court* (1630, English translation 1632). The "honest" (read "honorable") man treats "courtesy" like capital. A master economist, he can decide, based on the situation at hand, just how much of his courtesy to spend and how much to retain:

> His iudgment is so fit to finde tempers in all things, that without flattery & obsecuiousnesse, hee will obserue that rule of *Epictetus*, who doth aduise vs to yeeld without resistance, to the opinions and wils of great men, to consent as much as may bee to those of our equals, and to perswade those with mildnesse that are our inferiours. To these three maximes I will adde for the last generall precept, that hee must neuer attempt to entertaine any man to please him, vnlesse hee hath first duly considered of his humour, his inclinations, & of what temper his spirit is; to the end he goe not lower nor higher then hee ought, but to accompany him so neare, as all his discourses may be fitted to his carriage. (1632: 335–337)

A sociologist or a personnel management specialist could hardly say it better. Oratory differs from ordinary conversation not in kind but in degree: oratory belongs to a large forum—to what Della Casa calls "a public place"—and it entails a sense of occasion.

The standard types of speeches as distinguished in classical rhetoric—forensic, deliberative, epideictic—go a long way toward defining the range of occasions for oratory in early modern England. Debates in Parliament, speeches welcoming a dignitary's visit, funeral orations: on each of these occasions the orator attempts to speak, not just for himself, but for all present. He claims mastery over the entire auditory field, and must maintain that mastery to the end. Dramatized examples, with long odds and high stakes, include Sir Thomas More facing down the London mob in Add.II.D of *The Book of Sir Thomas More* and Menenius pacifying the Roman citizens in 1.1 of *Coriolanus*. It is at the beginning (*exordium*) and the end (*peroration*) of a speech, Cicero and Quintilian observe, that an orator needs to be most conscious of his audience. "The sole purpose of the *exordium*," Quintilian points out, "is to prepare our audience in such a way that they will be disposed to lend a ready ear to the rest of our speech. The majority of authors agree that this is best effected in three ways, by making the audience well-disposed, attentive and ready to receive instruction" (4.1.6 in 1921–1922, 2: 8–9). He proceeds to show just how to do that. At the end, the speaker will have been successful if he not only has maintained his hold on the audience but has moved them to action. The *per-oration*, a "through-speaking," belongs to the gestalt of force. A good speech, like a good play, ends in a physical act on the audience's part, an affirmation of their willing subjection to the force of the speaker's voice:

> if we have spoken well in the rest of our speech, we shall now have the judges on our side, and shall be in a position, now that we have emerged from the reefs and shoals, to spread all our canvas, while since the chief task of the peroration consists of amplification, we may legitimately make free use of words and reflexions that are magnificent and ornate. It is at the close of our drama that we must really stir the theatre, when we have reached the place for the phrase with which the old tragedies and comedies used to end, "Friends, give us your applause" [*Plodite*]. (6.1.52 in 1921–1922, 2: 414–415)

One can see such advice in action in transcriptions of Queen Elizabeth's public speeches. While it would be interesting to know just how moved the students in the last row may have been by Rainolds's lectures on Aristotle, contextual information is fully available for how successful Elizabeth was in dominating the auditory field in various public places. Printed accounts are political documents in themselves, of course, and, especially if laudatory, may exaggerate the queen's forensic power, but from the very beginning of her reign she seems to have known how to control the speech field with her voice no less than the presence chamber with her private conversations (see chapter 5 above). Leah Marcus, contrasting the terse, direct, and vivid language recorded by actual listeners to the queen's speeches with the more measured and periodic language of later official versions, suggests that Elizabeth customarily spoke without a prepared script (unpublished talk). Elizabeth's readiness to rise to the occasion rhetorically is evidenced as early as her coronation procession through London on January 15, 1558, when Elizabeth halted, among other places, at the cross in Cheapside, to receive from the City Recorder a purse of crimson silk, filled with a thousand marks in gold, on behalf the thousands of citizens crowded into what was London's largest public space—and ordinarily one of its noisiest. Her brief response, recorded in the souvenir pamphlet narrating the whole event, sets the pattern for Elizabeth's later public speeches in the way she casts the beginning and the end of the speech in second person. Her deployment of pronouns is all the more remarkable because of the way she grants the Lord Mayor and his fellows their third-person dignities: "I thanke my lord maior, his brethren, & you all. And wheras your request is that I should continue your good ladie & quene, be ye ensured, that I wil be as good vnto you, as euer quene was to her people. No wille in me can lacke, neither doe I trust shall ther lacke any power. And perswade your selues, that for the safetie and quietnes of you all, I will not spare, if nede be to spend my blood, Goad thanke you all." Elizabeth's deft way with a *peroration* is apparent in the powerful contrast between the abstractness of "safety and quietness" and the sudden physicality of "my blood." When the queen fell silent, the crowd filled the air with their cries: "if it moued a meruaylous showte and reioysing, it is nothyng to be meruayled at, since both the heartines thereof was so woonderfull, and the woordes so ioyntly knytte" (Osborn 1960: 46). In its testimony to the gestalt of force, the physical metaphor is telling.

On several occasions at Oxford and Cambridge, Elizabeth showed herself competent in the very language of Cicero and Quintilian as well as in their precepts. After several days of official entertainments at Cambridge in August 1564, the queen attended a disputation, in Latin, in St. Mary's church. When various auditors—courtiers, the chancellor, the Bishop of Ely—desired the queen herself to say something in Latin, she managed a brilliant *exordium* by pretending that she might be persuaded to speak in English but certainly not in Latin. She then astounded everyone with a three-hundred-word oration in the classical tongue. The reverberant vaulted space of St. Mary's, formerly echoing with the alternating voices of disputing scholars, was now filled with only one voice, speaking the *lingua franca* of male power and privilege. In the course of her speech the queen noted that learning is a matter of listening, of deferring to the words of one's superiors:

> Quod ad propagationem spectat, unum illud apud Demosthenem memini, "Superiorum verba apud inferiores librorum locum habent; et principum dicta legum authoritatem apud subditos retinent."
>
> [As to the increase of good letters, I remember that passage in Demosthenes "The words of superiors have the weight of books with inferiors; and the sayings of Princes retain the authority of laws with their subjects."]

For her part, she would have her listeners act as one body and, by pursuing good letters, show their subjection to her as magistrate:

> Hoc itaque unum vos omnes in memoria retinere velim, quod semita nulla rectior, nulla aptior erit, sive ad bona fortunae acquirenda, sive ad principis verstrae benevolentiam conciliandum, quam ut graviter studiis vestris incumbatis, ut coepistis. Quod ut faciatis, vos omnes oro, obsecroque.
>
> [This one thing then I would have you all remember, that there will be no directer, no fitter course, either to make your fortunes, or to procure the favor of your Prince, than, as you have begun, to ply your studies diligently. Which that you would do, I beg and beseech you all.]

A *peroration* of studied modesty—"Nunc tempus est, ut aures vestrae hoc barbaro orationis genere tam diu detentae, tedio liberentur" [It is time that your ears, which have been long detained by this barbarous sort of an Oration, should now be released from the pain of it]— produced its intended effect. After she had marked her end in a single word—"Dixi," "I have spoken"—the auditors acknowledged their subjection with the unanimity she had worked so hard to inspire:

> At this Speech of the Queen's, the auditors, all being marvelously astonied, and inwardly revising and revolving the sence of it, they presently spoke forth in open voice, "Vivat Regina." But the Queen's Majesty said on the other side, in respect of her Oration, "Taceat Regina," ["Let the queen be silent"]. And wished, "That they that heard her Oration had drunk of the flood of Lethe." (Nichols 1823, 1: 175–179)

In a successful oration, a first-person speaker controls second-person listeners to produce a desired third-person action. On other, more politically fraught occasions Elizabeth showed how clever she was in doing just that. Her speeches to Parliament concerning the fate of Mary Queen of Scots try to deflect responsibility for a death sentence from the queen as speaker to members of Parliament as listeners—and potential actors. When in November 1586 Parliament sent Elizabeth a petition requesting Mary's execution, Elizabeth replied in a speech that begins with "I" but ends, most emphatically, with "you": "And now for your petition, I pray you for this present to content yourselves with an answer without answer. Your judgement I condemn not, neither do I mistake your reasons, but pray you to accept my thankfulness, excuse my doubtfulness, and take in good part my answer answerless" (Rice 1951: 95).

DIALOGUE

"Conversation" in early modern English covered a much wider range of experience than the same word does in English today. When Sidney in *Arcadia* describes Philoclea as going to Pamela's chamber, "meaning to joy her thoughts with the sweet conversation of her sister," he has in mind just what we would: an exchange of thoughts and words. But Sidney is giving a new application (the earliest cited in the *Oxford English Dictionary*) to the older idea of conversation as social relations in general, the fact of "living or having one's being *in* a place or *among* persons" (OED, "conversation" 1, 7a). What sort of clothing one wears, how one carries one's body, where one goes, when one laughs and how much: these, for speakers of early modern English, were as much a part of conversation as what one might say with one's voice. As factors in early modern conceptions of conversation, place and persons were qualified by a third element: time. It is considerations of time, not patterns of speech, that give the Latin word *conversatio* its senses of frequenting a certain locality, dealing regularly with certain people, or using something frequently. Understood within these coordinates of place, persons, and time, "conversation" is not so much a set of actions as an existential condition: being with certain people in a certain place across a certain span of time. Castiglione catches the situation exactly when he has Federico Fregoso advise the reader of *Il Libro del Cortegiano* (1561) that all the prospective courtier's accomplishments will go for naught unless he constantly adapts his behavior to the people among whom he finds himself at a given moment: "among al the men in the world, there are not two to be found that in every point agree in mind together. Therfore he that must be pliable to be conversant with so many, oughte to guide himselfe with hys own judgement. And knowing the difference of one man and an other, every day alter facion and maner accordyng to the disposition of them he is conversant withall." Paying heed to the marginal note in Sir Thomas Hoby's translation (1561), the prospective courtier will see the social world as "So many men so many mindes" (1900: 123). There is one man, however, to which the courtier must pay primary attention: his prince. Castiglione devotes most of his advice to how the courtier ought to comport himself in the presence of his sovereign.

A much broader range of social relationships is considered in Stephano Guazzo's *Il Civile Conversazione* (1574). It was, perhaps, the wide applicability of Guazzo's book that caused it to be translated into English within half a dozen years of its publication in Italian. Down to the eighteenth century it was freely mined by writers of conduct books all over Europe. As one of two interlocutors in the book, Guazzo himself echoes Castiglione's doubts that any general rules can be given for conservation, "for that we are driven by diverse accidentes to have to deale with diverse persons, differing in sexe, in age, in degree, in conditions, in country, and in nation." No cause for despair, says his interlocutor Annibale Magnocavalli. Think of the people with whom you hold conversation as so many pipes on an organ: "You see in organes diverse pipes, whereof every one giveth a diverse sowne, yet they are all proportioned together, and make one onely bodie: in like sort, albeit there be diverse kindes of entertainement and conversation, yet in the end we shall perceive that they agree together in such sorte, that they seeme in a manner one onely sort, and perchaunce more easie then we thinke for." The conversationalist becomes, in Annibale's view, an organist; his conversants, pipes to be played upon. Hamlet, for one, would protest that advice.

Among so many men and so many minds one huge distinction can be made. Guazzo divides his attention between two primary sites of conversation: "publike, which is abrode with strangers: and private, which is at home in the house" (1967, 1: 114). Nicolas Faret in *The Honest Man* divides up his advice along the same lines.

To modern sensibilities, the distinction here might seem to be one between *staging* oneself in public and *being* oneself in private. On the public front, certainly, the conduct books devote page after page to advice on when a gentlemen should be deferential and when self-assertive. As noted by Castiglione, a prince presents a special case. "*Kinges* are like lightning," Daniel Tuvill observes in an essay collected in *Vade Mecum* (1629), "they neuer *hurt* but where they finde *resistance.*" Therefore, a courtier should never contradict his prince. "He must not *strive* or *contend* to goe beyond him in *apprehention,* in *iudgement* or *conceit,* but moderate his vnderstanding, and somewhat *abase* the value and estimation of his worth" (1629: 47).

History plays by Shakespeare and his contemporaries are full of scenes that put to the test the principles articulated by Castiglione, Guazzo, Faret, Tuvill, and other writers of conduct books. None of these scenes is more pointed than the opening of *Richard II.* The verbal authority established by King Richard in the play's very first lines is challenged by Bolingbroke and Mowbray, who enter fairly bursting with choler that demands to be vented. Their challenge to Richard's aural authority remains, at this point in the play, only implicit, since both of them observe a conversational decorum that recognizes Richard's superior power. The more punctilious of the two is Bolingbroke, who does not even presume to address Richard in second person:

> First, heauen be the record to my speech,
> In the deuotion of a subiects loue,
> Tendering the precious safetie of my Prince,
> And free from other misbegotten hate,
> Come I appealant to this Princely presence.
> (F1623: 1.1.30–34)

Mowbray, for his part, addresses Richard directly, but in terms that defer to the king's power: it is "the faire reverence of your Highnesse" that prevents Mowbray "from giuing reines and spurres to my free speech" (1.1.54–55). The tenuousness of Richard's power over these speakers is revealed, however, in their refusal to be reconciled by him. Richard acknowledges as much in the speech that closes the scene: "We were not borne to sue, but to command" (1.1.205).

If public discourse requires self-staging, then surely, we would think, domestic discourse allows one simply to be oneself. Faret contrasts the demands of speaking with one's superiors, on the one hand, with the relative ease of speaking with one's inferiors and equals, on the other. For the "honest" man, however, private conversation requires just as much diligence:

> it is more dangerous to commit errors then in the other, wheras his spirit is alwayes attentiue to those things whereof he vndertakes the discourse. This is especially obserued among priuate friends, whereas our mindes finding themselues freed from that constraint, which holds them in suspence in other companies, then giue way to all their naturall motions, with such negligence as it many times makes vs vnlike to that which we seeme in publique. Yet this liberty must neuer bee so neglected, but it must bee restrained within the rules of a sweet and honest respect, without offering violence to the minde, which suffers him to reape the contents of this pleasing kind of entertainment in their purity, and without any mixture or bitternesse.

A man can be so intent on self-advancement in public that he fails to cultivate the friends he has in private. Such people, Faret observes, may end up all alone (1632: 197–199). On the domestic front, Guazzo manages to fill an entire book of *Il Civile Conversazione* with "the orders to be observed in conversation, within doares, between the Husbande and the Wife, the Father and the sonne, Brother and Brother, the Mayster and the Servante" (1967, 2: 1).

Again, the early modern theater transcribes numerous examples of Guazzo's precepts (Magnusson 1992, 1998). Even the conversation of social equals—brother with brother, friend with friend—becomes, for Guazzo, an exquisitely balanced exchange of honor. A "modest respecte" must be used between brothers and between friends, "that grave and discreete maner, whereby we doe honour to others, and cause others to doe honoure to us" (1967, 2: 91). The vocatives exchanged between Palamon and Arcite—"Deere *Palamon*, deerer in love than Blood, / And our prime Cosen," "Cleere spirited Cozen," "Noble Cosen," "gentle Cosen"—are only the most rarified instances of a verbal decorum that obtains between all of Shakespeare's male/male and female/female friends (*TNK* Q1634: 1.2.1–2, 1.2.74, 2.2.1, 2.2.70). "Cousin," let it be noted, might indicate not only blood kin but chosen kin (Wrightson 1982: 46).

Self-awareness, restraint, calculation: these are, of course, universal factors in how human beings communicate with one another through speech. For speakers of early modern English the precepts of conduct books and the examples of playscripts suggest that those factors were unusually conscious and, by our standards, unusually rigorous. Their impact on individuals can be measured in several ways: physically, sociolinguistically, politically, and psychologically. With respect to the very sound of one's voice the conduct books emphasize the need for control. Taking his cue from the rhetoricians, Guazzo treats speech, first and foremost, as a physical feat: "the first part of action consisteth in the voice, which ought to measure its forces, and to moderate it selfe in suche sort, that though it straine it selfe somewhat, yet it offend not the eares by a rawe and harshe sownde, like as of stringes of instrumentes when they breake, or when they are yll striken." In using his own vocal chords Guazzo's speaker becomes a kind of instrumentalist, varying his technique "so the change of the voice, *like an instrument of strings,* is verie acceptable, and easeth both the hearer and the speaker" (1967, 1: 128–129).

Writers of conduct books submit the physical voice to the same three quantifiable measures as Cicero and Quintilian do: pitch, volume, and rhythm. With respect to pitch, Della Casa counsels moderation: "The voyce would be neither hoarse nor shrill. And, *when you laugh and sporte in any sorte: you must not crye out and criche like the Pullye of a well: nor yet speake in your yawning.*" Volume, too, should be modulated. *"It is an yll noyse to heare a man rayse his voyce highe, lyke to a common Cryer,"* Della Casa continues. "And yet I would not haue him speake so lowe and softly, that he that harkens, shall not heare him" (1576: 86–87). Guazzo's spokesman Annibale specifies some of the types of speakers a gentlemen, in terms of volume, is *not:* "the voyce must be neither fainte like one that is sicke, or like a begger: neither shrill nor loud like a crier, or like a schoolemaister, which doeth dictate or rehearse to his scholers some theame or epistle. For it would be saide, as it was saide to one, If thou singest, thou singest yll, if thou readest, thou singest" (1967, 1: 129). In rhythm, finally, a speaker must once again aim at a mean. Some of the most pointed advice on this particular measure of speech is to be found in a memorandum of "Short Notes for civill conversation" drawn up by Francis Bacon and printed posthumously among *The Remaines . . . being Essayes and severall Letters to severall great Personages, and other pieces of various and high concernment not heretofore published* (1648). The unsystematized punctuation of this document (commas, semicolons, colons, and periods are used interchangeably) and one reference to *"my* speeches" in a text otherwise cast in third person suggest that the object of Bacon's advice was, in the first instance, himself. For Bacon, rhythm provides a way of maintaining both self-possession and rhetorical control over one's listeners:

> In all kinds of speech, either pleasant, grave, severe, or ordinary, it is convenient to speak leisurely, and rather drawingly, then hastily, because hasty speech confounds the memory, and oftentimes (besides unseemlinesse) drives a man either to a *non-plus,* or unseemly stammering, harping upon that which should follow, wheras a slow speech

confirmeth the memory, addeth a conceit of wisdom to the hearers, besides a seemlinesse of speech and countenance. (1648: 6–7)

"Seemliness," as Bacon uses the term, is very much a matter of *seeming*.

Bellowing beggars, screeching criers, singsong schoolmasters, seemly gentlemen: however calculating or casual they may have been about the physical properties of their voices, all of these speakers of early modern English would have conformed to the laws of their own speech communities. Conduct books recognize the differences among these communities and give the would-be gentleman advice on how to place himself. After the Fall, "after the first confusion of tongues," Guazzo observes, "many sortes of languages have by the divine power of God remained in the worlde: whereby not onely one Nation was knowne from another, but also one Countrie, one Citie, one Village, and (which is more) one street from another." "Country," in George Pettie's translation of 1581, refers to a region, the West Country as opposed to Yorkshire, East Anglia as opposed to the Cotswolds. In general, Guazzo advises, a speaker should stay true to his national tongue, true to his variety, and true to his register. Garnishing one's speech with foreign words when no one understands them is vain affection. Unless circumstances make it absolutely necessary, Della Casa specifically forbids any attempt to speak a language one knows less well than one's own. People who do so "without any grace, thrust them selues in to Chat in their language with whome they talke, what so euer it be, and chop it out euery worde preposterously." Even if a tongue is reckoned to be better than one's own, a speaker should stick to the language he knows best, the language that is "natural" to him (1576: 79–80). Della Casa's example of a superior tongue is Tuscan as opposed to Lombard; the equivalent for early modern England might have been French as opposed to English. As far as Italy is concerned, Guazzo's spokesman Annibale insists that speakers of regional dialects like that of Guazzo's native Monferrato ought not to go around *speaking* Tuscan, even if they habitually *write* Tuscan. At the same time, a speaker must stay true to the register he speaks within a given variety. Since you won't let me speak Tuscan, Guazzo remonstrates with Annibale, "I will frame my selfe to speake as the common people doeth." Annibale replies:

So shall you commit a fault unseemely for so wise a Gentleman as you are, and you shall therein imitate some of our Citizens, who coveting to bee counted pleasant fellowes, take delight in counterfaiting to speake clownishly, whereof it commeth to passe, that comming afterwardes into the companies of grave persons, they are not able to refraine those follies, and so shewe themselves rusticall, and uncivile in their speach. (1967, 1: 140–142)

Prince Hal seems not to have read Guazzo.

Protocols of speaking and listening work their effects not only physically and sociolinguistically but politically. Guazzo's term "counterfeiting" suggests how strongly, how *essentially* social identity was identified with speech: a man can adopt another tongue, another variety, another register, but in doing so he presents himself to the world as other than he really is. For "really" read "naturally." According to Guazzo, it is as "natural" for a "Clown" to speak as he does as for a citizen and a gentlemen to speak as they do. Equally "natural" is the way speech functions politically. Conversation, as Guazzo's spokesman Annibale realizes, is at bottom a matter of rhetoric, of persuasion, of exerting power over other people: "And for so muche as all men naturally indevour them selves in speeche, to persuade and to moove, it is certaine that a sentence hath so much the more or lesse force and vigour, according to the difference of persons from whom it commeth, and of the words by which it is uttered." Speech, in more ways than one, is always strategic: it is a bid for power, and it operates within a range of limited possibilities. Aside from his mystification of "nature," Guazzo almost anticipates Lacan. "Our chiefe labour," Annibale continues, "must bee to moove the heartes of the hearers: and wee must weigh this, that nothing can enter into their hearts, which is not currantly spoken, and without offence to the eares: and therefore wee must labour to have (as Bias saide) a comely grace in holding our peace, and a lively force in speaking" (1967, 1: 125).

The politics of conversation are reflected in the way Guazzo conceives of the entire social world in terms of hierarchy. In offering advice on domestic conversation, he speaks in terms of "fathers" and "sons," but under those titles he wishes it to be understood that father is interchangeable with father-in-law, uncle, and tutor, and that son is interchangeable with daughter, son-in-law, daughter-in-law, and pupil (1967, 2: 3). In Guazzo's book, as in early modern England, everyone is assumed to be someone else's servant. Members of the body politic are linked to one another by chains. In the case of masters and their hired retainers, the chain connecting master and servant is made of iron; in the case of princes and the gentlemen who serve them it is made of gold. Ordinary servants may hate the chain; gentlemen love it. Guazzo himself complains when sickliness keeps him away from the court of his patron, the Duke of Mantua: "being about that Prince, I am in case every hower to pleasure a number of persons, to get friendes daylye, and to make my selfe honoured of the most honourable in the court: by reason whereof, I curse my infirmitye, whiche will not suffer mee long to be bound to this chayne of Gold, which I like above all thinges in the world." Knowing how to be a good servant to one's own master, Guazzo goes on to observe, teaches a man how to be a good master to his own servants (1967, 2: 94–97).

To give pleasure and to gain honor: the politics of civil conversation involve how much one says to whom, and under what circumstances. Thomas Wright offers precise advice in *The Passions of the Mind* (1630; first edition 1604). With respect to one's social superiors, reticence is called for: "Who talks much before his betters, cannot but be condemned of arrogancie, contempt, and lacke of prudence" (1630: 139). Daniel Tuvill in *Vade Mecum* copies Wright verbatim and adds this unforgettable tag: "*Loquacity* is the fistula of the mind euer running and almost incurable" (1629: 57). In the other direction, Wright continues, a gentleman needs to be just as careful. To converse too much with one's inferiors will only earn their contempt and make it appear that the speaker's "conceits" are no better than the people with whom he converses—unless, of course, these social inferiors can tell the speaker something he needs to know. Learned men, military experts, and "wise polititians" are exceptions to the general rule. The whole purpose of conversation, Wright assumes, is to keep social hierarchy in place: "familiaritie aspireth to equalitie" (1630: 139–140).

Early modern English was full of formulae for affirming social difference. When, for example, Autolychus poses as a courtier and confronts the Old Shepherd and the Clown as they carry proof of Perdita's noble birth to the king of Bohemia's palace in Act Four of *The Winter's Tale*, the two laborers are full of deference, as befits men at the bottom of the social hierarchy:

AUT[OLYCHUS]
How now (Rustiques) whither are you bound?
SHEP[HERD]
To th'Pallace (and it like your Worship). (F1623: 4.4.715–716)

That is to say, we, lowly creatures as we be, are bound for the palace—but only if that happens to please you as a person of superior station and freer volition. When, in Sicily two scenes later, the Old Shepherd and the Clown have been made gentlemen for the care they have shown to Perdita, the high-heeled shoes are suddenly on the other feet:

AUT[OLYCHUS]
I humbly beseech you (Sir) to pardon me all the faults I haue committed to your Worship, and to giue me your good report to the Prince my Master.

SHEP[HERD]
    'Prethee Sonne doe: for we must be gentle, now we are Gentlemen.
CLOW[N]
    Thou wilt mend thy life?
SHEP[HERD]
    I, and it like your good Worship. (5.2.147–153)*

Because they are coded in transcribed speech, the physical, socio-linguistic, and political constraints on early modern conversation are easier to mark than the psychological results of those constraints. Taken altogether, however, the books by Castiglione, Guazzo, Della Casa, Tuvill, and Faret suggest a relationship between the speaking "I" and other speakers that is downright *adversarial*. Every word, so these writers imply, has to be carefully calculated. The generally terse advice Sir Walter Raleigh wrote out for his son is not so terse when it comes to conversation. A man who talks too much, Raleigh observes, is like a city without walls:

> therefore if thou observest this rule in all assemblies thou shalt seldome erre, restraine thy choller, hearken much and speake little, for the tongue is the instrument of the greatest good, and greatest evill according to SALOMON; life, and death are in the power of the tongue: and as EURIPIDES truly affirmeth, every unbridled tongue in the end shall find it selfe unfortunate, for in all that ever I observed in the course of earthly things, I ever found, that Mens fortunes are oftner mard by their tongues then by their vices. (1632: 52–54)

From his own political career Raleigh knew whereof he spoke. In his "Short Notes for civill conversation," Bacon equates "civility" not, as we might expect, with politeness, but with statesmanship. *Civilitas*, after all, is an attribute of a *civis*, a citizen or royal subject. Bacon's reference to "*my* speeches" suggests that a conversationalist is a *public* spokesman—and a master of public deception—rather like himself: "To deceive mens expectations generally (which Cautell) argueth a staid mind, and unexpected constancie, *viz.* in matters of fear, anger, sudden joy or griefe, and all things which may effect or alter the mind in publique or sudden accidents, or such like" (Bacon 1648: 6–7).

With practical advice like Bacon's in mind, we should not be surprised when Della Casa in *Galateo* treats the whole system of honorific titles, polite phrases, uncovering one's head, bowing, and kissing hands as a form of deception or dreaming. "Ceremonies," as Della Casa calls such acts of reverence, once belonged only to religion. Carried over into social life, they become a polite form of lying. Like lies

---

*It was Susan Snyder who pointed out to me this elegantly compact instance.

and dreams, they have no connection with reality: "we honour me[n] to their face, whome we reuerence not indeede, but otherwhile co[n]temne. And neuertheles, because we may not go against custome, wee giue them these titles: *The most honorable Lord suche a one: the Noble Lord such a one.* And so otherwhile wee offer them our humble seruice: whome wee could better vnserue then serue, & co[m]maund then doe them any duety" (1576: 42). All of social life, that is to say, is a matter of service and command. To converse with other people is to enter an unending contest of power. Early modern conduct books confirm Mark Johnson's contention that speech is fundamentally a *physical* phenomenon. A speaker projects his person into the environment: he acts upon others and is acted upon in turn. Mark Johnson's term for this give-and-take is the gestalt of force (1987: 41–64). If compulsion is the structure force assumes in oratory, the structure of early modern conversation would seem to involve a conflicted combination of attraction, blockage, resistance, and diversion. Amid it all stands the speaking, listening subject. Acoustically, socially, psychologically, he maintains a strong sense of his own centrality and calculates carefully his interjections into the world of sound around him.

LITURGY

[W]hosoeuer shall call vpo[n] the Name of the Lord, shalbe saued. But how shal thei call on him, in whome they haue not beleued? and how shal they beleue in him, of whom they haue not heard? and how shal they heare without a preacher? . . . Then faith is by hearing, & hearing by the worde of God. (Geneva Bible, Romans 10:13–14, 17)

Paul's exhortation to the new Christian church at Rome became the rallying cry for new Protestant churches all over Europe in the sixteenth and seventeenth centuries. Apologists for the new religion like Cranmer cited the passage in Romans as the *locus classicus* for Reformation doctrine that gave precedence to hearing over seeing (Malette 1998: 24–32). In William Perkins's formulation, "The preaching of the word is indeed the most worthie instrument for the founding and confirmation of Gods Church" (1608–1609, 3: 243). Where the elevation of the Eucharist, viewed from afar by the congregants, had been the highlight of the Latin liturgy, Protestant theologians of all persuasions designed worship services to lead up to the sermon. Before the Counter Reformation it was usual for worshippers to come forward and drink from the cup only on special occasions, usually just once a year (Bossy 1983: 29–61). Contrast between the two liturgical ideas—one based on vision, the other on audition—is graphically apparent on the title page to Foxe's *Acts and Monuments* (1563 *et seq.*)

Figure 9.1. Title page (detail) from John Foxe, *Actes and Monumentes* (1570). Reproduced by permission of the Folger Shakespeare Library, Washington, D.C.

(fig. 9.1). Two vignettes at the bottom, boldly captioned in the second edition, offer a contrast between "The Image of the persecutyng Church" on the right and "The Image of the persecuted Church" on the left. The Roman homilist on the right speaks to listeners who are distracted by their rosaries (one female figure in the foreground is clearly asleep and another gestures with her arm as if talking), while in the distance a crowd follows a banner towards a roadside icon; meanwhile, the Protestant preacher on the left calls down the Word of God from heaven and broadcasts it among a circle of rapt listeners.

Theological justification for the Protestant cause was to be found in the Platonically inspired gospel according to John, with its emphasis on Christ as the embodied Word of God: "In the beginning was the Worde, and the Worde was with God and that Worde was God" (Geneva Bible, John 1:1). Saul was struck blind, so John Donne argues, the better to hear God's call and rise from the ground as Paul, the great preacher. As Dean of St. Paul's Cathedral, Donne preached a sermon to that effect on the Sunday after the feast of The Conversion of Paul in 1625. Donne's words resounded in a space dedicated to the very subject of his sermon:

Man hath a natural way to come to God, by the eie, by the creature; *So Visible things* shew the *Invisible God:* But then, God hath super-induced a supernaturall way, by the eare. For though hearing be naturall, yet that faith in God should come by hearing a man preach, is supernatural. God shut up the naturall way, in *Saul,* Seeing; He struke him blind; But he opened the super-naturall way, he inabled him to heare, and to heare him. God would have us beholden to grace, and not to nature, and to come for our salvation, to his Ordinances, to the preaching of his Word, and not to any other meanes. (1953–1962, 6: 216–217)

Preaching is an "ordinance," a ritual activity ordained by God, no less deserving of the adjective "holy" than communion is.

As the embodied Word of God, Christ was the *sounded* Word of God. Christ was not a written text; he *preached*. The written gospel (traditionally understood as *God-spell*, "God's tidings") records Christ's words, but the true power of those words is released when they are heard. An illustration of Pentecost in *A briefe summe of the Whole Bible* (1567) captures the theological point precisely (fig. 9.2). Christ passes a bound and labeled copy of the Gospel to the apostles, but at the same time he speaks the words that give readers of the Gospel the power to preach: "Receiue the holy ghost." Donne is not just being figurative when, in a sermon preached at Whitehall on March 4, 1625, he speaks of an echo:

> The Scriptures are Gods Voyce; The Church is his Eccho; a redoubling, a repeating of some particular syllables, and accents of the same voice. And as we harken with some earnestnesse, and some admiration at an Eccho, when perchance we doe not understand the voice that occasioned the Eccho; so doe the obedient children of God apply themselves to the Eccho of his Church, when perchance otherwise, they would lesse understand the voice of God, in his Scriptures, if that voice were not so redoubled unto them. (1953–1962, 6: 223)

As pronouncer of the Gospel, a preacher speaks *in loco Dei*.

John Gipkyn illustrates the point in his depiction of a sermon at Paul's Cross (1616) (fig. 9.3). Painted on commission from Henry Farley as part of Farley's one-man campaign for the refurbishment of the cathedral, Gipkyn's panel shows the preacher declaiming under a sounding-board large enough to serve as the roof to a pavilion. As the preacher speaks, a burst of light at the apex of the painting signals the inspiration of the Holy Ghost. That this light can be heard, rather than seen, is suggested by the birds who have been rousted from their roosts on the great tower, as if by the bonging of bells. In *St. Paules-Church Her Bill for the Parliament* (1621), a pamphlet Farley published to further his crusade, the cathedral herself speaks:

> As it is said that Ships doe fight,
> When tis the men that in them bee:
> So I (poore Church) pray, speake, and write,
> When tis my Friend doth all for mee.
>
> (1621: A4^v)

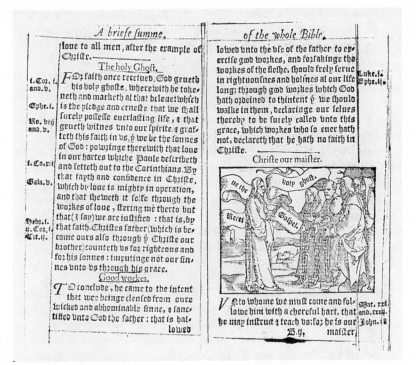

Figure 9.2. Pentecost. From *A briefe summe of the Whole Bible* (1567). Reproduced by permission of the Folger Shakespeare Library, Washington, D.C.

The authorized voice of the church is the preacher's. John the Baptist has great distinction in his name, Donne maintains in a sermon preached at Whitehall in 1619, but even greater are his names as the Prophet and the Preacher. John's self-description as *"Vox clamantis, The voyce of him that cryes in the wilderness,"* can stand for *all* preachers of God's word. John, according to Donne, claims "not A Voyce, Any voyce, but The voyce: in the Prophesie of *Esay,* in all the four Evangelists, constantly, The voyce. Christ is *verbum,* The word; not A word, but The word: the Minister is *Vox,* voyce; not a Voyce, but The voyce, the voyce of that word, and no other" (1953–1962, 2: 172).

The purpose of preaching is, then, to make Christian believers subject to God's voice, through the voice of the preacher. The liturgy of the Church of England was framed to accomplish that theological and political end. The three most commonly used orders of worship—Morning Prayer, Evening Prayer, and Holy Communion—are explicitly designed to turn the worshippers into a body that speaks in unison. Hooker takes great care in *The Laws of Ecclesiastical Polity*

Figure 9.3. John Gipkyn, Sermon at Paul's Cross, oil on panel (1616). Reproduced by permission of the Society of Antiquaries, London.

(1594–1597) to explain why recited prayers are central to Protestant worship:

> When wee publiquely make our prayers, it cannot be but that we do it
> with much more comfort than in priuat, for that the things we aske
> publiquely are approued as needefull and good in the iudgement of
> all, we hear them sought for and desired with common consent.
> Againe, thus much help and furtherance is more yeelded, in that if so

be our zeale and deuotion to Godward be slacke, the alacritie and fer-
uor of others serueth as a present spurre. *For euen prayer it selfe* (saith
St. Basill) *when it hath not the consort of many voices to strengthen it, is not
it selfe.* (5.24 in 1593–1597: 54)

Concerning the Lord's Prayer in particular Hooker reaches for a new
word to describe the power of mutually confessed sins. The worshippers become, he says, "eare-witnesses" of each other's assent to their
common indictment as sinners (5.36 in 1593–1597: 73). The orders for
Morning Prayer and for Evening Prayer reach a climax in such acts
of unanimous recitation.

The Order for Holy Communion, as revised in the 1559 *Book of
Common Prayer,* reveals a more complex strategy. It begins with a preface insisting that communion should be an act of community. If anyone who proposes to take communion is "an open and notorious
euyll lyuer, so that the congregacion by him is offended," he should
be called before the curate and enjoined openly to repent, "that the
congregacion may therby be satisfied, which afore were offended"
(M1). The opening prayers in the service itself are styled as "collects,"
prayers enunciated by the priest with responses to be spoken in unison by the congregants. The first of these collects requires the congregants periodically to respond "Lorde haue mercye vpon vs, and en-
cline our hartes to keep thys lawe" no fewer than ten times (M1ᵛ–M2).
Unified in voice by these prayers, the worshippers are ready to hear
the epistle and the Gospel read aloud. Next comes the Creed in first-
person singular ("I Beleue in one god, the Father almyghtye, maker
of heauen and earth . . ."), then the sermon (M2ᵛ–M3). Offerings collected after the sermon allow listeners to give visible proof of their
unanimity. "Let your light so shine before men, that they maye see
your good woorkes" (Matthew 5:16) goes the first of several sentences
among which the priest can choose as he announces the offertory
(M3–M4).

For the creed to *precede* the sermon, rather than to follow it, represents a particularly significant departure from the Roman liturgy.*
Acknowledgment of belief becomes a *condition* for hearing God's
word preached, not an *effect.* Conditions in the Latin rite, in the form
of prayers of confession, are set as preparations, not for the sermon,
but for the Eucharist. These, to be sure, are preserved in "The ordre
for the administration of the Lordes Supper, or holy Communion" in
the 1559 *Book of Common Prayer,* but they function there as confirma-
tions of a spiritual event that has already happened: the creation of

---

*For this information I am grateful to Scott Pilarz.

a Christian community through the ordinance of preaching. The effect is signalled by a shift in pronouns, from first-person singular in the creed before the sermon to first-person plural in all of the prayers that follow it. The political purpose of this shift from the individual to the group is apparent in three lengthy exhortations that can be read "at certayne tymes when the Curate shal see the people negligent to come to the holy Communion." "And whereas you offende God so sore in refusynge thys holye Banquet," the curate is instructed to say in the first of these speeches, "I admonishe, exhorte, and beseche you, that vnto this unkindenesse ye will not adde anye more. Whiche thing ye shall doo, yf ye stande by as gasers and lookers on them that doo Communicate, and be no partakers of the same your selues. For what thing can this be accompted els, then a further contempt and vnkindenes vnto God" (M5). No equivalent exists in the Roman rite for the post-Communion prayer that redefines the body of Christ as a "body" of believers: "Almighty and euerliuyng God, we moste hartelye thanke the . . . that we be uery members incorporate in thy mistical body, *whiche is the blessed company of all faythfull people*" (N1^v, emphasis added). In a final act of incorporation before the curate pronounces his benediction, the congregation affirms its first-person plurality in a *Gloria:* "We prayse thee, we blesse thee, we worship thee, we glorify thee, we geue thankes to the for thy great glory." In its direction that these words "shalbe sayd or song" (N1^v) the rite makes its only explicit mention of music, implicitly recognizing, perhaps, the totalizing power of music in an enclosed space. The psycho-acoustic effect of singing a familiar text along with other people has been called "unisonance" by Benedict Anderson, who identifies it as a particularly powerful way of affirming social identity: "the echoed physical realization of the imagined community" (1991: 145).

It was the physically sounded word of God that transformed Christ's "mystical body" from a wafer in the Roman rite into an assembly of people in the Protestant liturgy. A preacher, like an orator, accomplished that feat through the totalizing deployment of his voice. Unlike an orator, a preacher had divine sanction for what he did. The liturgy of the Church of England was designed to produce docile listeners, to turn the voices of the many into the voice of the one. Puritan doctrine put even greater emphasis on preaching—and on the submission of listeners. Among the many preachers John Manningham heard and recorded in his diary, Stephen Egerton of Peterhouse, Cambridge, is distinguished by his alterations in the order of worship. It was in December 1602 at the Blackfriars chapel—just steps from the Blackfriars theater—that Manningham joined "a great congregation, specially of women," to hear Egerton preach. The speaker's rhetorical mastery over his auditors is apparent at every point in Manningham's description. After prayers and readings from the scriptures, Egerton, "having sat a good tyme before in the pulpit, willed them to sing to the glorie of God and theire owne edifying, the 66 Psal. 2 part"—presumably the lines beginning "I wil go into thine House with burnt offrings, & wil paie thee my vowes, Which my lippes have promised, & my mouth hathe spoken in mine afflic-tio[n]" (Geneva Bible, Psalm 66: 13–14). Having secured this act of aural unaminity, Egerton "made a good prayer, then turnd the glas, and to his text, Acts 7.23, &c." The passage tells how the forty-year-old Moses, learned in the wisdom of his Egyptian captors, "mightie in wordes, and in dedes," comes to the rescue of one of his Israeli brethren, incorrectly supposing that the enslaved Israelites will see in that act a sign that God is ready to deliver them. Manningham fails to specify just what doctrines Egerton chose to read out of this text, but he makes special note of the fact that Egerton ended the sermon by catechising the congregation, putting questions to them and requiring them to respond. This, Manningham notes, "seemes an order which he hath but newely begun, for he was but in his exor-dium questions." Manningham's diary entry is concluded by a note on "the difference betwixt preaching and catechising, that that is a large continued course of speache, and may be performed onely by the minister" (1976: 152–153). Whatever Egerton may have found to say about the passage in Acts, he clearly cast his auditors as enslaved Israelites, himself as a Moses-figure, "mighty in words and deeds."

With respect to the ordinance of preaching the difference between the official Church of England and Puritan reformers was a matter of degree, not kind. Under both regimes, listeners were to be turned into obedient subjects. As William Perkins puts it, "In the right hearing of the word, two things are required. The first, that we yeelde ourselues in subiection to the word we heare: the second that we fixe our hearts vpon it. . . . Subiection to God must be yeilded in giuing subiection to his word: and our cleauing vnto God must be by fixing our hearts vpon his word" (1608–1609, 1: 695). The end of good preaching, in the formulation of James 1:22, is to turn "hearers" of the word into "*doers* of the worde" (Geneva Bible). Donne is careful to distinguish mere pleasure in listening from an inspiration to act. In a sermon preached at Whitehall on 12 February 1619, Donne took as his text Ezekiel 33:32: "And lo, thou art unto them as a very lovely song, of one that hath a pleasant voyce, and can play well on an instrument; for they hear thy words, but they doe them not." For a preacher to "play well" does *not* mean "to be the working upon the understanding and affections of the Auditory, that the congregation shall be his instrument; but as S. *Basil* says, *Corpus hominis, Organum Dei*, when the person acts that which the song says; when the words become works, this is a song to an instrument" (1953–1962, 2: 166–167).

That the liturgy did not always succeed in these goals is indicated by Donne himself in an undated sermon preached at St. Paul's on 2 Corinthians 5:20, "We pray yee in Christ's stead, be ye reconciled to God." Listeners in the early church, Donne observes, were apparently anything but passive: "all that had been formerly used in Theaters, *Acclamations* and *Plaudites*, was brought into the *Church*, and not onely the vulgar people, but learned hearers were as loud, and as profuse in those declarations, those vocall acclamations, and those plaudites in the passages and transitions, in Sermons, as ever they had been at the Stage, or other recitations *of their Poets, or Orators.*" The same custom still holds true in other places in the world, "where the people doe yet answer the Preacher, if his questions be applyable to them, and may induce an answer, with these vocall acclamations, *Sir, we will, Sir, we will not.*" Donne's own listeners come close to such practices "in those often periodicall murmurings, and noises, which you make, when the Preacher concludeth any point." Such exclamations take up valuable time—and turn the sermon into a performance, so that "many that were not within distance of hearing the Sermon, will give censure upon it, according to the frequencie, or paucitie of these acclamations" (1953–1962, 10: 132–134). Even in Gipkyn's idealized painting of a Paul's Cross sermon there are listeners who refuse to be drawn into the circle of aural subjection: a strolling couple, a man whipping a dog, two pairs enjoying conversations outside the cathedral's west door. These marginal figures belong to another scene of speaking entirely.

THEATER

If Della Casa makes no mention of theater in his typology of the scenes of speaking, it may be because theater is so difficult to place with respect to oratory, conversation, and worship. Theater is all three. In all sorts of ways play-acting declares its affinities with rhetoric. Actors, like orators, lay totalizing claim to the auditory field of a public place. Cicero and Quintilian gave early modern performers their readiest vocabulary for talking about the actor's art (Donawerth 1984: 13–55; Trousdale 1982: 22–38). As a mimesis of real-life speech situations, theater likewise resembles conversation. Anecdotes about the theatergoing public's behavior in early modern England suggest that the audience were far from passive listeners. Drayton alludes to "Showts and Claps at ev'ry little pawse, / When the proud Round on ev'ry side hath rung," and Dekker advises his playgoing Gallant, if bored, to "mewe at passionate speeches, blare at merrie, finde fault

with the musicke, whew at the childrens Action, whistle at the songs"
(Gurr 1992: 226, 228). To Puritan detractors, finally, theater presented
itself as a diabolical alternative to religious worship. John Stockwood,
in a sermon preached at Paul's Cross on August 24, 1578, fulminates
against the greater drawing power of play performances over sermons—though he does except Paul's Cross:

> Wyll not a fylthye playe, wyth the blast of a Trumpette, sooner call
> thyther a thousande, than an houres tolling of a Bell, bring to the Ser-
> mon a hundred? nay euen heere in the Citie, without it be at this place,
> and some other certaine ordinarie audience, where shall you finde a
> reasonable company? whereas, if you resorte to the Theatre, the Cur-
> tayne, and other places of Playes in the Citie, you shall on the Lords
> day haue these places, with many other than I can not recken, so full,
> as possible they can throng. (quoted in Chambers 1923, 4: 199–200)

The power of public theater in early modern England lies at the intersection of these three scenes of speaking: oratory, conversation, and
liturgy. Audiences liked it because it engendered, through sound, a
subjectivity that was far more exciting—and far more liberating—
than those created by oratory, conversation, and liturgy by themselves. Let us investigate how that effect comes about.

"O For a Muse of Fire": the very phoneme in which the Chorus
to *Henry V* first projects his voice into the space around him opens
up a circle. Visually and aurally, figuratively and physically, flatness
is "raised" into a sphere:

> But pardon, Gentles all:
> The flat unraysed Spirits, that hath dar'd,
> On this vnworthy Scaffold, to bring forth
> So great an Obiect.

Figuratively, the "object" in question is the story of Henry's epic exploits; physically, it is a circle:

> Can this Cock-pit hold
> The vastie fields of France? Or may we cramme
> Within this Woodden O, the very Casks
> That did affright the Ayre at Agincourt?
> O pardon: since a crooked Figure may
> Attest in little place a Million,
> And let us, Cyphers to this great Accompt,
> On your imaginary Forces worke.
>   (F1623: Pro.1, 9–19)

The circular shapes described in "this cock-pit," "this wooden O,"
and "the very casques," or helmets, find their human counterparts in

the "ciphers," or zeroes, in which the Chorus bodies forth himself and his fellow actors. In this acoustic context, [o:] is not only a shape but a sound, twice heard as an exclamation ("O for a Muse of fire," "O pardon"), once as an epithet for the theater ("this wooden O") that fuses shape and sound.

Instructed by phenomenologists of sound like Don Ihde, we can "see" the acoustic field by imagining the Prologue standing at the front edge of the stage, near the geometric center of The Globe. As he projects his voice in all directions, he defines a circle. Beyond the reach of his voice stretches a horizon of silence. Along with the speaker, the auditors stand well within the circle defined by that horizon. Actor and audience share the same field of sound. If the actor stands at the center of that shared acoustic space, each individual auditor stands nonetheless at the center of his or her own field of hearing—a field that includes the actor's voice but is not limited by it. The radius of sounds each auditor can hear is defined by its own encircling horizon of silence. For the space of the play, each individual auditor's radius of hearing is narrowed. As each of the "gentles all" focuses his or her attention on the speakers onstage, sounds outside the acoustic field of the play become, quite literally, peripheral (Ihde 1976: 170). The result is, or can be, a totalizing experience of sound that surrounds each hearer completely, penetrating his or her body through the ears, immersing him or her in the playful patterning of speech. When a listener heeds the Prologue's commands, the play fills that listener's entire auditory field. In Mark Johnson's terms, the playgoer willingly makes herself subject to the spoken word's gestalt of force (1987: 47–64).

In sending out a Prologue, Shakespeare was drawing on well established dramatic practice. The typical *incipit* of a medieval playtext is a command for silence. The Prolocutor ("the one who speaks before") in *The Pride of Life* marks out for the ensuing "game" a place that is bounded by sound as well as space:

> Pees, and herkynt hal ifer,
>> Ric and por, yong and hold,
> Men and wemen that bet her,
>> Bot lerit and leut, stout and bold.
>
> Lordinges and ladiis that beth hende,
>> Herkenith al with mylde mode
> How oure gam schal gyn and ende.
>> Lorde us wel spede that sched his blode!
>
> Now stondith stil and beth hende,
>> And teryith al for the weder,

And ye schal or ye hennis wende
Be glad that ye come hidir.
(1–12 in Davis 1970: 88)

The "here" of *The Pride of Life* (l. 3) is, to be sure, a physical place. Twice the lords and ladies are situated as being "at hand" ("hende," ll. 5 and 9). But the "here" of the play is also a field of sound, centered on the actors. The Prolocutor's job is to establish that field of sound before the other actors join in. The formation in which the lords and ladies stand "at hand" is, to judge from the evidence of this and other medieval playscripts, a circle. The messenger who is sent to end *The Pride of Life* with a moral warning from the King of Life is directed to "go [about] the place" ("*Et eat plateam*" [SD after 470]), and he talks to the audience all the while in terms just as direct as the Prolocutor has at the start ("Pes and listenith to my sawe, / Bothe yonge and olde" [471–472]). The *platea* of medieval theater-in-the-round coincides with an acoustic field whose shape is also circular (Southern 1975: 17–88; Wickham 1987: 100–102, 116–118). When a listener heeds the Prolocutor's commands, the play fills that listener's entire auditory field. "Pes, lordyngs, I prai yow pes, / And of your noys ye stynt and ses," cries the Prologue to an otherwise lost miracle play preserved at Durham Cathedral (1–2 in Davis 1970: 118–119). The Messenger at the start of *Everyman* voices a similar command: "I pray you all gyve your audyence, / And here this mater with reverence" (1–2 in Cooper and Wortham 1980: 3). "I have more tentreate youe of gentle Sufferaunce, / That this our matier may have quyet utteraunce," pleads the Prologue to *Respublica* (3–4 in Udall 1952: 1). Whatever else they may go on to say, figures like these serve to define an auditory circle that encompasses both actors and audience.

The personages who command silence in medieval drama tend to be major authority figures: a stand-in for the poet, a king, God Himself. The prologue to *Respublica* is specified in "The partes and names of the plaiers" to be "a Poet" (S.D. before 1). More commonly the prologue is not "*A Poet*" but "*The Poet*." Taking a cue from the prologues that have come down from antiquity with Terence's playtexts, schoolmasters and college dons felt compelled to write an introductory speech from the author, even if the script did not call for one, when they mounted productions of tragedies as well as comedies, Greek as well as Latin, in the sixteenth century. In thus giving voice to *Auctoritas*, these academic impresarios were realizing in an actual person the woodcut image of The Poet that presided as frontispiece over fifteenth- and sixteenth-century editions of Terence's plays (Smith 1988: 140–147, 199–202).

In scripts of the fourteenth and fifteenth centuries the figure who speaks first is often an Emperor, or someone like him. Sometimes preceded by a Prologue, sometimes not, these political authority figures impose an acoustical authority that silences the audience and reverberates through the entire play over which such figures preside. "I command sylyns in the peyn of forfetur, / to all myn audyeans present general," roars the "Inperator" (*sic*) who begins the Digby manuscript *Mary Magdelene* (S.D. before 1 in Furnivall 1896: 55). "I prey yow, lordyngys so hende, / No yangelyngys ye mak in this folde / To-day," commands Duke Moraud in the fragment that bears his name (7–9 in Davis 1970: 106). The "lordyngys" the Duke addresses here range beyond the listeners who are physically present. It extends to "Emperourys and kyngys," "Erlys and barunnys," "Bachelerys and knytys," "Sueyerys and yemen," "Knavys and pagys" (1–4). The effect of such catalogs is to extend auditory dominion not just over the listeners assembled within the compass of the speaker's voice but over the whole realm of England, if not the entire world.

Stand-ins for the estates of the realm are often made subject to the speaker's imperialism along with the audience. "Pes, now, ye princis of powere so prowde," commands the King who opens the narrative action of *The Pride of Life*—and goes on to call the roll of auditory subjects who are standing round about him:

> Ye kingis, ye kempis, ye knightis ikorne,
> Ye barons bolde, *that beith me obowte;*
> Sem schal yu my sawe, swaynis isworne.

To breakers of silence he promises physical punishments:

> Sqwieris stoute, stondit now stille,
>   And lestenith to my hestis, I hote yu now her,
> Or I schal wirch yu with werkis of wil
>   And doun schal ye drive, be ye never so dere.
>   (113–120, emphasis added)

Just as peremptory are the King who opens the Rickinghall fragment ("Lordinges / wytouten lesinge, / Ye weten wel that I am kinge / Her of al this lond" [1–4 in Davis 1970: 117]), Herod in the Digby manuscript *Slaughter of the Innocents* ("A-bove all kynges under the Clowdys Cristall / Royally I reigne in welthe with-out woo" [57–58 in Furnivall 1896: 3]), Mundus in *Mundus et Infans* ("Sirs, cease of your saws, what so befall, / And look ye bow bonerly to my bidding!" [1–2 in Lester 1981: 1]), and Cambises ("My Councell grave and sapient, with lords of legal train, / Attentive ears towards me bend, and mark what shal be sain" [1–2 in Creeth 1966: 449]). Such speeches not only serve the locutionary function of commanding the audience to be silent; in the process they give the audience an implicit fictional identity. Standing or seated within the same acoustic circle as the speaker/king, the audience become his political subjects as well as his auditory subjects. In effect, the audience become players in the fictional game. They become subjects to the speaker/king.

> Ego sum alpha et oo • principium et finis.

> Ego principium Alpha et O in altissimis habito. . . .

> Ego sum Alpha et O: vita, veritas, primus et novissimus.

> Ego sum alpha et o,
> I am the first, the last also. . . .

> Ego sum alpha et oo,
> primus et novissimus
> It is my will it shoulde be soe;
> hit is, yt was, it shalbe thus.

In the *Ludus Coventriae*, in the Norwich Grocers' Play, in the York cycle, in the Townley plays, in the Chester mystery cycle the voice that projects and defines the auditory field is God's (Block 1922: 16; Davis 1970: 8; Beadle 1982: 49; England 1897: 1; Lumiansky and Mills 1974: 1). He does so in a first-person indicative speech act that rings out in two resonant soundings of [o:]: "Ego . . . O." In creating himself acoustically God creates the world, and in creating the world God also creates the world of the play. Aurally as well as ontologically he gives voice to the scripture that has occasioned the entire cycle of plays:

> In principio erat Verbum,
> et Verbum erat apud Deum,
> et Deus erat Verbum.

> In the beginning was the Worde, and the Worde was with God and that Worde was God. (John 1:1–2 in *Biblia Sacra* and Geneva Bible)

It was at Corpus Christi, on the feastday of the Body of Christ, that the townsfolk of Norwich, York, Chester, and other cities gathered to play the history of the world from Creation till Doomsday. The events they enacted form a "cycle" because they begin and end with the same speaker, with God, *primus et novissimus*, the first and the last. In the first play of the cycle God creates; in the last he destroys. In both plays the words with which God heralds himself are the same. "*Ego sum alpha et omega*, I, *primus et novissimus*," God proclaims in the first words of the last play in the Chester cycle (24.1 in Lumiansky and Mills 1974, 1: 438). In the York judgment play God is even more forthright: he and he alone has been the creator of everything the audience has watched and heard:

Nowe is fulfillid all my forthoght,
   For endid is all erthely thyng.
All worldly wightis that I have wrought,
   Aftir ther werkis have nowe wonnyng.
        (47.373–376 in Beadle 1982: 415)

The Prologue to *Henry V* steps into the acoustic space created by these medieval predecessors. The fact that only six of Shakespeare's printed scripts happen to include speeches for a prologue should not be taken to mean that only in these six plays did the conventional trumpet-calls announcing the play give place to a single speaker, sent out to clear the air and command the stage before the play began. To judge from the endings at least, the printed texts fail to transcribe all that audiences heard and saw in actual performances. Eyewitnesses attest that jigs customarily concluded each play, tragedies and comedies alike. No verbal traces of these gestures of closure exist in any of Shakespeare's printed scripts, unless we accept the devices that round off some of the comedies: the dialogue of the owl and the cuckoo in *Love's Labor's Lost*, the wedding dance in *Much Ado*, Feste's song in *Twelfth Night*. Were there also prologues that, like jigs, never found their way into print? However distinctive they may be, Shakespeare's six surviving prologues are alike in casting the plays they precede as experiences to be *heard*. After declaring the entire plot, deaths and all, very much in the manner of medieval banns, the Chorus to *Romeo and Juliet* asks that the audience "with patient eares attend" the two hours' traffic of the stage (Q1597: Pro.13). "Painted full of Tongues," Rumor in *2 Henry IV* opens the play with a literally dumbfounding [o:]: "Open your eares" (Q1600: Pro.1). Not with explicit references to hearing does the armed Prologue to *Troilus and Cressida* establish the martial tone of the play but with the epic pitch of his rhetoric and the single resounding word that concludes his harangue: "war" (F1623: Pro.31). The whole of *Pericles* is cast as "a Song that old was sung," as a performed ballad with John Gower as the oral bard (Q1609: 1.1). The projected [o:] of Gower's first line, especially in an acoustic context of sibilants, serves to heighten the "old." The crier's [o:] in *Henry VIII* sounds out in "no more": "I come no more to make you laugh." The emphasis instead falls on "woe," "noble," and "flow":

        Things now,
That beare a Weighty, and a Serious Brow,
Sad, high, and working, full of State and Woe:
Such Noble Scenes, as draw the Eye to flow
We now present.

The audience to *Henry VIII* are flattered as "gentle Hearers," indeed "The First and Happiest Hearers of the Towne" (F1623: Pro.1–5, 17, 24).

With or without a prologue, all but a handful of Shakespeare's scripts display quite obvious devices for establishing the auditory field of the play within the first few moments. "A tempestuous noise of Thunder and Lightning heard": the opening stage direction to *The Tempest* is the last and most concentrated in a series of attention-focusing noises, clamors, and tumults that begins with *Titus Andronicus* ("Flourish. Enter the Tribunes and Senators aloft. And then enter below Saturninus and his Followers at one doore, and Bassianus and his Followers at the other, with Drum & Colours") and continues through *2 Henry VI* ("Flourish of Trumpets: Then Huboyes. Enter King, Duke Humphrey, Salisbury, Warwicke, and Beauford on the one side. The Queene, Suffolke, Yorke, Somerset, and Buckingham, on the other"), *3 Henry VI* ("Alarum. Enter Plantagenet, Edward, Richard, Norfolke, Mountague, Warwicke, and Souldiers"), *Coriolanus* ("Enter a Company of Mutinous Citizens, with Staves, Clubs, and other weapons"), and *Macbeth* ("Thunder and Lightning. Enter three Witches") (S.D.s in F1623). That all the plays that begin so are histories and tragedies is not happenstance. It is not just the boundaries of the auditory field that are established in the cacophony that opens these plays but the distinctive temper of that auditory field in volume (loud), pitch (masculine), and rhythmic energy (high). "Now is the winter of our discontent, / Made glorious summer by this sun of Yorke" (*R3* Q1597: 1.1.1–2): the voices heard first in Shakespeare's history plays and tragedies repeat an aural device as old as the earliest surviving playscript in English, the Durham Prologue of 1300.

To the end of an authority figure's first speech a play works, in effect, as an oration. But plays are staged conversations: they happen through dialogue. The effect of introducing a second speaker, then a third, then a fourth, then a fifth is to *decenter* the aural field, to set up a competition for mastery over that field—and hence over the listeners' subjectivity. Sometimes the competition is quite explicit. The "flourish" of trumpets and drums that inaugurates *Titus* is followed at once by orations from Saturninus, Bassanius, and Marcus Andronicus that recall, in substance and in dramatic effect, the speeches of the emperors and kings that inaugurate scripts like *The Pride of Life*, *Mundus et Infans*, and *Cambises*. Like their medieval and Tudor predecessors, Shakespeare's rivalrous Romans attempt to impose their aural authority not only on the estates ranked on the stage but on the audience standing and seated about the playing place. "Noble Patricians, Patrons of my right, / Defend the iustice of my Cause with Armes,"

Saturinus demands; "And Countrey-men, my louing Followers, / Pleade my Successiue Title with your Swords." Bassanius counters with a similar hierarchical catalog: "Romaines, Friends, Followers, / Fauourers of my Right. . . ." When Marcus enters aloft, he too attempts to speak for all:

> Princes, that striue by Factions, and by Friends,
> Ambitiously for Rule and Empery:
> Know, that the people of Rome for whom we stand
> A speciall Party, haue by Common voyce
> In Election for the Romane Emperie
> Chosen *Andronicus*. . . .

(F1623: 1.1.1–4, 9–10, 18–23)

The ensuing tragedy can be heard as a competition among voices for dominance of the auditory field. In plays like *Titus*, the auditory field and the political arena are one and the same place. Politics in such plays can be not only seen but heard. Connecting breath control, humoral psychology, and politics, Lois Potter finds in history plays a particularly patent instance of the competitive nature of speech: in such a world the only character who can "control his own air space" is the king (Potter 1999). Or at least he *should* be such. With its confrontation, rhetorical and physical, of Bolingbroke and Mowbray, Act One, scene one of *Richard II* plays out an aural competition in which Richard is the loser. His failure to establish verbal authority in this opening scene—in rhetorical effectiveness, Bolingbroke clearly carries the day—predicts the silencing of Richard's voice in the end.

In many of Shakespeare's most famous scripts the decentering happens immediately. The "private" conversations that begin *Othello*, *Hamlet*, *King Lear*, *Antony and Cleopatra*, *Cymbeline*, and *The Winter's Tale* give way in each case to a "public" scene presided over by an authority figure, but each of these plays begins, as it were, *in medias res*, with conversations that seem already to be in progress. "Tush, neuer tell me . . . ": the words Actor A speaks to Actor B as they enter the stage together (someone to whom Actor A obligingly gives the name "Iago") imply an offstage reality to which the audience is now being made privy. Never tell him *what*? Whatever it was, it must have been shocking to earn the exclamation "Tush." And just who is Actor A? The audience is not told the first speaker's name until fifty lines later—a good five minutes into the play: "sir, / It is as sure as you are *Roderigo*, / Were I the Moore, I would not be *Iago*" (Q1622: 1.1.1, 55–57). No play approaches *Hamlet* in the disturbingly direct way it poses such questions of identity. The 1603 quarto is particularly stark in the way it presents the speakers as empty ciphers:

1.
>   Stand: who is that?

2.
>   Tis I. (Q1603: 1.1–2)

When the actor speaking Hamlet eventually joins the action in 1.2, a good ten minutes into the play, he enters an aural field that has already been radically decentered. He establishes mastery over that field by degrees: first in an aside ("A little more then kin, and lesse then kinde"), then in a series of short speeches ("Not so my Lord, I am too much i'th'Sun"), finally in an oration after everyone else has vacated the field ("Oh that this too too solid Flesh, would melt") (F1623: 1.2.65, 67, 129ff). Once he has achieved acoustic control, however, Hamlet maintains it until his stage death in 5.2, until the moment when he relinquishes control over language in a series of preverbal (or is it postverbal?) interjections: "O, o, o, o" (F1623: 5.2.311).

In the meantime, Hamlet has demonstrated the way in which character in early modern theater happens through sound, through spoken language. In the circumstances of early modern performance—circumstances in the literal, physical sense of the word—"character" is an achievement of actors, not of scriptwriters. Among allusions to Shakespeare during and just after his career "character" is conspicuous by its absence. Ben Jonson is altogether typical of Shakespeare's contemporaries in praising his natural facility, his style, his flair for writing great lines:

> . . . Looke how the fathers face
>   Liues in his issue, euen so, the race
> Of *Shakespeares* minde, and manners brightly shines
>   In his well torned and true filed lines.
>
>   (F1623: 10)

What makes Shakespeare "not of an age, but for all time" is his genius as an artist of sounds, not his ability to create memorable characters. That is the province of actors. Even in character-books modeled on Theophrastus (where, if anywhere, "character" would seem to be something conjured up by words on a page), the ability to *do* characters is assigned to actors, not to playwrights. Take, for example, Sir Thomas Overbury's character "Of an Excellent Actor": "All men haue beene of his occupation: and indeed, what hee doth fainedly, that doe others essentially: this day one playes a Monarch, the next a priuate person. Heere one Acts a Tyrant, on the morrow an Exile: a Parasite this man too night, to morrow a Precisian, and so of diuers others" (M2ᵛ–M3). The character of "A Player" in R. M.'s *Micrologia* is drawn on similar lines: "He is much like the counters in arithmetic, and may stand one while for a king, another while for a beggar, many times as a mute or cipher. Sometimes he represents that which in his life he scarce practises—to be an honest man" (reprinted in Morley 1891: 285–286). It is an actor's ability to create the illusion of reality, to seem to be someone else, that distinguishes him from the soldiers, the lovers, the melancholics, and all the other men caricatured by strokes of the character-writers' pens. Actors, not writers, are the great impersonators.

However clever Shakespeare's fellow actors may have been in the art of impersonation, not all the characters in Shakespeare's scripts enjoy the same *kind* of reality. Direct address is also what the audience heard when an actor spoke in soliloquy. As J. L. Styan remarks, soliloquies are the primary occasion when both modes of drama (Styan's own terms are "ritualistic" and "naturalistic") seem to be operating simultaneously. To the degree that he is thinking out loud in private, the speaker of a soliloquy is representing an event within the fictional play-as-object; to the degree that he is standing center stage and playing to the house, he is participating in the play-as-game (1967: 36–40, 72–75). As different as they may be ontologically, acoustically these two actions achieve the same effect: the solitary speaker fills the entire auditory field with his voice. For the temporal space of the soliloquy, the physical space of the stage and the acoustic space of the theater are his alone. In soliloquies, if anywhere, we come closest to Ihde's observation that "dramaturgical voice" can heard by the audience in the same way they might hear music, as a totalizing volume of sound that shuts out all other sounds and fills the listener's very body.

The voice of a speaker in soliloquy creates a very different subject-position than does the voice of a Prolocutor or a King. Instead of being subject *to* the speaker, the listener becomes a subject *with* the speaker. The shift from one subject position to the other is encouraged by the speaker's situation within the fiction, but it is facilitated also by the speaker's temporary position with respect to the other actors. The aloneness of a speaker onstage, as Styan and others have observed, is one of the most dramatically powerful opportunities early modern stage practice offered playwrights and actors. In acoustic terms, some of that power may derive from the fact that the speaker replicates in his own isolation vis-à-vis the other speakers the very situation of isolation in which each listener exists vis-à-vis other listeners. Especially in soliloquy, when the speaker's solitary voice fills the entire auditory field of the theater and fills each listener's ears, the subject in the fiction might sound, from within, like the listener's own self. It was, perhaps, the *heard* dimension of dramatic impersonation, rather than the actor's visible presence, that most powerfully caused early modern audiences to "interiorize" characters. Character is a function of performance in general, but of voice in particular. It is located not in the actor onstage, or even in the audience's imagination, but somewhere between the two—in the air, within the wooden [o:].

Instead of "characters" we might more accurately talk about the "persons" of the play. Not only is"person" the term used by early modern witnesses themselves; it also captures the double sense of person as both a body (the actor's "person" [OED 3]) and a voice (sound-through-a-mask, from *per-sonare*, "to sound through"). Together the im*person*ators of early modern theater create one single interpersonal circle, but that circle has a "center." The "peripheral" figures have an existence by virtue of their relationship to the central figure or figures. Hamlet, who commands 38 percent of the script's lines, offers perhaps the clearest example of all. Against Hamlet's visual and aural centrality Gertrude, Claudius, Horatio, and the others play out their parts as Mother, Stepfather, and Reasonable Friend. Ophelia's identity, by contrast, seems anything but fixed. The knowing interlocutor with her brother in 1.3, the thoroughly frightened reporter of Hamlet's apparent madness to her father in 2.1, the bait to draw out Hamlet in 3.1, the "straight man" for Hamlet's lewd remarks during *The Murder of Gonzago* in 3.2, the mad singer of ballads in 4.5, the corpse that precipitates Hamlet's belated leap into action in 5.1: the radical discontinuities in Ophelia's dramatic identity all make sense, not with reference to herself, but with reference to Hamlet as the figure standing and speaking at center stage. What we witness in such a centralized design is, of course, the structural principle of medieval morality plays. The dark moral universe of *Hamlet* may be worlds away from the morally just universe of rewards and punishments that governs *Everyman*, but the centeredness of the protagonist is in both cases the same (Bevington 1962: 114–127; Margeson 1967: 29–59). At the middle of the interpersonal circle stands and speaks a single figure who has choices to make—choices that generate all that happens within the circle of the play. Other figures within the circle can claim verisimilitude only through that central figure's personhood.

In comedy, as in history and tragedy, Shakespeare's scripts often begin with the voice of authority:

> Merchant of *Siracusa* plead no more. . . .

> Now faire Hippolita, our nuptiall houre
> Drawes on apace. . . .

> Of Gouernment, the properties to vnfold,
> Would seeme in me t'affect speech & discourse. . . .

> If Musicke be the food of Loue, play on. . . .
> (F1623: *CE* 1.1.3, *MND* 1.1.1–2, *MM* 1.1.2–3, *TN* 1.1.1)

Each of these speakers, let it be noted, is a Duke. It is altogether typical of comedy, however, that these oratorical figures are quickly displaced. And the voices that drown them out often belong to persons whose social station is anywhere but the top. Bottom in *A Midsummer Night's Dream*, Fluellen in *Henry V*, Helena in *All's Well That Ends Well*: each of these socially marginal figures lays claim to the center of the acoustic field and to a considerable share of the play's air space, whether that be measured in cubic feet, in minutes, or (save Helena perhaps) in decibels. With their intellect the audience may consign Bottom to the bottom of the Great Chain of (Social) Being, but with their ears they hear a voice that compels attention, first in booming [u] and [o:] ("I will mooue stormes; I will condole in some measure"), then in roaring [r] and blustery [ʃ]:

> To the *rest* yet, my chiefe humou*r* is fo*r* a tyrant. I could play *Ercles*
> *r*arely, or a pa*r*t to tea*r*e a Cat in, to make all split the *r*aging *R*ocks;
> and *sh*iuering *sh*ocks *sh*all break the locks of prison gates, and *Phibbus*
> carre *sh*all *sh*ine from fa*rr*e, and make and ma*rr*e the fooli*sh* Fates. This
> was lofty. (F1623: 1.2.22–35, emphasis added)

In more ways than one. Fluellen enters *Henry V* in a barrage of fisticuffs against Nim, Bardolph, Pistol, and Boy—and in an assault of monosyllables impelled by dentals and labials: "Go*dd*es *pl*u*d* v*p* to the *b*reaches / You rascals, will you not v*p* to the *b*reaches?" (Q1600: 3.2.21–22). From that moment on he shares dominion over the aural field with Henry and the Chorus, yielding power only in the play's last scene. Even Helena, with her piping boy's voice, emerges as the dominant sound in *All's Well*. After a decorously restrained dialogue with the Countess (1.1.1–77), she asserts her rhetorical power in one of the play's few soliloquies (1.1.79–104), confirms it in an astonishing series of bawdy exchanges with that master-of-words Paroles (1.1.105–181), and in a final soliloquy (1.1.212–225) assures that her voice is still ringing in the audience's ears when the King of France opens the next scene. What the audience hears in all of these instances is dissonance, a physical contra*diction* between social/political status on the one hand and acoustic status on the other. Through the sound of their voices alone, such characters establish a subject position the auditors have no choice but to share. When such decentering occurs in more serious contexts—as it does, for example, when Richard Gloucester's speeches reverberate in the audience's ears and guts—the effect can be very unsettling indeed.

With Bottom and the rude mechanicals, with Fluellen and the common soldiers, divergent speech communities are brought together onstage just as they were in the arena and in the galleries. As far as the audience is concerned, the prologues to both *Henry V* and

Henry VIII do all in their rhetorical power to counter this diversity: the listeners in each case are flattered as all being gentry. "Pardon, Gentles all," entreats the Chorus to *Henry V* (F1623: Pro. 8); "gentle Hearers . . . The First and Happiest Hearers of the Towne," the Prologue to *Henry VIII* styles the audience (F1623: Pro. 17, 24). Concerning the persons of the play, every single one of Shakespeare's scripts ends with an act of acoustic centering. Sometimes that takes the form of music and dancing. Duke Theseus exerts auricular power over Bottom, after all. Henry V silences Fluellen. The King of France has the last word in *All's Well* twice over—first as King, then as speaker of the Epilogue. Richmond concludes *Richard III* with a forensic oration that verges on liturgy: "Now Ciuill wounds are stopp'd, Peace liues agen; / That she may long liue heere, God say, Amen" (F1623: 5.8.40–41). That is, let the *audience* say "Amen."

These oracular appeals to audience unanimity were customarily succeeded by the even stronger seductions of music and dancing. At least through the 1590s on the South Bank and even later in theaters north of the city, danced jigs brought closure to serious plays and festive plays alike, as a few internal clues and several earwitnesses attest (see chapter 7 above). The Epilogue to *2 Henry IV* promises as his parting gesture a dance: "My Tongue is wearie, when my Legs are too, I will bid you good night" (F1623: Ep.30–32). Dancing ended the performance of *Julius Caesar* the Swiss traveler Thomas Platter saw on the South Bank in 1599. Elements of jigs seem apparent in the devices that round off several of Shakespeare's comedies: the dialogue of the owl and the cuckoo in *Love's Labor's Lost*, the wedding dance in *Much Ado*, Feste's song in *Twelfth Night*. Scripted or not, plays ended in London's public theaters with acts of enchantment. The singing and dancing that close *A Midsummer Night's Dream* are cast by the fairies themselves as a piece of magic-making that will bring blessings upon the newly married couples and ward off from their progeny such "blots of Nature's hand" as moles, harelips, and scars (F1623: 5.1.36–42). Though the connection with sympathetic magic may be less striking, the dances that are scripted to conclude *Much Ado* and *As You Like It* serve to close the liminal circle of the play in just way the fairies close *Dream*. In the absence of dance, music alone completes the compass of sound that began with three trumpet blasts. The songs that end *Love's Labor's Lost* and *Twelfth Night* fill the auricular field one last time before the audience deconjures the circle with the clapping of their hands. The music that closes these comedies has its tragic counterpart in the "flourishes" of trumpets and/or the thudding of drums that are specifically called for to close *2* and *3 Henry VI*, *Macbeth*, and *The Two Noble Kinsmen*, just as the comic dances have their tragic counterparts in the "dead marches" that conduct corpses and survivors out of the liminal circle in the folio *Lear,* in *Macbeth,* and in the folio *Hamlet.*

Whatever the scripted sounds happen to be, all plays end with a bid for "unisonance." The question remains, however, whether such gestures of acoustic closure actually succeed in turning three thousand listeners into one body of clapping enthusiasts. *Can* Richmond's 27-line *peroration* damp the echo of Richard's 1,124 lines earlier in the play? *Can* the wedding dance in *As You Like It* drown out the parting words of Jacques, who has held forth in such memorable speeches as "All the world's a stage"? The effect of decentering the acoustic field of the play is surely to decenter the acoustic field of the audience as well. Within the wooden O assembled an audience that was neither the unified congregation assumed in oratory and the liturgy nor the self-possessed adversaries implied by conduct-book discussions of conversation. Rather, the audience for public theater was both. Theater inserted listeners into a scene of speaking that was unlike any other in early modern culture. It subjected them, not to the voice of authority, but to the voices of diverse persons, every one of them competing for air space. And it permitted them to carry on conversations with persons above and below them in social station, something not allowed in conduct books.

Through the distinctive penetrating properties of sound, theater made those competing voices, for a time at least, each listener's own. As part of his consideration of the ears in *Microcosmographia* (1616), Helkiah Crooke stops to consider "How it comes to passe that *we are more recreated with Hearing then with Reading:* For we are wonderfully delighted in the hearing of fables and playes acted vpon a Stage, much more then if wee learned them out of written bookes." After considering several quick explanations—good actors are rarer than good books, it is easier to listen than to read—Crooke locates the reason in the phenomenology of hearing. A live actor's voice is more affecting "by reason of his inflexion and insinuation into our Sense" than the "dumbe Actor" one encounters in reading. Hence, "those things which be heard, take a deeper impression in our minds, which is made by the appulsion or ariuall of a reall voyce. But those things which are seene are always intentionally imprinted, & therfore the Act of Seeing is sooner ended and passeth more lightly by the Sense then the Act of Hearing. Whence it followes necessarily that things seene do not sticke so fast vnto vs." Then too, we experience voices *as part of a social group:* "there is a kinde of society in narration and acting, which is very agreeable to the nature of man, but reading is more solitary." Furthermore, in live voice events, we have the possibility of talking back, of changing the direction of the event. Certainly that is true in conversation; Crooke suggests that it is also true in the theater. When the speaker is right in front of us we pay closer attention than we do when reading:

> a certaine shamefastnesse and obseruancie doth cause vs to apply our eares to him that vttereth an thing by voyce, but in reading there is a kinde of remission in the minde and security from any blame of not profiting. Now wee conceiue more pleasure in a diligent and curious acting, then in a negligent and careless.

The result of such close attention, Crooke implies, is *ad lib* on the part of actors:

> Bookes cannot digresse from their discourse for the better explication of a thing, as those may which teach by their voyce. For in changing of words or mutuall conference, many pleasant passages are brought in by accident, as the Interlocuters list to aduance themselues; as we see in Comedies it is very ordinary. And by these sauces, as it were, of discourse, is the Hearing more sumptuously feasted, but the vniformity of the stile in things written and the continuity of the sentences causeth the Reader to loathe it. (1616: 698)

All these factors taken together—the penetrative power of speech, the hold of heard voices on the imagination, the sociability of listening to plays in the theater, the excitation to interaction between speakers and listeners—help to explain the privileged position of theater in the formation of early modern subjects. In the theater was to be found a degree of resonance unmatched elsewhere among body, society, psyche, and voice.

PART
THREE

Beyond

10 Listen, Otherwise

This time, thank God, the Indians did not respond with the irrelevant questions they had asked on a similar occasion six weeks before—questions about what causes thunder, the ebbing and flowing of the sea, the wind. Instead, the Algonkian-speaking Indians gathered in Waaubon's wigwam on October 18, 1646, seemed really to have understood "all the principall matter of religion" as Eliot and other missionaries from Plymouth Plantation preached it to them in their own tongue for an hour and a quarter. In his report of the event in *The Day-Breaking, If Not the Sun-Rising of the Gospell with the Indians in New-England* (1647)—he describes himself there as "an eare witnesse"—Eliot *says* he and his fellows were after "discourse" with the Massachusetts Indians—in effect, conversation with them—but the Englishmen *began* with an oration, with what Eliot calls "set speech." It was after the Englishmen had finished that the Indians were invited to ask questions.

Only one of their responses struck Eliot as naive—something only an Indian would think of, a question "sounding just like themselves." One of the Indians related to the Englishmen how he was praying to Jesus Christ in his wigwam, when one of his fellows interrupted him and told him he was praying in vain, "because Jesus Christ understood not what *Indians* speake in prayer, he had bin used to heare *English* men pray and so could well enough understand them, but *Indian* language in prayer hee thought hee was not acquainted with it, but was a stranger to it, and therefore could not understand them." Could God understand Indian prayers or not? Oh yes, the English brethren assured him. Since God made all men in the first place, he can understand all the languages men speak. In such doubts Eliot and his associates had to confront among their Native American subjects an idea of divinity very different from their own. One great obstacle to the Indians' acceptance of Christianity, Eliot notes, was the idea that one God can be in many places at the same time (1834: 2–7). Divinity, to Waaubon and his friends, resided in particular places, in animals, in plants; divinity made itself known to them in sounds, in thunder, sea tides, the wind.

To these aural intimations of immortality the Plymouth colonists were stone deaf. They had their own ideas about sound and space.

"The *Indians* inhabiting within our bounds"—one notes the first-person plural possessive—needed to be convinced of truths that, to the English, seemed self-evident, incontrovertible, universal. In their insistence on these truths the Plymouth colonists were being true to their sense of what John Gillies has called "poetic geography," a mapping of the world according to clear distinctions between an "us" and a "them." Such a geography accepts the spectator's own culture as a central reference point and proceeds to mark frontiers from there, presuming that *terminus* and *fines*—"our bounds"—have an objective reality that anyone can recognize. Even for the Greeks, language functioned as a major factor in this boundary-marking process: the word *barbarian* has its origins in babble, in the [br] [br] [br] noises that came out of the mouths of non-Greek-speakers (Gillies 1994: 6–12, 17). The exact shape of the space created by poetic geography varies according to the lay of the land—mountains, rivers, seas—and the human habitations built within it, but two features seem essential: extension in all directions and continuity. There is no room for "islands" of other cultures between center and terminus, no permanent possibility for outposts of civility beyond the frontier. Hence, the Romans extended their empire league by league in all directions—south, east, west, north—but they systematically ignored evidence of a rival center of civilization beyond the frontier in China (1994: 36).

In its circularity, continuity, and directionality, the shape of empire replicates the shape of sound. In acoustic terms as well as geographic, one might conceive of the Roman world as a huge circle with Rome as its effective center. The sounds that historically gave that acoustic field its coherence were the phonemes of Latin. However varied the sounds in a given spot within the circle—the roar of the north wind over Hadrian's Wall, the blasts of hunting horns in German forests, the babble of traders' tongues in Tyre, the crackle of earth under foot on the *llanos* of Spain—the Latin language made the huge circle of sound a single cultural space. It subsumed—or presumed to subsume—local soundscapes all around the Mediterranean into a single entity. Fynes Moryson approves such ambitions when he argues in material collected for his *Itinerary* (1617) that English-speakers ought to be establishing the same kind of aural empire in Ireland:

> this communion or difference of language, hath allwayes beene obser-
> ued, a spetiall motiue to vnite or allienate the myndes of all nations, so
> as the wise Romans as they inlarged theire Conquests, so they did
> spreade their language, with theire lawes, and the diuine seruice all in
> the lattene tounge, and by rewardes and preferments inuited men to
> speake it, as also the Normans in England brought in the vse of the

French tounge, in our Common lawe, and all wordes of art in hawking, hunting, and like pastymes. And in generall all nations haue thought nothing more powerfull to vnite myndes then the Community of language. (1967: 213–214)

What the Romans succeeded in imposing on their conquered lands— and what the English were failing to impose on Ireland—was a distinctive "acoustemology," a world view centered on sound. In the case of the Romans, as Fynes recognizes, that acoustemology was organized around the sounds of Latin.

If each culture has its own distinctive ways of understanding the world through sound, the borders between cultures become, potentially at least, sites of noise, confusion, pandemonium. The American composer Charles Ives delights in inhabiting such liminal places. One of the pieces in his "Three Places in New England" (composed 1903– 1914) renders the odd effect of standing in one place during an Independence Day parade as marching bands, playing different tunes in different keys, pass by and fade into one another. In the merging of two soundscapes, in the clash of two acoustemologies, there are always political factors at work—preoccupations that prevent the playful, open listening of someone like Ives.

John Brinsley, for example, positions himself solidly within the acoustemology of imperial-minded England when he proclaims on the title page to his pedagogical manual *A Consolation for Our Grammar Schooles* (1622) that he has written the book for schoolmasters,

> *More specially for all those of the inferiour*
> sort, and all ruder countries and places; namely,
> for *Ireland, Wales, Virginia,* with the *Sommer*
> Ilands, and for their more speedie attaining of our
> *English tongue by the same labour, that all*
> may speake one and the same
> *Language.*

That Brinsley expected practical political results is registered in his dedication of the book to the Lord Deputy for Ireland, the Lord Lieutenant of Wales, the Council of the Virginia Company, officials of the Somer's Island Company, and the Governors of Gurnsey and Jersey. (French, apparently, was as much beyond the pale for Brinsley as Welsh, Irish, and Algonkian.) Brinsley sees the ability of subjects to speak "one and the same language"—namely, English—as being essential not only to civilization but to spiritual salvation. His main purpose, he says, is "to make the way of knowledge more easie vnto them, not onely to the attaining of the Latin tongue, but also that hereby they may much more easily learne our English tongue, to helpe to reduce the barbarous to more ciuilitie, and so to plant Gods

true religion there, that Iesus Christ may reigne amongst them, Sa-
thans kingdome fall, and they be saued eternally, if the Lord vouchsafe them that mercie" (1622: 14–15). Especially among "our louing
countrie-men in *Virginia*," living as they do among the Indians, Brinsley is anxious that the English language be maintained, "when as
there are in the same [place] so manifold perils, and especially of
falling away from God to Sathan, and that themselues, or their posterity should become vtterly sauage, as they [the Indians] are"
(1622: A3).

By extending his apocalyptic vision to Somer's Island (later
known as Bermuda) as well as to Virginia, Brinsley may be including
African slaves among the peoples subject to "the slauerie of Satan."
A ship named the *Edwin* had brought "an Indian and a negar" from
the Savage Islands (the Bahamas) to Somer's Island in 1616; "twenty
Negars" were sold ashore by a Dutch ship at Jamestown three years
later (Wilkinson 1933: 115; Catterall 1926–1937, 1: 54–55). They joined
thirty-two "Negros in the service of planters" who were inventoried
a few months before as already present in Jamestown (Murphy 1998).
Brinsley's hopes of salvation for the Welsh, the Irish, the Powhatan
Indians, African slaves, and the French speakers of Gurnsey and
Jersey—and his safeguards against going native in Virginia—are
founded on just two skills: facility in reading and writing English
and an ability to translate from another language into English—not
from Welsh, Irish, Algonkian, African creole, or even French but from
the master language that rules them all: *Latin*. The bulk of *A Consolation* is made up of precepts cribbed from Edmund Coote's *The English
Schoole-Maister* (1596), tips on teaching translation from Latin to English, a list of approved English translations of approved Greek and
Latin authors (including Chapman's Homer, Thomas Newton's Cicero, Golding's Ovid, and Lodge's Seneca), advice on using dictionar-
ies, and a commercial plug for copy books on penmanship published
by the same printer as *A Consolation for Our Grammar Schools*.

Wales, Ireland, Virginia, Somer's Island: in naming these places
Brinsley is singling out four localities in which the dominion of the
English language was, in 1622, not yet certain. Furthermore, the
speech communities situated in these places were by and large free
from centralized control over communication in the form of writing
and the printing press. Welsh, Irish, Algonkian, and the creole speech
of African slaves existed in an uneasy relationship not only to the
English language but to writing as a means of political control. Each
of these languages was situated in its own geographical space—a
space to which English-speakers in the seventeenth century were attempting to lay claim. Each was composed of its own diverse speech
communities, most of them quite remote from English-speaking com-

munities. Each possessed its own distinctive acoustemology. Each presented, in Niklas Luhmann's terms, its own distinctive "zones of consensuality" among body, society, psyche, and sound as a medium. The convergence of these acoustic fields is, in Caliban's words, a place "full of noises, / Sounds, and sweet airs" (*Tempest*, F1623: 3.2.138–139). Whether those sounds "delight and hurt not" is something else again. Exploring the acoustemology of early modern England and the crisis of authority presented by its contacts with the acoustemologies of other cultures is, at bottom, an *ecological* project. As Luhmann observes, ecology is a study of borders, of "horizons" where self-maintaining systems touch one another and "resonate" (1989: 15–21). For Luhmann resonance is a figurative phenomenon, but in the case of spoken languages it is a physical fact. Let us listen for the resonances at four sites: within the self-contained acoustemology of English-speaking Britain, at the borders of English speech with Welsh and Irish, on the plantations of Virginia and New England, and at the intersection of early modern culture with our own.

FROM ABOVE

If the range of sounds that human beings everywhere make and apprehend can be visualized as a circle, cultures vary in how they use that compass of sounds to get their bearings in the world. The move from primal [o:] to speech to music to ambient sound and back to primal [o:] offers a range of possible reference points. Speech might seem to be the obvious, indeed inevitable, equivalent of due north, but cultures vary in how closely attuned they are to other points on the compass, particularly toward naturally occuring ambient sounds. Caliban, like the native peoples English-speakers encountered in the Caribbean and in North America, professes to *hear* an entirely different world than his European masters do. By contrast, the acoustemology of early modern England was solidly anchored in certain ideas about language and music.

With respect to language, there was a deeply entrenched belief that aural signifiers matched—or ought to match—actual signifieds. Francis Bacon and others may have argued the arbitrariness of human languages, but the power of realism over nominalism was hard to dislodge (Hudson 1994: 49). After all, it accorded with common sense, especially if one happened to speak only English. Richard Brathwait makes fun of such suppositions in his character of a word-bound Puritan under the guise of "A Zealous Brother": "Hee houlds his Mother tongue to be the Originall tongue; and in that only he is constant, for he hath none to change it withall. Hee wonders how *Babel* should have such a confused variety of *tongues*, and hee understand but one" (I12ᵛ). Here is logocentricism with a vengeance. Once, of course, things had been different: in the beginning, according to Genesis, "the whole earth was of one language and one speache." Not only did everyone speak the same language, but the sounds they made when they spoke were the same. Language was, so to speak, transparent: sound and meaning were one. The effect was spoiled, according to the Genesis account, by two things: the people decided to give themselves a name "lest we be scatred vpon yᵉ whole earth," and they proposed to build a city with a tower that reached up to heaven. Speaking one language, living in one place, acting on the same transcendental ambitions, all of mankind shared the same acoustemology. God realized as much—"the people is one . . . nether can thei now be stopped from whatsoeuer thei haue imagined to do"—and proceeded to punish what the gloss in the Geneva Bible terms "mans pride and vain glorie" by confounding human speech and scattering humankind upon all the earth (Geneva Bible, Genesis 11:1–9).

To Thomas Dekker's imagination of the event in his cony-catching pamphlet *O per se O*, the world before the construction of the tower was all one place: "A man could trauell in those dayes neyther by Sea nor Land, but he met his Countreymen, & none others" (1612: A3). The destruction of that unity is something Dekker not only sees but hears. God's divine "We" in Genesis ("Come on, let vs go downe, and there confounde their language") becomes, in Dekker's account, God and a messenger—a female messenger—named Confusion. Bred in Chaos, wild in her appearance, loose in her particolored attire, Confusion begins her attack by destroying the boundaries between human language and ambient sound:

> In one hand she grip'd an heape of stormes, with which (at her plea-sure) she could trouble the waters: in the other she held a whip, to make three *Spirits* that drew her to gallop fast before her: the *Spirits* names were *Treason, Sedition,* and *Warre,* who at euery time when they went abroad, were ready to set *Kingdomes* in an vproare. She roade vpon a Chariot of clowdes, which was alwayes furnished with *Thunder, Lightning, Windes, Raine, Haile-stones, Snow,* and all the other Artillerie belonging to the seruice of *Diuine Vengeance:* and when she spake, her *voyce* sounded like the roaring of many *Torrents,* boysterously strugling together, for betweene her iawes did she carry 1000000 tongues.

Stepping up to the workmen on the tower one by one, she whispers something into each one's ear. Suddenly, no one can understand anyone else. Furious and frustrated, the workmen give up their building and begin to disperse, each seeking out others who can understand him, and all go their separate ways to the ends of the earth (1612:

A3^v–A4^v). Since Babel, the history of languages has been one of confusion, dispersal, and degeneration. In place of the original *Ur*-language we now have different languages—and different acoustemologies.

To Renaissance Humanists, however, the memory of originary unity remained vivid and inspired idealistic linguistic projects like Brinsley's *Consolation*. Humanists never quite gave up the dream of building a tower that would allow them to seize transcendental authority for human signifiers. Cicero and his Humanist disciples may have understood style as a dynamic process, a performance that takes place between a speaker and an audience, but Aristotle and *his* disciples understood style quite differently. The laws that govern language, they affirmed, are not dynamic but fixed. What Latin, and most European languages after Latin, render as two separate words, *ratio* and *sermo*, Aristotle encapsulates in a single word, *logos*, suggesting a fundamental connection between speech and reason as two forms of discourse, mental and verbal. It was just this nexus between speech and reason that inspired medieval philosophers to trace out in grammar a (literally) mirror image of Aristotle's logic in *grammatica speculativa*, from the Latin word *speculum* or "mirror" (Bursill-Hall 1972). Two of the major speculative grammars of the thirteenth century were printed in the Renaissance: the *Grammatica Speculativa* attributed to Duns Scotus (Venice, 1499 and later editions) and the treatise *De Modis Significandi* attributed to Albertus Magnus (London, 1496 and 1515). While shifting the grounds of *logos* from theological faith to scientific observation, seventeenth-century empiricists pursued the same aspirations with respect to language. In *The Advancement of Learning*, Book VI, Bacon distinguishes between two kinds of grammar, "literary" and "philosophical." "Literary" grammar is simply the practical rules and formulae used for learning foreign languages. "Philosophical" grammar, on the other hand, "may diligently enquire, not the *Analogie of words one* with one another, but the *Analogy* between Words and Things, or Reason" (1640: 261). "Words" and "things": Bacon's speculations about a "philosophical" grammar were part of a Europe-wide interest in equating logical categories with grammatical categories. The great monument of the movement is the Port Royal *Grammar* of 1660. In its attempt to arrive at the universal principles of grammatical structure the Port Royal *Grammar* seems in hindsight to mark the beginning of modern linguistics (Salmon 1979).

To speculative grammarians, the universal principles of grammar are not arbitrary; they reflect the logical principles that govern the whole cosmos. The syntax of human speech, in this view, mirrors the structure of the cosmos in three essential ways: (1) as a direct reflection of things in words, (2) as a creative act proceeding from the same four causes that govern the physical universe, and (3) as a process that articulates space and time. "Are the modes of signifying in the word as subject or in the thing itself?" asks the writer of *De Modis Significandi* as printed by Wynkyn de Worde. Not in the word, he answers, "because as the mode of being is absolutely to the thing, and the mode of understanding to the thing understood, so is the mode of signifying to the thing signified" (Albertus Magnus 1977: 21). The most extreme form of such thinking would insist that even syntax replicates the cosmic order. *Grammatica speculativa*, inspired by Priscian, is founded on the conviction that things in the world dictate language, and not the other way around. For example, Thomas of Erfurt, a fourteenth-century grammarian whose work was printed in the sixteenth century, argues that "since ... active modes of signifying are not fictions, it follows necessarily that every active mode of signifying must originate basically from some property of the thing"—and goes on to apply that principle not only to nouns but to pronouns, verbs, participles, adverbs, conjunctions, prepositions, interjections, and the construction of whole sentences (1972: 137–139).

It was just such an understanding of language that invested writing with transcendental powers. To John Wilkins, summing up received opinion in *An Essay Towards a Real Character and a Philosophical Language* (printed 1668), the invention of the alphabet stands as testimony to "the divinity and spirituality of the humane soul," inasmuch as "it must needs be of a farr more excellent and abstracted Essence then mere Matter or Body, in that it was able to reduce all articulate sounds to 24 *Letters*" (quoted in Hudson 1994: 37). Far from being merely a record of speech sounds, however, writing became a way of getting in touch with Ideas, with the essences of meaning themselves. Hebrew in particular was regarded by many Renaissance scholars as a kind of *Ur*-language that directly transcribed timeless truth. For most Humanists, Hebrew remained a *visual* language—all the more since many scholars gave lip service (or rather, pen service) to Hebrew's status without actually attempting to learn the language themselves. Nicholas Hudson's history of *Writing and European Thought, 1600–1830*, casts the seventeenth century as a transitional period, in which such ideas of writing were being demystified, but failure to possess a writing system remained a major criterion by which the Irish, Native Americans, and the peoples of Africa could be adjudged barbaric (1994: 32–54).

Within the acoustemology of early modern England music, even more than language, could be heard as audible proof of cosmic order. The title to Thomas Morley's treatise *A Plain and Easy Introduction to Practical Music* (1597) points to the existence of "*impractical*" music in the form of philosophical ideas going back to Plato. As gymnastics train the body, Socrates insists, so music trains the mind. Why music has such power is explained in the *Timaeus:*

> so much of music as is adapted to the sound of the voice and to the sense of hearing is granted to us for the sake of harmony; and harmony, which has motions akin to the revolutions of our souls, is not regarded by the intelligent votary of the Muses as given by them with a view to irrational pleasure, which is deemed to be the purpose of it in our day, but as meant to correct any discord which may have arisen in the courses of the soul, and to be our ally in bringing her into harmony and agreement with herself; and rhythm too was given by them for the same reason.(Plato 1937, 2: 28)

The qualifying phrase "as is adapted to the sound of the voice and to the sense of hearing" locates the Real power of music in Ideas. Categories logically separate to modern ways of thinking—harmony and rhythm (musical concepts), revolutions (a physical concept derived from astronomy), and soul (a theological or perhaps psychological concept)—are here fused. Earthly music is, or ought to be, a sensible translation of the music of the spheres.

By Pythagoras's reckoning, that could physically be the case, since the differences in the lengths of lyre-strings necessary to produce harmonic intervals are proportionate to the distances between the planets (Iamblichus 1926: 62). Find the right proportions—or so goes the theory—and you can intercept cosmic harmony. Hence the early modern belief in the physical power of music to penetrate the body and to work physiological effects. The music that awakens Lear from his madness has a historical counterpart in the music English adventurers used to charm the wary people who awaited them on the shores of the New World. The number of musicians aboard the ships—from a trumpeter or two and a drummer to a full consort—varied according to the social rank of the head of the expedition (Woodfield 1995: 5–37). In addition to providing music for worship and festive occasions aboard ship, the musicians might be sent out as agents of civilization. John Davis describes an occasion when the Native Americans greeted the arriving English with cries that sounded like "the howling of wolves." Davis and his companions, in their turn, "made a great noise, partly to allure them to us, and partly to warne our company of them. Whereupon M. Bruton and the Master of his shippe, with others of their company, made great haste towards us, and brought our Musicians with them from our shippe, purposing either by force to rescue us, if need should so require, or with courtesie to allure the people." Music makes all the difference.

"When they came unto us, we caused our Musicians to play, our selves dancing, and making many signes of friendship" (Woodfield 1995: 109). The Indians responded in kind.

To work as Socrates and Pythagoras say it does, music needs to be monophonic (composed of a single melodic line, as Greek and Roman music was) and the language sung to such music needs to be organized by the quantity of vowels (as Greek and Latin poetry was). By the late sixteenth century, however, art music was already characterized by the polyphony that still distinguishes Western music from music elsewhere in the world, and English poets, despite the best efforts of Sidney and others, had accepted the fact that rhythm in English is a matter of stress and not vowel quantities. One can read early modern writing about music in general, and Morley's treatise in particular, as a sustained attempt to reconcile Platonic theory with latter-day realities (Smith 1979). Nonetheless, the old ideas kept their affective power. If no longer a physical reality, the translation of cosmic order into sensible sound remained a psychological factor in how people listened to music. One of the commonplaces transcribed into Folger MS V.a.381 attests to the persistence of Plato's ideas: "Musicke hath a certein secret passage into mens Soules & worketh so diversely in the mynde that it elevateth the harte myraculously & resembleth in a certein manner the voyces & Harmonye of Heaven, & questionless there is nothinge in this lief, wch so sensibly discourseth unto us the pleasures of Paradyce, as a sweete consorte of Musicke." The passage goes on to question the source of such power, whether it is a property of "the shakinge or arteficiall crispinge of the Ayre" or "a certein Sympathie, Correspondence or Proportion betwyxte oure Soules & Musicke" or "Gods generall provydence" or the quick passage of sound through the ears to the heart (112–113).

In the last song in *A Booke of Ayres* (1601) Thomas Campion attempts to make cosmic descriptions of music more than a metaphor by contriving that the rhythm of the tune exactly replicates the classical sapphic meter of the words (musical quotation 10.1). The fact that this and other songs in the collection are, according to the title page, "*Set foorth to be song to the Lute, Orpherian, and Base Violl*" casts the singer and his accompanists in the role of Orpheus, shaping the physical world with voice and lyre. Humanist philosophy kept alive the ancient contrast between *logos*-inspired song and the merely sensuous appeal of music played on pipe or flute. A performer's manipulation of the strings, making them now longer, now shorter, served as a visual reminder of the origins of harmony in the varying distances between the planets (Smith 1979: 85–87). Ultimate authority for the power Campion celebrates comes, of course, from God, "author of number," whose act of creation the song replicates.

Musical quotation 10.1

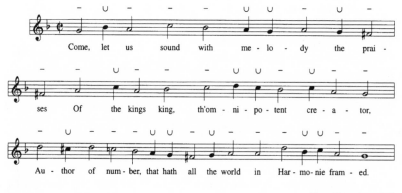

(from Campion 1601)

The acoustemology of early modern England can be considered from two vantage points: internal and external. On the internal front, inherited ideas about sound from Plato, Aristotle, Pythagoras, Priscian, and other authorities served, quite literally, to *harmonize* body, society, psyche, and media as self-maintaining systems. A conviction that sounds came with a cosmic guarantee of meaning positioned the listening body vis-à-vis ambient sounds in quite distinctive ways. Evidence of that conviction is to be found in high-culture hostility to gests, jigs, ballads, and stage plays. On the external front, the acoustemology of early modern England maintained a strong sense of its own boundaries. Brinsley imagines a world unified in terms of language, but the language he has in mind is *ours*. Sounds from the world beyond make sense only in terms of the English language. Thomas Gainsford, for one, is particularly suspicious of music that is imbued with the sounds of foreign tongues. "Poetry animated by musicke," he observes, "are dangerous companions amongst working spirits, and barbarous nations: witnesse the bardes and rimers of *Ireland,* and *Wales,* whose Siren songs haue excited such hellish treasons, and horrible tumults" (1616: 112ᵛ). On both fronts, the acoustemology of early modern England was defended by a formidable battery of philosophical ideas. Some of those ideas were as old as Plato, but the especially potent defenses were new, a product of Protestant theology. "In the beginning was the word": the acoustemology of early modern England was, from the top at least, radically logocentric. Even music, or so the theory demanded, should be imbued with the power of *logos*. Along the compass of sound from [oː] to [oː], the official acoustemology of early modern England was firmly centered on language.

AT THE BORDERS

Well *Eccho* tell me yet,
    howe might I come to see:
This comely Queene of whom we talke?
    oh were she nowe by thee.

   Eccho.   By thee
By me? oh were that true,
    howe might I see her face?
Howe might I knowe her from the rest,
    or judge her by her grace?
   Eccho.   Her grace. . . .

Amid the Kenilworth revels in 1575 it is Echo who directs the Savage Man into the presence of the queen, into the civilizing center of the eminently civilized soundscape realized within the grounds of Kenilworth Castle (Gascoigne 1907–1910: A6). Courtiers, city-dwellers, country folk: all present are invited to revel in what Benedict Anderson describes as "unisonality . . . the echoed physical realization of the imagined community" (1991: 145). The voice of Echo sounds the circumference of that social space and returns reassuring reverberations of sameness. Some fifty miles west of Kenilworth, however, the acoustic horizon would have sent back, to English ears, the sounds of cultural difference. To cross the border into Wales was to enter a soundscape distinguished from the soundscapes of England by the very language its inhabitants spoke, by their distinctive ways of *using* language, by their styles of singing and making music, by the mountains that acted as a natural sounding board to all these acoustic events. To cross the sea into Ireland was to enter a soundscape more different still. And to cross the ocean to America was to venture, aurally as well as visually, into *terra incognita*. Wales, Ireland, and North America presented three ever more distant horizons not only geographically but acoustemologically.

The Act of Union between England and Wales in 1535 gave the inhabitants of Wales a voice in Parliament, but the language they were required to speak there was English. At home, too, English became the official language. One result was a decline in the volume and the quality of Welsh poetry as traditional patrons lifted their cultural sights eastward beyond the Marches (Hughes 1924: 11–12). Despite the prominence of many Welshmen in English public life in the sixteenth and seventeenth centuries, Wales maintained a reputation for difference—for *audible* difference—that found its most influential description in the *Itinerarium* of Giraldus Cambrensis. Four hundred years after he had put pen to parchment Gerald of Pembroke, a Welshman himself, was still providing the topographical groundwork for early modern conceptions of Wales. An abridgement of Giraldus's *Itinerarium*, in Latin, was published in London in 1585 as part of Edmund Bollifont's edition of Ludovico Pontico Virunio's *Britannicae Historiae*. With or without a nod to Bollifant's edition, Fluellen's forthrightness with Henry V on the stage of the Globe in 1599 bespeaks a distinctively Welsh "boldness and confidence in speaking and answering, even in the presence of their princes and chieftains" that Giraldus attributes to the Welsh nation's ancestry in the warm climes of Asia Minor. One mark of their descent from the Trojan refugee Brutus is the presence of Greek words—or what Giraldus takes to be Greek words—in the Welsh language (1892: 500–501). Along with forthrightness comes a distinctively Welsh facetiousness, which Giraldus delights to illustrate. For example, hinting at the stinginess of his hostess, a Welshman is reported to have remarked, "I only find fault with our hostess for putting too little butter to her salt" (1892: 498–499). Above all, Giraldus admires the "natural rhetoric" of the Welsh: "In their rhymed songs and set speeches they are so subtile and ingenious, that they produce, in their native tongue, ornaments of wonderful and exquisite invention both in the words and sentences. Hence arise those poets whom they call Bards, of whom you will find many in this nation, endowed with the above faculty" (1892: 496).

It is not to verbal arts, however, but to music that Giraldus devotes his lengthiest attention. Unlike the inhabitants of other countries, who sing in unison a single melodic line against a rhythmic accompaniment, the Welsh like to sing in multiple parts, "so that in a company of singers, which one very frequently meets with in Wales, you will hear as many different parts and voices are there are performers, who all at length unite, with organic melody, in one consonance" (1892: 498). In a world of musical monophony the Welsh produce polyphony. Their instrumental music is no less affecting. The three instruments the Welsh like to play are the harp, the pipe, and a violin-like instrument called the crwth. On all three the Welsh are virtuoso performers:

> It is astonishing that in so complex and rapid a movement of the fingers, the musical proportions can be preserved, and that throughout the difficult modulations on their various instruments, the harmony is completed with such a sweet velocity, so unequal an equality, so discordant a concord, as if the chords sounded together fourths or fifths. . . .
> They enter into a movement, and conclude it in so delicate a manner,

and play the little notes so sportively under the blunter sounds of the base strings, enlivening with wanton levity, or communicating a deeper internal sensation of pleasure, so that the perfection of their art appears in the concealment of it. (1892: 493)

In the fineness of discrimination he devotes to music, as well as in the quantity of ink, Giraldus implicitly locates the aural sensibilities of the Welsh at a point along the compass of sound somewhere beyond language itself, at a point where language verges into music. He pushes that point still further when he remarks among the Welsh a gift of prophecy going back to Calchas and Cassandra: "There are certain persons in Cambria, whom you will find nowhere else, called Awenydhyon, or people inspired; when consulted upon any doubtful event, they roar out violently, are rendered beside themselves, and become, as it were, possessed by a spirit" (1892: 501).

All told, the ancient Britons of Wales occupied an aural universe different from that of the Anglo-Saxon-Normans in England. Communicating with one another in a language incomprehensible to English speakers, singing in polphony, performing on harp, pipe, and crwth with astonishing speed and deep "internal sensation," given to roaring in prophecy, the Welsh presented themselves to the English across the border not just as a cultural Other, but as a distinctly *aural* Other. Garrulousness and musicality were, to English ears, the very qualities that made the Welsh *Welsh*. Two diverse ways of responding to those differences are suggested by the comment Giraldus makes on the effects music—Celtic music in particular—is likely to work on listeners: "those very strains which afford deep and unspeakable mental delight to those who have skilfully penetrated into the mysteries of the art, fatigue rather than gratify the ears of others, who seeing, do not perceive, and hearing, do not understand; and by whom the finest music is esteemed no better than a confused and disorderly noise, and will be heard with unwillingness and disgust" (1892: 495). "Deep and unspeakable mental delight" or "noise": the difference consists in how well one listens otherwise.

The persistence of the aural topography first laid out by Giraldus can be witnessed in Michael Drayton's *Poly-Olbion* (1612, 1622), in which Wales is introduced via a musical contest with England for possession of an island in the Severn River that separates the two territories. Wales pleads its case in a markedly traditional way, through the voices of bards

> that in their sacred rage
> Recorded the Descents, and acts of everie Age.
> Some with their nimbler joynts that strooke the warbling string;
> In fingering some unskild, but onelie us'd to sing

Unto the others Harpe: of which you both might find
Great plentie, and of both excelling in their kind. . . .
(Drayton 1961: 73)

John Selden's learned commentary on the passage explains the bardic tradition and quotes Giraldus on the distinctive music of Wales (Drayton 1961: 83–84). England, on the other hand, underscores its claims with the sophisticated music of a broken consort—just the ensemble of instruments that regularly played for performances at the Blackfriars Theater in the years Drayton was completing his poem. Lute, viol, gamba, cithern, pandore, therbo, gittern, cornet, fife, hoboy, sackbut, recorder, flute, and shawm demonstrate the sonic sophistication of the people living on the English side of the Severn. Instruments of the sort favored by people living on the Welsh side are tamed and civilized by being orchestrated into the artful music of the larger consort: "Some blowe the Bagpipe up, that plaies the Country-round: / The Taber and the Pipe, some take delight to sound" (Drayton 1961: 78). How these consorted musicians are able to "sing" the ancient English history Drayton goes on to relate is not altogether clear—unless Drayton is implicitly imaginging performance in the theater.

Popular representatives of English ideas of Welsh aurality do, in fact, walk and talk onstage in the persons of Fluellen in *Henry V* and Morgan, Earl of Angelsey, in *The Valiant Welshman*. Acted by the Prince's Men, probably in 1612, *The Valiant Welshman* sounds all the traditional notes of Welshness loud and clear. The presenter of the action, much in the manner of Gower in *Pericles*, is a "Bardh" who makes his first entrance by rising from his tomb while harpers play (R. A. 1615: A4^v). Harp music likewise accompanies a wedding masque performed in the course of the play, complete with a song in Welsh (1615: C2^v). It is Morgan, however, who most forcefully speaks Wales into dramatic presence. When he encounters a group of officious rustics right out of *Much Ado About Nothing*, Morgan no sooner hears them speak than he rushes to embrace them as linguistic kinsmen: "Harke you, harke you Cousin, he speakes Brittish, by shesu, I not strike him now, if he call mee three kanues more. God plesse vs, if he do not speake as good Brittish, as any is in Troy walles. Giue me both your right hands, I pray you, let vs be friends for euer and euer" (1615: G2). (This despite the fact that the rustics are scripted to speak perfectly normal Stage Rustic English.) In his irrepressible spirit, his redundancies, his occasionally dubious word order, his regular substitutions of [p] for [b] and [sh] for [ʤ] Morgan declares himself to be fellow countryman to Fluellen and other stage Welshmen of the early seventeenth century.

Overbury's character of "A Braggadochio Welchman" fits Morgan to a T—or perhaps to an LL: "The first note of his familiaritie is the confession of his valour; and so hee prevents quarrels. He voutcheth Welch, a pure and vnconquered language, and courts Ladies with the story of their Chronicle" (1615: D8–D8ᵛ). The acoustic effect of such figures in the theater is suggested by the illustration accompanying "The Welshmans praise of Wales," one of many "folk" items collected in *Recreation for Ingenious Head-peeces. Or, A Pleasant Grove for their Wits to walke in* (1650) (fig. 10.1). The speaker's irrepressible braggadoccio on behalf of his native land—"For hur will tudge your eares, / With the priase of her thirteen Seers"—finds its visual counterpart in his huge round hat, his jutting left leg, and his tilted pikes, not to mention his thrust diaphragm and outsized mouth. The graphemes of feather, foot, and pikes extend the figure to the very margins of the frame, filling the visual space just as Fluellen and Morgan filled the acoustic space of the theater with their voices.

Fancies and Fantasticks.

Fancies and Fantasticks.

But hark you me now, for a liddell tales
Sall make a gread deal to the creddit of VVales.
    For hur will tudge your eares,
    VVith the praife of hur thirteen Seers;
    And make you as clad and merry,
    As fourteen pot of Perry.

'Tis true, was weare him Sherkin freize,
But what is that? we have ftore of feize;
And Got is plenty of Coats milk
That fell him well, will buy him filk
Inough, to make him fine to quarrell
At *Herford* Sizes in new apparell;
And get him as much green Melmet perhap,
Sall give it a face to his Momouth Cap.
    But then the ore of *Lemfter*,
    Py Got is uver a Sempfter;
    That when he is fpun, or did
    Yet match him with hir thrid.

Aull this the backs now, let us tell yee,
Of fome provifions for the bellie:
As Cid and Goat, and great Goats mother,
And Runt, and Cow, and good Cows uther.
And once but taft on the VVelfe Mutton;
Your *Englis* Seeps not worth a button.
And then for your Fifs, fall fhoofe it your difs,
Look but about, and there is a Trout.

A

The Welfhmans praife of Wales.

I's not come here to tauke of *Prut*,
From whence the *Welfe* dos take hur root;
Nor tell long Pedegree of Prince *Camber*,
Whofe linage woud fill full a Chamber,
Nor fing the deeds of ould Saint *Davie*,
The Vrfip of which would fill a Navie.

But

Figure 10.1. "The Welshmans praise of Wales." From *Recreation for Ingenious Head-peeces. Or, A Pleasant Grove for their Wits to walke in* (1650). Reproduced by permission of the Folger Shakespeare Library, Washington, D.C.

Aural assault on English sensibilities was even more acute in Ireland. Christopher Highley has called attention to the English impulse to merge all its borderlands—Wales, Scotland, and Ireland—"into a single territorial and ethnographic zone, with common linguistic and cultural ties, and with a shared hostility to the English" (1997: 68). Giraldus, for his part, is just as full of praise for Irish music-making as he is for Welsh. (He in fact uses some of the same wording to describe musical practices in the two places, with full acknowledgment that he is doing so.) On most other aspects of Irish culture, however, Giraldus anticipates later observers in taking a much harsher view. "In the common course of things," Giraldus remarks, "mankind progresses from the forest to the field, from the field to the town, and to the social condition of citizens; but this nation, holding agricultural labour in contempt, and little coveting the wealth of towns, as well as being exceedingly averse to civil institutions—lead the same life their fathers did in the woods and open pastures, neither willing to abandon their old habits or learn anything new" (1892: 124).

Early modern commentators locate the Irish of their day in the same physical circumstances. As Edmund Spenser's spokesman Irenius observes in *A vewe of the present state of Irelande*, most of the Irish do not live in towns but move about the land in search of fresh pasturage for their herds. Out in the open, Irenius claims, the people can live "more licentiouslye" than they would in towns. Having once known such freedom, they "doe like a steare that hathe bene longe out of his yoke grudge and repine ever after to Come vnder rule againe" (E. S. 1949: 98).* It is partly the geographical distribution of the people, Sir John Davies argues in *A Discovery of the True Causes Why Ireland Was Never Entirely Subdued . . .* (1612), that explains the failure of the English to impose law and order. The first English "adventurers" chose the choicest land for themselves, pushing the native Irish out into the woods and mountains—out of sight and out of hearing. Had the English done just the reverse and erected their castles in the mountains, they would have forced the Irish to live in open country, "where they might haue had an eye and obseruation vpon the[m]" (1612: 161). Customs of land-holding and tillage discourage fixed habitations, as Moryson points out in one of the unpublished parts of his *Itinerary*. Irish landlords typically distribute their lands among their tenants for only one, two, or three years at a time, so the people build no houses, but, "like Nomades living in Cabins,

---

*Although Jean Brink has pointed out that no evidence connects Edmund Spenser's name with *A vewe* until Sir James Ware's edition of 1633, I follow Highley, Baker, and most other students of the tract in accepting it as Spenser's work.

remoue from one place to an other with their Cowes, and commonly retyre them within thick woods not to be entred without a guide delighting in this Rogish life, as more free from the hand of Justice and more fitt to commit rapines" (Moryson 1967: 198).

Moryson goes on to connect the musical nature of the Irish with their freedom to move about in the open land. It is their idleness, Moryson says, that "makes them to loue liberty aboue all thinges, and likewise naturally to delight in musick, so as the Irish Harpers are excelent, and their solemne musicke is much liked of strangers" (1967: 483–484). Such civil institutions as the Irish possess are located outdoors. Richard Stanyhurst in "A Treatise Conteining a plaine and perfect description of Ireland," printed in the 1587 edition of Holinshed's *Chronicles*, specifies that when one of the traditional dispensers of justice holds court, he "sitteth on a banke, the lords and gentlemen at variance round about him, and then they proceed" (1586, 2: 45). Spenser seems to have such a place in mind when he describes "raths" or artificial mounds where people come together for "folkmotes" or talks. These traditional sites for coming together and talking out the differences "betwene Towneshipp and Towneshippe or one private persone and another" seem dangerous to Spenser's Irenius because they are beyond the English government's surveillance, aurally as well as geographically: "well I wote and trewe it hathe bene often times approued that in these metinges manye mischiefs haue bene bothe practised and wroughte ffor to them doe Comonlye resorte all the sum of lose people wheare they maye frelye mete and Conferr of what they liste which else they Coulde not doe without suspicion of knowledge of others" (E. S. 1949: 128–129). Spenser and Moryson both remark on the Irish eagerness for news, carried from place to place by professional jesters (E. S. 1949: 126–127; Moryson 1967: 199). Outdoors, unbounded, beyond the hearing of centralized authority: the soundscape of Ireland outside the pale around Dublin presented the English colonizers with circumstances in which it was difficult to get their acoustic bearings.

Not surprisingly, English intruders into the Irish soundscape followed the lead of Giraldus and pronounced what they heard brutish and uncivilized. All three of the "evils" considered by Spenser's Irenius—law, religion, and customs—bear aural marks of savagery. Laws there are unwritten. The *brehon* who dispenses justice among men gathered around him on a bank speaks according to "a certaine rule of righte vnwritten but deliuered by tradicion from one to another" (E. S. 1949: 47). As the *brehon* pleases, murder may be atoned by the payment of blood money. Land-holding is likewise grounded in oral agreement. When a lord dies, his allies come together and choose a successor (or *tanist*) from among the surviving relatives. Often this successor is not the eldest son, the rightful heir according to English sights, but the man who can best defend his holdings. His succession is confirmed by his taking an oath while he stands on a special stone (E. S. 1949: 49–40). Brehon law and tanistry are likewise condemned by Davies (1612: 166–168).

The terms in which Spenser has Irenius cast his condemnation are especially telling: "it is daungerous to leave the sence of a law vnto the reasone or will of the Iudge whoe are men and maye be miscaried by affeccions and manye other meanes But the lawes oughte to be like stonye tables playne stedfaste and vnmoveable" (E. S. 1949: 78). Law should be visible, legible, touchable. Implicit in Spenser's turn of phrase are not only the Ten Commandments engraved on stone tablets by the hand of God but the material solidity of early modern law books. Irish religion is likewise an oral affair— and hence an unreasonable affair. "Not one amongest a hundred," Spenser claims, "knowethe anye grounde of religion anie article of his faithe but Cane perhaps saie his pater noster or his *Ave marye* without anie knowledge or venderstandinge what one worde theareof meanethe" (E. S. 1949: 136). Irish religion, in fine, is a matter of "saying," not "knowing." A religion founded on meaningless sound, it lacks the *logos* of Protestant faith. Elizabeth attempted to correct that situation by giving £6 13s 4d to have cast a set of type in Irish characters for printing the Old Testament and the catechism of the Church of England (Salmon 1985: 335). Not for nothing does Spenser title his treatise "A *Vewe.*"

Aural savagery, to the ears of Spenser's Irenius, sounds out loudest of all not in law or religion but in Irish "customs," particularly the Irish [o:]. In battle or "at other troblesome times of vprore" the Irish raise a cry that Irenius connects with their supposed Scythian ancestors, as written up by Herodotus and Diodorus Siculus (E. S. 1949: 103). The Irish, according to Moryson, "are by nature very Clamorous, vpon euery small occasion raysing the hobou (that is a dolefull outcrye) which they take one from anothers mouthe till they putt the whole towne in tumult" (1967: 484). English speakers have the Irish to thank for the word *hubbub:* "That which in *England* we doe call the Hue and Cry," Barnabe Rich explains from forty years of firsthand (or first-ear) experience, "in *Ireland*, they doe call the *Hubbub*" (1619: 1). Stanyhurst, Moryson, Rich, and other witnesses note that the same kind of cry was used in mourning. As Edmund Campion describes the custom in his history of Ireland, "They follow the dead corpes to the grave with howlings and barbarous out-cryes, pittyfull in apparance"—a most curious word in this context—"whereof grew (as I suppose) the Proverbe, to weepe Irish" (1633: 13–14). Stanyhurst repeats Campion verbatim (1586, 2: 44). According to Rich, all the shrieking and howling is done by women hired for the purpose, so that "to weep Irish" is really "To weepe at pleasure, without cause, or griefe" (1610: 13). What's more, the Irish will raise the hubbub at the drop of a hat. If a couple of drunks get into a fight, Rich reports, if a husband strikes his wife, if a master or mistress beats a servant, the victim raises an outcry taken up by all the town. "Of these Alarmes and Outcries, we haue sometimes three or foure in a weeke, and that in *Dublin* it selfe, among the base and rascall sort of people, and as these *Hubbubs* are thus raised in cases of anger and discontent, so they vse to giue the *Hubbubs* againe in matters of sport and merriment" (1619: 1–2). If the Welsh, in English imagination, are centered on music, the Irish are centered on [oː].

English estimations of the Irish language classify it as one step removed from noise. As the Irish themselves are, according to Spenser's Irenius, descended from Scythians and "another nacion Comminge out of spaine"—whether native Spaniards, Gauls, Goths, or Africans is uncertain—so the Irish language is a hodgepodge made up of scraps of other tongues (E. S. 1949: 84). Stanyhurst cites Giraldus to the effect that one Gathelus or Gaidelus "deuised the Irish language out of all other toongs then extant in the world" (1586, 2: 12). Stanyhurst himself is of the opinion that the Irish language was forged out the debris of Babel. One reason for the English hostility may have been the fact that Irish was notoriously difficult for a non-native to learn. Stanyhurst tells the story of a woman in Rome who

> was possessed with a babling spirit, that could haue chatted anie language sauing the Irish; and that it was so difficult, as the verie diuell was grauelled therewith. A gentleman that stood by answered, that he tooke the speech to be so sacred and holie, that no damned feend had the power to speake it; no more than they are able to saie (as the report goeth) the verse of saint Iohn the euangelist, *Et verbum caro factum est.* Naie by God his mercie man (quoth the other) I stand in doubt (I tell you) whether the apostles in ther copious mart of languages at Ierusalem could haue spoken Irish, if they were apposed: whereat the companie heartilie laughed. (1586, 2: 12)

To English ears, Irish sounded whiny. Like the slaves in Roman comedy, Davies says, "the Common people haue a whyning tune or Accent in their speech, as if they did still smart or suffer some oppression" (1612: 176). (Thanks to Davies and his ilk, indeed they did.) Stanyhurst attributes to Irish women of all social classes a similar way of speaking "peeuishlie and faintlie, as though they were halfe sicke, and readie to call for a posset" (1586, 2: 11). James Howell hears the same cadences in all speakers of Irish. Comparing Welsh to Irish,

Howell notes that "the tones of both the Nations is consonant," but with this difference: "the *Irish* is a little more querulous and whining" (1650, 2: 78–79). But who speaks proper Irish? One thing that helped English speakers keep their distance from the consonant clusters flying all about them was the conviction that no one was speaking the ancient language correctly. Stanyhurst's judgment is carefully considered. "The toong is sharpe and sententious, & offereth great occasion to quicke apophthegms and proper allusions," he will concede. "But the true Irish indeed differeth so much from that they commonlie speake, that scarse one in fiue hundred can either read, write, or vnderstand it. Therefore it is preserued among certeine of their poets and antiquaries" (1586, 2: 12). That is to say, "true Irish" exists only in writing, only in a form that can be studied, censored, and supervised. Moryson sums up the English appraisal of Irish as a spoken language: "if no such tongue were in the world I think it would never be missed either for pleasure or necessity" (1904: 317).

Ireland in the late sixteenth century must, in fact, have sounded very much like Babel before the dispersal of the workmen. As Vivian Salmon has pointed out, there was not just one language being spoken in Ireland at the time but *four:* Irish, English, Latin, and a creole of Irish-English (1985: 331). Since the great age of the Irish missionaries, Latin had operated for the Irish as a *lingua franca* in communicating with speakers of other languages. Stanyhurst confirms that in the sixteenth century Latin was widely spoken "like a vulgar language, learned in their common schooles of leachcraft and law" (1586, 2: 45). It was not so much Irish or Latin but the Irish-English creole that most irritated the Anglophone interlopers. Stanyhurst reports the experience of an English nobleman, recently arrived as a commissioner, who was surprised at how many words he could recognize in "the rude complaints of the countrie clowns." Happy that he could understand so much, the nobleman "told one of his familiar friends, that he stood in verie great hope to become shortlie a well spoken man in the Irish, supposing that the blunt people had pratled Irish, all the while they iangled English" (1586, 2: 11). Stanyhurst proceeds to offer examples of how English is being mangled. Irish speakers persist, he complains, in putting the accent on exactly the wrong syllable— "markeat" for "market," "baskeat" for "basket," "gossoupe" for "gossip," "Robart" for "Robert," "Niclase" for "Nicholas"—rather like the British of a certain social class today, who, when they go to the "balley" (ballet), always like to visit the "buffey" (buffet). It was not a "posset" that peevish Irish women called for, but a "possoat" (1586, 2: 11). Stanyhurst describes the everyday language of English-occupied Ireland as "a mingle mangle or gallamaulfrey of both the languages

... [they have] so crabbedly iumbled them both togyther, as commonly the inhabitants of the meaner sort speake neyther good English nor good Irishe" (1586, 2: 11–12).

Given their conviction that spoken Irish was a confused babel, given their distrust of oral communication in the absence of writing, given their visceral fear of [o:], the English were especially suspicious of Irish bards. Moryson speaks for most of the English censors in condemning the way Irish bards choose to praise licentious men (read Irish rebels) instead of moral exemplars (read collaborators with the English): "Alas how vnlike to Orpheus, who with his sweete harpe and wholesome precepts of Poetry laboured to reduce the rude and barbarous people from liuing in woods, to dwell Ciuilly in Townes and Cittyes, and from wilde ryott to morall Conuersation" (1967: 199). Bards and their performances belong, clearly enough, to the woods and wide open spaces, not to towns. The oral improvisations of Irish bards are dangerous to listen to: they appeal to the body, to sensuous perception, not to the mind. Stanyhurst connects the bards specifically with feasting, with music, with carnival: "Their noble men, and noble mens tenants, now and then make a set feast, which they call coshering, wherto flocke all their reteiners, whom they name followers, their rithmours, their bards, their harpers that feed them with musicke: and when the harper twangeth or singeth a song, all the companie must be whist, or else he chaseth like a cutpurse, by reason his harmonie is not had in better price" (1586, 2: 45). Listening to the bard's recitations becomes a form of "feeding." John Derricke's diatribe against Irish kerns in *The Image Irelande* (1581) illustrates the kind of feast Stanyhurst is talking about (fig. 10.2). First the feasters fill their bellies.

> Now when their gutts be full,
>   then comes the pastyme in:
> The Barde and Harper mellodie,
>   vnto them doe beginne.
> The Barde he doeth report,
>   the noble conquestes done,
> And eke in Rimes shewes forth at large,
>   their glorie thereby wonne.
> Thus he at randome ronneth,
>   he pricks the Rebells on. . . .
>   (1581: 55–56)

The concatenation here of food-drink-music-bardolry-rebellion positions bards and their performances solidly within the range of [o:].

Spenser's attitude to the bards, as both David Baker and Christopher Highley have argued, is highly conflicted. Clearly enough

A spice spectator sic me docuere parentes

Me quoque maiores omnes, virtute carentes

3   Who play'th in Romish toyes the Ape, by counterfetting Paull:
     For which they doe award him then, the highest roome of all.
     Who being set, because the cheere, is deemed little worth:
     Except the same be intermixt, and lac'de with Irish myrth.

D   Both Barde, and Harper, is preparde, which by their cunning art,
     Doe strike and cheare vp all the gestes, with comfort at the hart.

Figure 10.2. Irish bard (detail). From John Derricke, *The Image of Ireland* (1581). Reproduced from a copy in the Library of Congress, Washington, D.C.

Spenser realized the threat oral bards posed to English authority; at the same time he saw in the bards a counterpart to himself as a poet with political aspirations. Irenius confesses to his interlocutor Eudoxus that he has gone so far as to have some of the bards' poems translated into English (and, in the process, presumably written down) and finds that "surelye they savored of swete witt and good

invencion but skilled not of the goodlie ornamentes of Poetrye yet
weare they sprinkled with some prettie flowers of theire owne naturall devise which gaue good grace and Comlinesse vnto them" (E. S.
1949: 127). Both Baker and Highley contrast Eudoxus' intolerant An-
glocentrism with Irenius' more nuanced sense of the Irish situation.
(Does Irenius play the *Ironist* as well as the *Eire-nist*?) Baker finds
in Irenius' sympathies evidence that Spenser wished to critique the
government's Irish policy even as he rehearsed some of its major
premises (1997: 66–123). Highley finds in Irenius' treatment of the
bards nothing less than a prototype for Spenser's own career as a
political poet. From *A View* through the dedicatory sonnet to Ormond
in the first printing of *The Faerie Queene* through *The Teares of the Muses*
to *Colin Clouts Come Home Againe* Highley traces a "movement up the
hierarchy of bardic orders, from a kind of wandering bard on his trip
to England, to an ollamh, the modern descendant of the ancient dru-
ids and seerers of Celtic society" (1997: 36). If so, Spenser stands out
as an anomaly among English observers who tried to get the Irish
down in writing.

Amid so many barbaric practices of orality, the next-to-most infuriating thing about the Irish to English ears is their absolute refusal,
unlike the Welsh, to learn to speak English. The most infuriating
thing of all, however, is the way they have seduced English settlers
into abandoning the English language and joining them in their linguistic savagery. Davies compares the Englishmen's corruption to
drinking from Circe's cup and turning into beasts—that is to say, giving up language for bellows and grunts (1612: 181). The two main
causes for going native Irenius locates in intermarriage and in "foster-
ing," in sending infants out to suck with Irish-speaking women. Once
again, the Irish language is associated with carnality:

> the Childe that suckethe the milke of the nurse muste of necessitye
> learne his first speache of her, the which beinge the firste that is envred
> to his tounge is ever after moste pleasinge vnto him In so muche as
> thoughe he afterwardes be taughte Englishe yeat the smacke of the
> firste will allwaies abide with him . . . ffor the minde followethe muche
> the Temparature of the bodye and allsoe the wordes are the Image of
> the minde So as they procedinge from the minde the minde must be
> nedes affected with the wordes So that the speache beinge Irishe the
> harte muste nedes be Irishe for out of the abundance of the harte the
> tonge speakethe. (E. S. 1949: 119)

Moryson, for his part, reports encounters in which residents who
knew how to speak English flatly refused to do so (1967: 213–214).
Stanyhurst, wondering why "we must embrace their language, and
they detest ours," tells the story of one man who "demanded merilie

whie O[']neile that last was would not frame himselfe to speake English? What (quoth the other) in a rage, thinkest thou that it standeth with O[']neile his honor to writh [*sic*] his mouth in clattering English? and yet forsooth we must gag our iawes in gibbrishing Irish?" (1586, 2: 11–12). One reason for English touchiness on this point was the recognition, acknowledged by Spenser, that the Irish were, once upon a time, literate and cultivated when the English were still savages (1949: 88). As Irenius has to confess, England at the time of the first Irish conquest, *temp.* Henry II, was itself a place less civilized than it is now: "it is but even the other daye since Englande grewe Civill" (1949: 118).

What is more, "British" had been spoken on English soil before the invasions of the Saxons and the Normans. In Wales and in Ireland, England confronted an acoustic image of its earlier self—and the prospect was unnerving. Brathwait's "Zealous Brother" may have thought English above the confusions of Babel, but that did not seem the case to speakers of languages other than English. French-speaking Etienne Perin, for example, in a description of England and Scotland written in 1558, treats the English language as a grab-bag of fragments from other languages, in just the way English writers were apt to treat Irish:

> Caesar, in his Commentaries, says, That England is an island in the sea, serving as a retreat for thieves and robbers; for it being uninhabited, many banished persons and vagabonds fixed their residence there; insomuch that the emperor Julius Caesar says, their language is an assemblage or jumble of all tongues; which in truth, I find to be true, for their language partakes as well of the German as of divers others; on which account, the poets of past times have despised their pretences to antiquity, and have always estimated them as a strange and barbarous people, not deducing their origin from the aborigines, but from strangers, barbarians, and runaways; as when any country has been destroyed by war, the inhabitants have come by sea and settled in Great Britain. (1809: 504)

Even the Dutch could feel superior. Emanuel van Meteren in his *History of the Netherlands* (1559) describes the English language as "broken German, mixed with French and British terms, and words, and pronunciation, from which they have also gained a lighter pronunciation, not speaking out of their heart as the Germans, but only prattling with the tongue." When a word is wanting, English speakers resort to Latin and sometimes to German and Flemish. The real language of the place, he points out, is not English at all: "In Cornwall—England's furthest boundary westward—and in Wales, they speak the old British language, which they call in their own language *Cym-*

raeg, and which the English call Welsh, as the Germans do" (1865: 71). In their commentaries on Wales and Ireland, English writers resort to three strategies for countering such claims: (1) hearing the other languages as only so much noise, (2) asserting the power of English as a written, printed language, and (3) refining those languages by appropriating them, by redacting them into English texts like *Henry V* and *The Valiant Welshman* in what Michael Neill rightly calls an "act of translation" (1994: 19).

It is highly ironic that the sounding board to speech in several strategic locations in England was made out of Irish wood. Foreign travelers were inevitably told that the hammer-beam roof of Westminster Hall was made of Irish wood. As Paul Hentzner explains: "In the chamber where the Parliament is usually held, the seats and wainscot are made of wood, the growth of Ireland; said to have that occult quality, that all poisonous animals are driven away by it; and it is affirmed for certain, that in Ireland there are neither serpents, toads, nor any other venomous creature to be found" (1901: 31). Other locations supposed to be covered with the same reverberant material included St. George's Chapel at Windsor, the summer house at Woodstock, and the great hall at Hampton Court (Gerschow 1892: 45, 49, 53). English aspirations to subsume Irish laws, Irish religion, and Irish customs into an imperial unity are given a distinctively aural description by Davies, who praises the reforms wrought by James I:

> The strings of this Irish Harpe, which the Ciuill Magistrate doth finger, are all in tune (for I omit to speak of the State Ecclesiasticall) and make a good Harmony in this Commonweale: So as we may well conceiue a hope, that *Ireland* (which heertofore might properly be called the *Land of Ire*, because the *Irascible* power was predominant there, for the space of 400. yeares together) will from henceforth prooue a Land of *Peace* and *Concorde*. (1612: 284)

So sudden and sweeping a transformation of a people's acoustemology may happen within the charmed confines of Jonson's *Irish Masque at Court*, danced in 1613, but in the world at large such things take longer. The prophecy of harmony Davies optimistically voiced in 1612 continues to be shattered, nearly four hundred years later, by the explosion of bombs in Belfast.

OUT THERE

In the land that came to be known as Virginia the first encounter between English speakers and the native inhabitants was an event more keenly heard than seen: "saylinge further, wee came vnto a Good bigg yland, the Inhabitants therof as soone as they saw vs began to make a great and horrible crye, as people which neuer befoer had seene men apparelled like vs, and camme a way makinge out crys like wild beasts or men out of their wyts" (Hakluyt 1893: 54–55). Hakluyt's vignette sets the stage for later linguistic encounters between English writers and the inhabitants of North America: a writing "we," positioned at a comfortable perspective distance, confronts a roaring "them," caught up in the blind intensity of their own cries. How the Native Americans on their side of the water may have seen—and heard—this encounter is something else again. In several important respects the acoustemology of Native Americans differed from that of the English speakers who suddenly appeared one day on the shore of the river, at the edge of the forest. To begin with, most Native American speech communities identified themselves with a certain geographical place and hence with one particular soundscape. The creation story of most Native American cultures locates the people where they are not because they walked across a land bridge from Asia tens of thousands of years ago but because they emerged into the sunlight from a hole in the ground or up through the waters of a lake or stream. Roger Williams, after surveying learned opinion about the Indians' origins, observes, "They say themselves, that they have *sprung* and *growne* vp in that very place, like the very *trees* of the *wildernesse*" (1643: A4). However peripatetic a people may have been, however far they may have wandered from their mythic point of origin, Native American peoples moved as a single social unit, and they moved through the same kind of landscape: forest (the Powhatans and other peoples of the East Coast and the South), river valley (the pueblo peoples of the Southwest), open plains (the Apaches, the Kiowas, the Arapahos, and other peoples of the Great Plains), mountains (the Cherokees of the Appalachians and the Utes of the Rockies). Moving within one kind of terrain, each migrating people continued to inhabit its own distinctive soundscape. Sounds of the same human activities, in the same rhythm of days, seasons, and years, were interwoven with the same sounds in the ambient world, again in the same rhythm of days, seasons, and years. For forest-dwellers in particular hearing might well be the most sharply attuned sense: in a forest a great deal more can be heard than seen (Feld 1996: 98). By contrast, the soundscapes inhabited by English speakers were, as we witnessed in chapter 3, far more varied and dispersed.

In addition to being focused on a single place, the acoustemology of Native American peoples gave mythic importance to sounds coming from beyond the sphere of human speech and human activity. Most Native American people believed that there once had been a time when animals spoke and when humans and animals could understand one another. While staying with the English at Jamestown during Christmas 1610, for example, Iopassus, Powhatan's brother, offered to tell the English the Indians' creation story. He did so, William Strachey reports, through the interpretations of a boy named Spilman, who had lived among the Indians and had learned their language. "We haue (said he) 5. gods in all our chief god appeares often vnto vs in the likewise of a mightie great Hare, the other 4. haue no visible shape, but are (indeed) the 4. wyndes, which keep the 4. Corners of the earth (and then with his hand he seemed to quarter out the syctuation of the world) . . . " (1953: 102). Four of the people's five gods exist only as sounds, as the four winds; even the great hare, in Iopassus's telling, exists primarily as a voice. The hare creates two men and two women but stores them in a great bag. Spirits in the shape of giants come and ask to eat the men and women, but the hare replies by sending the cannibals away. The hare-god feasts the other gods on the first deer he has created, then scatters the deer's hairs over the land to produce other deer, before freeing the men and women and setting them upon the land. As Iopassus's story testifies, even the elements—lightning, rain, the wind—could speak through their appropriate spirit-gods. The result of such beliefs was to focus aural attention on the ambient sound-world in a way quite foreign to the English speakers who abruptly thrust themselves into the Native American soundscape. If the acoustemology of English speakers was declared in speech and guaranteed by music, the acoustemology of Native Americans traced the entire compass of sound.

What the Native Americans, some of them at least, thought about the invaders' way of speaking is reported by the merchant William Wood in *New Englands Prospect. A true, liuely, and experimentall description of that part of* America *commonly called* NEW ENGLAND . . . (1634). For a start, the Indians thought the English talked too much. "Garrulitie is much condemned of them," Wood observes, "for they utter not many words, speake seldome, and then with such gravitie as is pleasing to the eare." Among the Algonkian phrases collected by Roger Williams in *A Key into the Language of America* are "Neesquttónck-quffu" ("A babler, or prater") and "Cunnecsquttonckqussímmin" ("You prate"), the latter derived from the English word "cock" (1643: 45). Clucking chickens, Williams notes, were among the animals introduced to North America by the English. According to Strachey's account, the boy who interpreted Iopassus's story at Jamestown in 1610 had evidently learned Indian protocols of speaking, along with their language: "Nowe yf the boy had asked him of what he made those men and women and what those spirritts more particularly had bene and so had proceeded in some order, they should haue made yt hang togither the better, but the boy was vnwilling to question him so many things lest he should offend him" (1953: 102).

As obnoxious to the Indians as too much talk were the loud, angry outbursts to which the English were prone. Wood observes of the Massachusetts Indians:

> such is the milde temper of their spirits that they cannot endure objurgations, or scoldings. An *Indian Sagomore* once hearing an *English* woman scold with her husband, her quicke utterance exceeding his apprehension, her active lungs thundering in his eares, expelled him the house; from whence he went to the next neighbour, where he related the unseemelinesse of her behaviour; her language being strange to him, hee expressed it as strangely, telling them how she cryed Nannana Nannana Nannana Nan, saying he was a great foole to give her audience, and no correction for usurping his charter, and abusing him by her tongue.

It was not just from women, however, that the "Aberginians" objected to such outcries. "I have beene amongst diverse of them," Wood continues, "yet did I never see any falling out amongst them, not so much as crosse words, or reviling speeches, which might provoke to blowes." Even in games of chance they are calm when luck goes against them, never crying out or cursing. They are just as calm when they play football: "I never heard yet of that *Indian* that was his neighbours homicide or vexation by his malepart, saucy, or uncivill tongue: laughter in them is not common, seldome exceeding a smile, never breaking out into such a lowd laughter, as doe many of our *English*" (1634: L1–L1ᵛ). Strachey confirms that the Powhatan Algonquians behaved the same way. After the Indians had killed several Englishmen and taken two others prisoner, Strachey reports, they made up a "scornefull song . . . in manner of Tryumph" (Strachey transliterates the entire Algonkian text), complete with a repeated chorus that makes fun of the English community's lamentations, "namely saying how they would cry whe whe, etc., which they mock't vs for and cryed agayne to vs Yah, ha ha, Tewittaw, Tewittawa, Tewittawa: for yt is true they never bemoane themselues, nor cry out, giving vp so much as a groane for any death how cruell soever and full of Torment" (1953: 85–86).

All in all, Native Americans were great appreciators of silence. Williams is not the only European commentator to be impressed by the behavior of gathered listeners when someone spoke:

> Their manner is upon any tidings to sit round double or treble or more, as their numbers be; I have seene neer a thousand in a round, where *English* could not well neere halfe so many have sitten: Every

man hath his pipe of their *Tobacco,* and a deepe silence they make, an attention give to him that speaketh; and many of them will deliver themselves, either in a relation of news, or in a consultation with very emphaticall speech and great action, commonly an houre, and sometimes two houres together. (1643: 55)

John Ogilby confirms the same behavior among the Tuscarora and Catawba peoples of Carolina: "After their Salutation they sit down; and it is usual with them to sit still almost a quarter of an hour before they speak, which is not an effect of stupidity or sullenness, but the accustom'd Gravity of their Countrey" (1671: 209). To the protocols of few words and careful listening the English were not well attuned, as Ogilby unwittingly implies in one of the anecdotes he collects in *America: Being the Latest, and Most Accurate Description of the New World* (1671). Summarizing Woods and other sixteenth-century sources, Ogilby affirms that "in Affairs of concern" the natives of North America "are very considerate, and use few words in declaring their intentions." As an instance he cites a dialogue between Leonard Calvert, newly arrived in Maryland, and the leader of the Pascatoway Indians whose lands the English were hoping to occupy: "the *Werowance* of *Pascatoway* being ask'd by him, *Whether he would be content, that the English should sit down in his Countrey?* return'd this answer, *That he would not bid him go, neither would he bid him stay, but that he might use his own discretion.*" Lord Calvert of course takes this reply to mean yes, when it might very well have meant "Custom forbids me to withhold hospitality" or "When *will* you be leaving?" or "What choice do you leave me?" Thanks to Lord Calvert's "discreet Demeanor" towards the Indians at first, and "friendly usage of them" afterwards, the Pascatoway Indians, Ogilby concludes, "are now become, not only civil, but serviceable to the *English* there upon all occasions" (1671: 191).

Displeased or baffled by the invaders' ways with the human voice, the natives of North America are nonetheless, Wood notes, "not a little proud that they can speake the *English* tongue, using it as much as their owne, when they meete with such as can understand it, puzling stranger *Indians,* which sometimes visite them from more remote places, with an unheard language" (1634: N2ᵛ). John Brereton, in his firsthand account of life in the Roanoke colony (1602), likewise testifies to the way the Indians "pronounce our language with great facilitie." On one occasion, Brereton records, he playfully taunted an Indian who was sitting beside him: "How now (sirrha) are you so saucie with my *Tabacco?* which words (without any further repetition) he suddenly spake so plaine and distinctly, as if he had beene a long scholar in the language" (1983: 158).

The English, for their part, entertained a similarly ambivalent attitude toward the speech of Native Americans. On the one hand, the English were highly impressed by the Indians' "natural" gifts of oratory. John Josselyn, traveling twice to New England in 1634 and 1663, terms the native speakers "Poets . . . as may be ghessed by their formal speeches": "Their speeches in their Assemblies are very gravely delivered, commonly in perfect *Hexamiter* Verse, with great silence and attention, and answered again *ex tempore* after the same manner" (1988: 97). On more than one occasion English writers take pains to transcribe—or to reconstruct from memory—speeches of the sort that Josselyn praises. John Smith in the third book of *The Generall Historie of Virginia* (1624) goes so far as to provide full English texts of the speeches he exchanged with Powhatan, the Algonkian-speaking chief who at first welcomed the English to his peoples' lands at the mouth of the James River but then maneuvered to get rid of them when the English made it clear that they wished not merely to trade but to take over the Indians' lands. The fact that Powhatan's speeches do not appear in earlier versions of the narrative like Smith's *A True Relation of such occurrences and accidents of noate as hath hapned in Virginia* (1608) casts some doubt on how close the printed speeches may be to what Powhatan actually said. Indeed, the speeches sound more than a little like the set speeches Herodotus is always putting into the mouths of barbarian Others (Hartog 1988: 269–273). The point, however, is that Smith grants Powhatan a voice and reports the gist of Powhatan's words. It is unlikely that Smith would have invented Powhatan's perfectly logical argument that, by using guns and swords to secure food, the English are cutting off their noses to spite their faces:

> What will it availe you to take that by force you may quickly have
> by love, or to destroy them that provide you food. What can you get
> by warre, when we can hide our provisions and fly to the woods?
> whereby you must famish by wronging us your friends. And why are
> you thus jealous of our loves seeing us unarmed, and both doe, and
> are willing still to feede you, with that you cannot get but by our la-
> bours?

The concreteness of Powhatan's images, their rootedness in "nature," became a standard feature of reported Native American speech in later periods:

> Thinke you I am so simple, not to know it is better to eate good meate,
> lye well, and sleepe quietly with my women and children, laugh, and
> be merry with you, have copper, hatchets, or what I want being your
> friend: then be forced to flie from all, to lie cold in the woods, feede
> upon Acornes, rootes, and such trash, and be so tyred men must

watch, and if a twig but breake, every one cryeth there commeth Cap-
taine Smith. . . .

In addition to its evocative pastoralism, David Murray points out,
there was a second feature of reported Indian speech that assured a
hearing among Europeans: the fact that the speaker was always already doomed to defeat (1991: 36). Powhatan continues in Smith's

> then must I fly I know not whether: and thus with miserable feare, end
> my miserable life, leaving my pleasures to such youths as you, which
> through your rash unadvisednesse may quickly as miserably end, for
> want of that, you never know where to finde. Let this therefore assure
> you of our loves, and every yeare our friendly trade shall furnish you
> with Corne; and now also, if you would come in friendly manner to
> see us, and not thus with your guns and swords as to invade your foes.

The reported speech in which Smith responds to this "subtill discourse" is a none-too-subtle reminder that the English have firearms
and can use them at any time to get what they want (1986, 2: 196).

A similar contrast between oratorical power and political frailty
marks a speech by Chuatawback, sachem of Passonagessit, that
Thomas Morton transcribes in *The New English Canaan . . . Containing
an Abstract of New England* (1637). Wittingly or unwittingly, the Plymouth planters had defaced the tomb of the sachem's mother. Chuatawback was enraged. How Morton or any other Englishman might
have heard the speech the sachem delivered to his men is unclear, but
Morton delights to report it *verbatim*—Copernican astronomy and all:

> When last the glorious light of all the sky was underneath this globe,
> and Birds grew silent, I began to settle as my (custome is) to take re-
> pose; before mine eies were fast closed, mee thought I saw a vision, (at
> which my) spirit was much troubled, & trembling at that dolefull sight,
> a spirit cried aloude. . . .

The voice that then speaks sounds very much like that of someone
who has spent time in the theaters of the South Bank:

> (behold my sonne) whom I have cherisht, see the papps that gave thee
> suck, the hands that lappd thee warme and fed thee oft, canst thou for-
> get to take revenge of those uild people, that hath my monument de-
> faced in despitefull manner, disdaining our ancient antiquities, and
> honourable Customes. . . .

Chuatawback's mother betrays a thoroughly English sense of social
propriety:

> See now the Sachems grave lies like unto the common people, of igno-
> ble race defaced: thy mother doth complaine, implores thy aide against

this theevish people, new come hether[.] if this this be suffered, I shall not rest in quiet within my everlasting habitation. (1637: 106–107)

As with Powhatan, it is all well and good for Chuatawback (and his mother) to go on at such length, because the "vild" and "theevish" English quickly prove their military superiority when the sachem and his men proceed to wage war.

English mastery comes not only through guns but through quill pens. Woods and other commentators may invest the Indians with a thrilling orality, but writing is patently presented as the superior means of communication. Thomas Hariot in *A Briefe and True Report of the New Found Land of Virginia* (1588) makes a list of all the devices the English possess that the Indians do not: "Mathematicall instruments, sea compasses, the vertue of the loadstone in drawing yron, a perspectiue glasse whereby was shewed manie strange sightes, burning glasses, wildefire woorkes, gunnes, bookes, writing and reading, spring clocks that seeme to goe of themselues" (1588: E4). In this list "writing and reading" figure as material objects, as tools that can be taken in hand and used. According to Morton, one "Master Bubble" was sent over from England as a "Master of Ceremonies, betweene the Natives, and the Planters: for hee applied himselfe cheifly to pen the language downe in Stenography" (1588: 122). "Master Bubble" certainly had his work cut out for him. Williams in his *Key into the Language of America* explains just how hard it was to transliterate Indian speech, how he had to resort to the diacritical marks of Greek to indicate the complicated system of vowel values. As matters fell out, there was little call for "Master Bubble's" services, so he pursued a number of merchandizing schemes.

It was a signal irony that the native peoples of Massachusetts Bay found themselves invaded by one of the most literacy-conscious cultures ever to exist. Believing salvation to be founded on a personal understanding of the word of God as recorded in the Bible, the Puritans who emigrated to Massachusetts Bay boasted a literacy rate notably higher than that of early modern England at large (Lockridge 1974: 43–47). The effect of bringing such highly literate people face to face with people who, to all appearances, lacked any writing system whatsoever was to enhance "orality" as a state of being. Back in England, where only 20 to 25 percent of the people could sign their own names, the borderline between literacy and illiteracy was not fixed. There were various kinds and degrees of literacy, ranging from the functional literacy of a tradesman or an artisan within his own sphere of activity up to the highly sophisticated literacy of someone who could read not only English but Latin and various European languages (Cressy 1980: 118–141; Thomas 1986: 98). In New England,

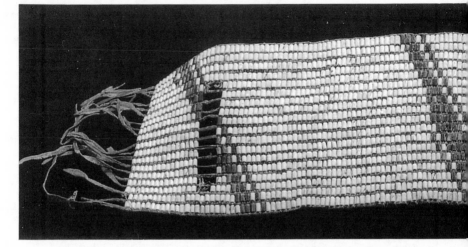

Figure 10.3. Wampum belt remembering a treaty between Delaware Indians and Quaker colonists of Pennsylvania (1682). Reproduced by permission of The Historical Society of Pennsylvania, Philadelphia.

however, the distinction was absolute: a highly literate "us" confronted a totally illiterate—or so it appeared to the English—"them." "Orality" as a concept depends, after all, on a dominant literacy. A completely oral culture isn't "oral"; it simply *is*. In fact, Native Americans did have a form of writing in the form of *wampumpeag*, shells or beads strung together in certain distinctive patterns to aid memory of important speech events. Spoken words in their languages were remembered, not in marks on a piece of paper, but in muscles of the chest, throat, and mouth (fig. 10.3). Against Jack Goody, who insists on the difference between mnemonic cues, which *prompt* linguistic statements, and "writing proper," which *reproduces* them, Walter D. Mignolo argues forcefully for recognizing wampum as a form of writing (Goody 1977: 25; Mignolo 1995: 78). Keepers of collective memory were instructed how to "read" wampum. In a speech to the governor of New France in 1684, for example, Otreouti, an Onondaga orator, repeatedly cites a piece of wampum as his "text" as he calls to mind the broken promises of the French: "This Belt Preserves my Words" (Calloway 1994: 118–120).

Missionaries like Williams and Eliot set out to provide Native Americans with more durable means to salvation in the form of bibles printed in Algonkian. In one of the dialogues he invented to help Indian converts proselytize among their brethren, Eliot registers an acute sense of the differences between written authority and oral ignorance. A literate convert is being instructed how to finesse the

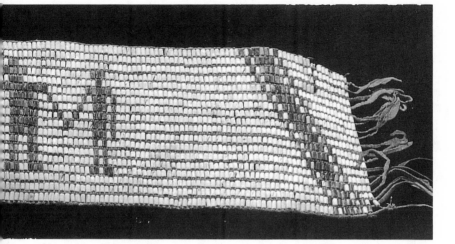

questions likely to be raised by a skeptical, illiterate kinsman. "What do you mean by scripture?" one of the kinsmen is scripted to ask. The convert's reply shifts the locus of meaning from the human body to the printed page: "a word spoken is soon gone, and nothing retaineth it but our memory, and that impression which it made upon our mind and heart. But when this word is written in a book, there it will abide, though we have forgotten it. And we may read it over a thousand times, and help our weak memories, so that it shall never be forgotten" (1980: 139–140). In teaching the Indians to read, Eliot and other missionaries were in fact engineering a shift in the site of philosophical, moral, and political authority. The Indians were expected to transfer reverence from the human voice to a physical object—and in that act to forget thousands of years of tradition. The Indians responded by revering the object just as they had formerly revered the voice. Hariot tells how, despite his protestations, the Virginia Indians would take the Bible and try "to touch it, to embrace it, to kisse it, to hold it to their brests and heades, and stroke ouer all their bodie with it; to shew their hungrie desire of that knowledge which was spoken of" (1588: E4–E4ᵛ). In effect, the English missionaries were casting doubt on oral tradition as a way of knowing the world—and in that act they were helping to consolidate the political power of the English. The Indians are written *about*, the Indians are written *to*, but the Indians themselves do not write *back*. The lavishly reported speeches of Powhatan and Chuatawback are, then, hedged about with ironies. The power of such reported speeches lies in the way politics is allied with philosophy, psychology, and phenomenology. Military might, reason, selfhood, and the technological proficiency of writing are aligned in ways that, to the colonizer, are supremely satisfying.

It should come as no suprise, therefore, that English writers could entertain, within the pages of the same book, admiration for Native American rhetorical power on the one hand and contempt for Native American languages on the other. Prompted by the biblical account of the Tower of Babel, Europeans were predisposed to hear in Native American languages evidence of confusion, dispersal, and degeneration. Robert Johnson, arguing the case for further colonization in his pamphlet *The New Life of Virginea* (1612), cites the barbarity of the Algonkian language as evidence of its speakers' depravity—and of their aptness for being pulled up and discarded like so many weeds. For their presumption in erecting the Tower of Babel, God dispersed the builders "so that from this scattering and casting them out like unprofitable seed upon the dust of the earth, did spring up (as weeds in solitarie places) such a barbarous and unfruitfull race of mankinde, that even to this day (as is very probable) many huge and spactious Countries and corners of the world unknowne, doe still swarme and abound with innumerable languages of this dispersed crue, with their inhumane behaviour and brutish conditions." A signal case in point is provided by "the sundrie nations of *America*," made up as they are "of infinite confused tongues and people." Until the arrival of the English on those shores, "God did never vouchsafe the hand of the weeder, to clense and give redresse to so desolate and outgrowne a wildernesse of humaine nature" (1612: B1–B2). Into that wilderness the English stepped forth as linguistic gardeners, as improvers of nature.

Reasons for the English attitude to native speech need not be fetched from so far afield. James Axtell catches the feel of Algonkian in the English throat, the clamor of Algonkian in English ears, when he observes, "Words, including names, were often long and seemingly undifferientiated, full of throaty glottals and short of defining labials" (1981: 54). Amid the aural confusion, European listeners were certain they could hear evidence of dispersal. Algonkian and other Native American languages seemed to be living proof of the claim in Genesis that, after destroying the Tower of Babel, God "did there confounde the language of all the earth: from thence then did the Lord scater them vpon all the earth" (Geneva Bible, Genesis 11:9). Having studied the dialects spoken in the lower Connecticut Valley, Williams notes affinities with Hebrew, Greek, and Turkish (1643: A4^v–A5). Opposite the marginal heading "The Natives have a mixed Language" Morton likewise lays claim to firsthand knowledge of North American languages—and proof of the disperal theory: "by continuance & conversation amongst them, I attaned to so much of their language, as by all probable conjecture may make the same manifest, for it hath been found by divers, and those of good judgement that the Natives of this Country, doe use very many wordes both of Greeke and Latine, to the same signification that the Latins and Greekes have done, as *en animia,* when an Indian expresseth, that hee doth any thing with a good will" (1637: 18). Morton takes this to be evidence that Native Americans derive from refugees from Troy. They are descendants of Brutus, who left Latium after Aeneas had settled it: "this people were dispersed there is no question, but the people that lived with him, by reason of their conversation with the Graecians and Latines, had a mixed language that participated of both, whatsoever was that which was proper to their own nation at first" (1637: 20). Others heard evidence of Welsh. Sir George Peckham's *True Reporte* of Sir Humphrey Gilbert's voyage to Newfoundland (1583) notes the Indian names of several plants and birds, as well as "divers other welch wordes at this daie in use," and takes them to be vestiges of a visit by Madoc ap Owen Gwyneth, Prince of Wales, in 1170 (Peckham 1940: 459). Even Wood can quote, without exactly approving, the opinion of some that the Massachusetts Indians' language "might be neare unto the *Hebrew;* but by the same rule they may conclude them to be some of the gleanings of all Nations, because they have words which sound after the *Greeke, Latine, French,* and other tongues." Whatever its origins, the Indians' language "is onely peculiar to themselves, not inclining to any of the *refined* tongues" (1634: N2, emphasis added).

Opinions like those reported by Wood and endorsed by Williams kept English speakers assured of their own linguistic superiority well into the twentieth century (Murray 1991: 14–33). Along with confusion and dispersal, Native American languages seem to demonstrate degeneration from the *Ur*-language spoken before Babel. Commenting on the tongues of Indians living in New York, Ogilby makes the truly wild claim that such languages lack syntax altogether: "The Language of this Countrey is very various, yet it is divided into onely four principal Tongues, as the *Manhattans, Wappanoo, Siavanoo,* and *Minqua's,* which are very difficult for Strangers to learn, because they are spoken without any Grounds or Rules" (1671: 178). The colonial French and English encountered only four of the major language groups of North America, Axtell points out, but they found them all wanting: rich in proper nouns for concrete objects, in metaphorical expressions, and in variety of ways to signify the same thing in different states or relationships, but poor in abstract words, universals, and words like "salt," "sin," "gold," "prison," "candle," "king," "shepherd," and "flock" (1981: 76–77).

Sounding like a cacophony of diverse tongues, lacking a developed syntax, innocent of letters: Indian languages were suspect in many ways. At bottom, however, the problem with the acoustemology of Native American culture was simple: it constantly threatened to devolve into [oː]. Perhaps the reason Strachey takes such pains to transliterate the Powhatan Indians' song of triumph is the satisfaction of exerting literate control over an oral gesture that turned the English mourners and their dead comrades into nonsense:

> Mattanerew shashashewaw crawango pechecoma
> Whe Tassantassa inoshashaw yehockan pocosack
> Whe, whe, yah, ha, ha, ne, he, wittowa wittowa.

And so on for three more stanzas (1953: 85–86). Other accounts of Indian music are just as suspicious. In *A Map of Virginia* (1612) John Smith describes flutes made from hollow pieces of cane—instruments similar to recorders and hence familiar to English readers. The Indians' "chiefe instruments," however, are not tunable flutes but percussive drums and rattles.

> These mingled with their voyces sometimes twenty or thirtie together, make such a terrible noise as would rather affright, then delight any man.

Even the Indians' speeches approach the frontier of noise. Smith presumably speaks from personal experience when he describes the welcome a European receives when he visits a Native American leader:

> If any great commander arrive at the habitation of a Werowance, they spread a Mat as the Turkes doe a Carpet for him to sit upon. Upon another right opposite they sit themselves. Then doe all with a tunable voice bid him welcome. After this doe two or more of their chiefest men make an Oration, testifying their love. Which they doe with such vehemency, and so great passions, that they sweat till they drop, and are so out of breath they can scarce speake. So that a man would take them to be exceeding angry, or stark mad. (1986, 2: 121)

Wood, who in general offers a sympathetic portrait of the Massachusetts Indians in *New Englands Prospect*, is horrified at the sounds of Native American healing ceremonies. Human speech is left behind as the voices of the participants range out, first into music, then into the cries of animals:

> The parties that are sick or lame being brought before them, the Powwow sitting downe, the rest of the *Indians* giving attentive audience to

his imprecations and invocations, and after the violent expression of many a hideous bellowing and groaning, he makes a stop, and then all the auditors with one voice utter a short *Canto;* which done the Pow-wow still proceeds in his invocations, sometimes roaring like a Beare, other times groaning like a dying horse, foaming at the mouth like a chafed bore, smiting on his naked brest and thighs with such violence, as if he were madde. (1634: M2)

The Algonkian-speakers near Roanoke utter similar cries in their ceremonies involving tobacco. When danger presents itself, Hariot reports, the Indians cast tobacco into the air, into the water, or into the fire, "all done with strange gestures, stamping, sometime dauncing, clapping of hands, holding vp of hands, & staring vp into the heauens, vttering therewithal and chattering strange words & noises" (C3ᵛ). Williams seconds such impressions in observing of the Narragansetts that "Howling and shouting is their Alarme; they having no Drums nor Trumpets: but whether an enemie approach, or fire breake out, this Alarme passeth from house to house; yea, commonly, if any *English* or *Dutch* come amongst them, they give notice of strangers by this signe: yet I have knowne them buy and use a *Dutch* Trumpet, and knowne a *Native* make a good Drum in imitation of the *English*" (1643: 18–19). Josselyn explicitly connects the Indians' power as orators with their propensity for music: "Their learning is very little or none, Poets they are as may be ghessed by their formal speeches, sometimes an hour long, the last word of a line riming with the last word of the following line, and the whole doth *Constare ex pedibus* [stand together in metrical feet]. Musical too they be, having many pretty odd barbarous tunes which they make use of vocally at marriages and feastings" (1988: 97).

Such dilations around the full compass of sound sent the literate imagination running toward horizons more immediate than those of North America. In describing the music and the dancing of the Indians of Carolina, Ogilby finds a comparison ready at hand: "they are in their Tempers a merry, frollick, gay People, and so given to Jollity, that they will Dance whole Nights together, the Women sitting by and Singing, whilest the Men Dance to their Ayrs, which though not like ours, are not harsh or unpleasing, but are something like the Tunes of the *Irish*" (1671: 209). Vivian Salmon (1985) has charted the multiple ways in which Native American languages were read in terms of Irish and Irish in terms of Native American languages. Comparison with the Irish on a number of other fronts seems to have been irresistible. Morton likens the timber-framed wigwams of the New England Indians to the huts of "the wild Irish" (1637: 24). Wood, for his part, compares the mantles of the Massachusetts Bay Indians to those of the Irish and pronounces their stewed food to be like "Irish Bonniclapper" (1634: K1, K2ᵛ). His most extended comparison, however, concerns Indian ways of mourning the dead: "The glut of their griefe being past, they commit the corpses of their diceased friends to the ground, over whose grave is for a long time spent many a briny teare, deepe groane, and Irish-like howlings" (1634: N3).

Even closer on the domestic front to the music and the speeches of Native Americans were gests and ballads. Richard Brathwait makes the connection explicit in his character of an itinerant ballad-monger and his singer. Once they have arrived in a village, "you shall see them (if both their stockes aspire to that strength) droppe into some blinde Alehouse, where these two naked Virginians will call for a great potte, a toast, and a pipe." The poetry and music this duo produces, fortified with ale, is all blather and bluster: "But now they are parted: and Stentor ha's fitted his Batillus with a Subject: whereon hee vowes to bestow better Lines than ever stucke in the Garland of good will. By this time with botches and old ends, this Ballad-Bard ha's expressed the Quintessence of his Genius, extracted from the muddie spirit of Bottle-Ale and froth" (1631: 10–11). Within the acoustemology of early modern England, ballads and barbarian speeches occupy similar positions. Both things came together when the colonists at Passonagessit rechristened their domicile "Merrymount" and erected an eighty-foot maypole. Their Puritan neighbors were, of course, duly outraged. To the usual abominations deplored by the Puritans at Wells in 1607 was added a new, distinctively North American abomination. Morton's gleeful account of the affair in The New English Canaan includes the detail that the Indians joined hands, and possibly voices, with the English: "And upon Mayday they brought the Maypole to the place appointed, with drumes, gunnes, pistols, and other fitting instruments, for that purpose; and there erected it with the help of Salvages, that came thether of purpose to see the manner of our Revels" (1637: 132).

To the English speakers' linguistic self-assurance the acoustemology of African slave communities posed even more of a challenge. To begin with, there was the reputation of the African continent itself for incredible linguistic diversity. The travel literature on Africa was unanimous in making that clear. So many languages and dialects were spoken there, European voyagers testified, that people a few miles apart could not understand one another. More than two hundred contact languages were reported by European visitors to Africa's west coast in the sixteenth, seventeenth, and eighteenth centuries (Hancock 1986: 88). One of the authorities collected in John Ogilby's Africa: Being an Accurate Description of the Regions of AEgypt, Barbary, Lybia, and Billedulgerid, The Land of Negroes, Guinee, AEthiopia, and the

Abyssines . . . sums up the European view of African languages in general when he describes the specific situation in Insoko: "The People of *Akara, Ningo,* and *Sinko,* though near Neighbors, yet understand not one anothers Tongue; but all that dwell upon the Shore, or Trade with the *Whites,* speak most of them a mixture of broken *Portugese, Dutch,* and *French:* yet in all this variety of Tongues, they all are ignorant of Writing, not knowing indeed what a Letter means" (1670: 458).

Such modern impressions were backed up by classical precedent. Pliny and Diodorus Siculus could be cited as authority that certain peoples in Africa have, in fact, no language at all. An account of two sixteenth-century English voyages to Guinea, appended to Richard Eden's translation of Peter Martyr's *The Decades of the newe worlde or west India, Conteyning the nauigations and conquestes of the Spanyardes* . . ., assures English readers that the inhabitants of "Trogloditica" in Ethiopia not only "dwell in caves and dennes" but eat snakes and communicate with one another in animal-like fashion: "For these are theyr houses, and the desshe of serpentes theyr meate, as wryteth Plinie and Diodorus Siculus. They haue no speache, but rather a grynnynge and chatterynge" (1555: 356ᵛ). For all that, evidence offered elsewhere in the *Decades* indicates that traders were usually able to communicate with Africans via some version of Portuguese, Dutch, or French—unless, like some of the peoples to the south, the inhabitants refused language altogether: "In some of these parts the people are so sullen and brutishly inclined, that they will neither speak, be sociable, nor appear to any; and in case one of them be taken, he will rather starve to death, than open his mouth and speak" (Ogilby 1670: 30).

The kind of animal Pliny, Diodorus, and the English chroniclers have in mind is made explicit in Thomas Herbert's firsthand *Discription of the Persian Monarchy Now beinge: The Orientall Indyes, Iles & other part's of the Greater Asia, and Africa* (1634). Concerning the inhabitants around the Cape of Good Hope—a region the seventeenth century mapped as "Nether Ethiopia"—Herbert declares: "Their words are sounded rather like that of Apes, then men, whereby its very hard to sound their Dialect, the antiquitie of it whither from *Babell* or no. The qualitie, whither beneficiall or no, I argue not." To "sound" the Africans' dialect is both to probe its historical depths and to attempt to reproduce its cacophony. As Woods and Williams attempt with Algonkian, Herbert proceeds to offer a brief pronouncing dictionary of useful words and phrases: the numbers one through ten, the words for objects pertinent to merchants (knife, quill, sword, book, ship), the terms for male and female body parts (breasts, penis, womb, testicles). The only verb he includes is *Quoy,* "Giue me" (1634: 16). Transliterating the sounds of these words and phrases is not easy, Herbert notes—and makes the inevitable comparison: "their pronunciation is like the Irish." Herbert takes the full measure of his interactions with the people of Nether Ethiopia when he concludes: "comparing their imitations, speech and visages, I doubt many of them haue no better Predecessors then Monkeys: which I haue seene there of great stature" (1634: 16–17).

Like the Irish, like the Powhatan Indians in Virginia, like the Indians of Massachusetts Bay, Africans are described as inhabiting a sound-world in which language keeps turning into noise. Human communication merges with sounds in the ambient world to create an acoustemology that frightened the English speakers who found themselves situated within it. Knives, quills, swords, even breasts, penises, and testicles: these visual components of another culture remain "out there" as objects in space. The *sounds* of another culture, however, penetrate "in here" and threaten to overwhelm the listener's sense of well-being. Cries of war and wails of mourning seem to have been two signal occasions when English speakers felt their hold on language to be slipping. The Irish and their ululations inspire the marginal note "After the *Irish* manner and wont, with a loud cry, or *ou la loo*," next to this account of mourning in Ogilby's *Africa:*

> When one dyes, the Wife or next Neighbour goes out of the Tent, howling in a strange manner with a loud cry, or *Ou-la-loo;* by which Summons the Women start out from their Tents, and joyning their sad notes, make a hideous and doleful harmony: others mean while repeating as it were in a Song, his Eulogies, chanting forth his Praises and Vertues, till at last they bring him to the Grave, according to the custom of the Mahumetans. (1670: 29)

It was presumably the threat of [oː] that prompted the Virginia authorities in 1680 to pass a law banning gatherings where slaves "played on their Negroe drums" (Thornton 1992: 228).

The "otherness" of African acoustemology was different from that of native North Americans not only in degree but in kind. The slaves imported to Somer's Island in 1616 and to Jamestown in 1619 did not, after all, come directly from Africa but from the West Indes. In all likelihood, they or their parents did not all derive from the same speech community in Africa but from any number of places along the west coast—regions Ogilby identifies as "Negro-Land," Senegal, and Guinea (Thornton 1992: 211–218; Ogilby 1670: 318, 347, 379). Importation of slaves into Virginia and Maryland via the Caribbean and New Amsterdam continued until the 1670s, when slaves began to be imported directly from Africa (Rickford 1997: 315–336). What language did these early African Americans speak among themselves?

The Indians in Virginia and Massachusetts Bay found their cultural identity in dialects that firmly established the "here-ness" of the places they inhabited; slaves, uprooted from their speech communities in Africa, communicated through makeshift creole dialects. That, according to the authorities collected in Ogilby's *Africa*, was just what they were used to doing at home. Along the Hens River in Guinea, Ogilby relates, the people "speak a particular Language that seems harsh and unpleasant; but when they go to *Quoya*, or *Cabo Monte* to traffick, they express their meanings significantly in another Tongue that runs smooth and easie, either to be learnt or understood" (1670: 379). This "other tongue," to judge from descriptions elsewhere in Ogilby's *Africa*, was likely a creole based on Portuguese. As early as the 1620s, however, there is evidence that a creole form of English was being spoken in the Gambia River area. Richard Jobson in *The golden trade: or, a discovery of the river Gambra, and the golden trade of the Aethiopians* (1623) makes note of "a pretty youth called *Samgulley*, who ... had alwayes lived among the English, and followed their affairs, so as hee was come to speake our tongue, very handsomely" (113, quoted in Hancock 1986: 75).

Evidence assembled by Ian Hancock suggests that it was an Anglo-African creole that supplied the basis for the creole spoken by slaves in the Caribbean and hence by slaves in the Chesapeake Bay region. In the first instance, this creole had been devised to permit Africans of diverse native tongues to talk to each other in African trading ports, but it evolved to serve the needs of cross-racial households, of Africans who worked as wage-earners for white men living in Africa, of polyglot slaves being held for shipment, of slaves already established in the New World who were called upon to acculturate new arrivals. In all these transactions English provided the structure; local African languages, its phonology and semantics (Hancock 1986: 71–102). While that may have been true for slaves shipped from the Gambia River region, the Christian Spanish names borne by most of the black-skinned people landed at Jamestown in 1619 suggest that it was Spanish that supplied their speaking vocabulary (Catterall 1926–1937, 1: 54–55). Though Spanish words would gradually have yielded to English words in the Africans' new environment, it must have sounded to the Jamestown settlers as if the Tower of Babel had toppled over in their very midst.

Something of the strangeness of the newcomers' speech survives in a court case in St. Mary's County, Maryland, in the 1650s, involving a runaway slave named Tony. When Tony refused to work, his owner Oversee first beat Tony with twigs, then poured hot lard over his body, and finally tied him to a ladder, in which posture after a few hours Tony died. Oversee was tried for murder. In the course of the trial a neighbor, Job Chandler, told the court about an earlier encounter with the runaway Tony. Chandler apprehended Tony stealing food and noticed that one of his hands was so badly wounded that gangrene was setting in: "I examined him how it came, but could not with all the words and signes I would imagine understand from him how it came, For of all humane Creatures that ever I saw, I never knew such a Brute: for I could not perceive any speech or language hee had, only an ugly yelling Brute beast like" (Catterall 1926–1937, 4: 12). Oversee was acquitted.

All in all, the earliest African Americans disrupted the acoustemology of English speakers in fundamental, frightening ways: they chattered like monkeys, they bellowed like beasts, they mourned in chants of *ou la loo,* they delighted in drumming, they spoke a language that was no language. The dubious linguistic status of Africans in seventeenth-century Virginia and New England stands as an index of their dubious legal status. In both places the word "slave" had not yet acquired its modern meaning of chattel for life. In the 1630s and 1640s the courts in New England punished Englishmen and Indians alike by ordering them to be handed over to certain individuals as "slaves"—presumably for fixed periods of unpaid labor (Catterall 1926–1937, 4: 469–471). In Virginia meanwhile the "twenty Negars" bought from the Dutch were clearly purchased as laborers, though whether for life or for a term of indenture is uncertain (Catterall 1926–1937, 1: 55–59). It was not until 1659 that a Virginia statute referred to African inhabitants of the colony as slaves in the modern sense of the word. Their legal status in the meanwhile seems to have been variable from individual to individual if not ambiguous as a group (Higginbotham 1978: 19–60).

Equally variable and ambiguous was their status as members of the local speech community. Court records provide a constant reminder that, for Indians and Africans alike, the *legitimacy* of non-English-speaking voices was always in question in seventeenth-century America. Among civil institutions, courts of law provide perhaps the clearest test case of who counts, and who does not, as a citizen. "The law of the land" presumably applies to all who live there. Among civil institutions, courts of law likewise provide a distinctively vocal form of social exchange. Legal cases are, after all, "heard," not "seen." The participation of Native Americans in these oral rites of community was always equivocal. On the one hand, colonial authorities in New England and Virginia alike were quick to insist that the law of the land applied to Indians as well as Englishmen. The difficulty of making that proclamation work in practice is indicated by a case from 1636 at Accomack on Virginia's Eastern Shore. In a series of depositions witnesses testified how "certine Indyans . . .

from the laughing kinge" arrived at the house of Daniel Cugley with "roanoke," or payment in wampum, which they wished to leave in recompense for the murder of an Englishman—or at least that is what the Indians *seemed* to be saying. No one present at Cugley's house, the deponents testified, "could well understand the Indyans [but as] far as they did interpret it did appeere that it was the death of some Englishman" (Ames 1954: 56–58). Voluntary recompense was a traditional way of atoning for murder in Native American communities (Calloway 1997: 118). To accept the wampum would be, for the English, to accept Indian justice, and the court at Accomack ordered Cugley to be held prisoner until the roanoke he was accused of having accepted be handed over to the English authorities.

Compelled to honor English laws, Native Americans were, nonetheless, usually denied a direct voice in court proceedings. In part at least, that was due to language differences. A case heard by the Quarterly Court at Ipswich, Massachusetts, in 1660 catches some of the inconveniences of living in the shadow of the Tower of Babel. When John Bishop, claiming to have purchased the labor of "Indian Mall," sued to secure the runaway squaw's return, the sixteen-year-old defendant entered the records of the court, not in a transcription of her own voice, but in papers filed on her behalf by a bonafide citizen, John Hawthorne. Hawthorne explicitly calls the court's attention to the problem of voice and justice: "the indian is extremely discontent to liue with John bishop . . . the testimony in court doeth not witnes from her mouth (which if they did it is nothing for she is to be at her parents disposing) but by an interpreter[.] now the witnes canot say that he did report her words unto them or ther words true to her. . . ." If the court has any doubt about the legitimacy of Mall's case, Hawthorne writes, let the judge recall that "the law is undeniable that the indian may haue the same distribusion of Justice with our selues: ther is as I humbly conseiue not the same argument as amongst the negroes for the light of the gospell is a begineing to appeare amongst them—that is the indians" (Dow 1912, 2: 240).

Hawthorne's careful specification of just who "they" are—Indians, not Negroes—calls attention to the even more dubious place of African voices in court proceedings. Among the few such voices in the Virginia records is that of the defiant "Anthony the negro," who enters the written papers of the Accomack court in the 1640s for the smart remark he gave to Edwin Conaway, clerk, when Conaway encountered Anthony and his master Robert Wyard Taylor on their way home from the fields. Conaway testifies that he

did see Capt. Taylor and Anthony the negro goeing into the Corne Feild And when they return'd from the said Corne Feild the said Ne-

gro told this deponent saying now Mr. Taylor and I have devided our
Corne And I am very glad of it now I know myne owne, hee finds
fault with mee that I doe not worke but now I know myne owne
ground I will worke when I please and play when I please, and the
said Capt. Taylor asked the said Negro saying are you content with
what you have And the negro answered saying I am very well content
with what I have or words to that effect. . . . (Ames 1973: 457)

"Or words to that effect": Indian voices and African voices ultimately defied the surveillance of writing, just as Irish voices did. The
extreme otherness of African voices is registered in the fact that creole
dialects, unlike Algonkian, did not attract the systematic attention of
people like William Wood, Roger Williams, and John Eliot. The few
African American voices that enter the written record in the first hundred years do so with markers of radical difference. Among the
women examined in the Salem witch trials in 1691–1692 was one
"Candy," described in the court papers as "A Negro Woman Servant
to Margarett Hawkes" and a native of Barbados (Boyer and Nissenbaum 1977, 3: 180). The language she speaks is clearly a creole:

Q. Candy! are you a witch? A. Candy no witch in her country. Candy's
mother no witch. Candy no witch, Barbados. This country, mistress
give Candy witch. Q. Did your mistress make you a witch in this country? A. Yes, in this country mistress give Candy witch. (1977, 3: 179)

The contrast with one of the other examinees, Tituba, described as
"an Indian woman" belonging to Samuel Parish, is sharp—at least
on the page. In two written versions of the same examination Tituba
is recorded as speaking something very close to standard English.
After getting Tituba to confess that she pinched certain women and
children by command of an apparition in the shape of a man, the
examiner presses for more:

Q. w't Other likenesses besides a man hath appeared to you? A. Sometimes like a hogge Sometimes like a great black dogge, foure tymes. Q.
but w't did they Say unto you? A. they tould me Serve him & that was
a good way; that was the black dogge I tould him I was afrayd, he
tould me he would be worse then to me. Q. w't did you say to him
after that? A. I answer I will Serve you noe Longer he tould me he
woulde doe me hurt then. (1977, 3: 750–751)

The two records of Tituba's examination are variable enough in narrative sequence and in diction to make us realize we are dealing here,
at least in part, with attributed speech and not with actual speech.
Tituba seems to have been *expected* to talk in standard English; Candy,
in pidgin. (In the event, Candy was acquitted and Tituba convicted.)
    African American speech presented challenges to orthography as

well as orthodoxy. Even so competent a rhetorician as Cotton Mather is clearly at a loss in his tract *Some Account of What Is Said of Inoculating or Transplanting the Small Pox* . . . (1721) as to how to tell his learned readers that certain African slaves in Boston know, from their own experience, a cure for smallpox. "The more plainly, brokenly, and blunderingly, and like Ideots, they tell their Story," Mather hazards,

> it will be with reasonable Men, but the much more credible. For *that these* all agree in *one Story;* "That abundance of poor Negro's die of the *Small Pox,* till they learn this *Way;* that People take the Juice of the *Small Pox,* and *Cut the Skin,* and put in a drop; then by'nd by a little *Sick,* then few *Small Pox;* and no body dye of it: no body have *Small Pox* any more." (quoted in Jordan 1968: 202)

Mather's attempt to reproduce the syntax of African American creole speech is, Winthrop D. Jordan claims, the first such attempt in print. A manuscript version of the same passage is even closer to the speech rhythms of Mather's African informants: "in their Country *grandy-many* dy of the *Small-Pox*" before they "*Cutty-skin,* and Putt in a Drop; then by'nd by a little *Sicky, Sicky*" (1968: 204). At about the same time Daniel Defoe in his novel *The Life of Colonel Jacque* (1722) makes a point of distinguishing the register of a Virginia slave's speeches ("Yes, yes . . . me know, but me want speak, me tell something. O! me no let him makee de great master angry") from that of white inden-tured servants (quoted in Rickford 1997: 315–336). By the 1720s, then, African American speech had been listened to, transcribed, and codi-fied as a communication system operating within its own horizons of audible difference.

A hundred years earlier no one would have been interested enough in the sounds of African speech to attempt a reproduction like Mather's or Defoe's. Or would they? Dissonances between the acoustemology of early modern England and the acoustemologies of Native American and African American communities should make us freshly attentive to differences in syntax, rhythm, and diction of fictional speakers like the Prince of Morocco, Othello, and Caliban. Since the 1550s Africans had been transported to England to learn English and then sent back to act as interpreters (Hancock 1986: 74). Whether William Shakespeare actually encountered African creole speech from the son of "Caddi-biah," an African trade-factor who was baptized at St. Mildred Poultry in 1611 (Knutson 1994: 110–126), or from other immigrants from Africa or the New World is less im-portant, perhaps, than the effect of acoustic difference in scripts like *The Merchant of Venice, Othello,* and *The Tempest.* If the acoustemolo-gies of Africa and North America were more open to nonverbal sounds than the acoustemology of early modern England, how might

Shakespeare's scripts register that difference? Morocco, Othello, and Caliban provide three test cases.

The voices of all three characters, let it be noted, are heard within authoritarian if not explicitly legalistic frames of reference. These outsiders do not just speak; they are specifically *licensed* to speak. Morocco's appearance as Portia's first suitor in *The Merchant of Venice* 2.1 sounds the tension between the African's ardent rhetoric and the cold restraint of Portia's father's will. The lines in which the oxymoronic "Moor of Venice" speaks himself into dramatic presence in *Othello* 1.3 are delivered as self-defenses in what amounts to a court of law. Concerning his adversary Desdemona's father, Iago has given Othello due warning in the previous scene:

> Be assur'd of this,
> That the Magnifico is much belov'd,
> And hath in his effect a voice potentiall
> As double as the Dukes.

Not to worry, Othello replies, continuing Iago's forensic image:

> My Seruices, which I haue done the Signorie
> Shall out-tongue his Complaints.
> (F1623: 1.2.11–14, 18–19)

Othello's confrontation with Brabantio in the next scene is a clash not just of wills but of voices, and that clash takes place in a judicial setting. Caliban, referred to by Prospero as "slave," first speaks on command: "What hoa: slaue: *Caliban*: Thou Earth, thou: speake." Replying from offstage, Caliban is, for the audience, a voice before he is a body: "There's wood enough within." Only after repeated commands—"Come forth I say; there's other busines for thee: / Come thou Toroys, when? . . . Thou poysonous slaue . . . come forth"—does the voice assume a body at the stage direction *"Enter Caliban"* seven lines after the audience has first heard from the indigenous inhabitant of the story's "vn-inhabited Island" (F1623: 1.2.315–318, 321–322, heading before dramatis personae).

In distinct ways each of these three characters—Morocco, Othello, and Caliban—declares his affinity with a sound-world different from that of his European interlocutors. Numerous modern critics have heard echoes of Morocco's verbal extravagance in Othello's speeches in 1.3, suggesting an aural mark of African identity every bit as strong as quick-wittedness was for Welsh characters (see, for example, Hecht 1986: 60–62). Othello positively invites such a hearing in his description of places out of travel books like Peter Martyr's *Decades*, places inhabited by "Canibals that each others eate, / The *Anthropophegue*, and men whose heads / Grew beneath their shoulders" (F1623: 1.3.142–144). Less explicitly, Morocco invokes a similar habitation for himself in his opening boast to Portia:

> Bring me the fayrest creature North-ward borne,
> Where *Phoebus* fire scarce thawes the ysicles,
> And let vs make incyzion for your loue,
> To prouve whose blood is reddest, his or mine.

Coming only moments after Shylock has set as forfeit "a pound of mans flesh taken from a man," Morocco's speech conjures up images of that ultimate mark of barbarity, cannibalism.

The immediate inspiration for Morocco's speeches, and perhaps Othello's in 1.3, is not geography books like Peter Martyr's *Decades*— what use would such books be, since none of them grants Africans the subjecthood of speech?—but the plays of Christopher Marlowe.

> By this Symitare
> That slewe the Sophy, and a Persian Prince
> That wone three fields of Sultan Solyman,
> I would ore-stare the sternest eyes that looke:
> Out-braue the hart most daring on the earth:
> Pluck the young sucking Cubs from the she-Beare,
> Yea, mock the Lyon when a rores for pray
> To win the Lady.
> (Q1600: 2.1.24–31)

In geographic sweep, in expansive periodic syntax, in images of violence, in assertive alliteration, Morocco sounds for all the world here like Tamburlaine. Othello's rhetoric may be more temperate, but he nonetheless joins Morocco in laying claim to an eloquence that, in context, sounds distinctly foreign. The Duke acknowledges as much when he confesses after Othello has concluded his defense: "I thinke this tale would win my Daughter too" (F1623: 1.1.170). In their aural foreignness, both speakers bring to the public stage gifts of speech granted on the printed page to Powhatan and the sachem of Passonagessit. Like these Native American orators, Morocco and Othello perhaps have license to speak so compellingly precisely because their English-speaking listeners know they are doomed before they ever open their mouths. If, as Anthony Hecht has argued, Othello achieves "a painful but undoubted nobility" in his death speech (1986: 83), the price of tragic sympathy is learning to speak as Europeans speak.

Morocco and Othello share speech traits with the *miles gloriosus* of Western dramatic tradition, but Caliban inhabits a sound-world that seems distinctively alien. Before the arrival of Prospero and Miranda, Caliban was ignorant, not just of Arabic, Italian, or English, but of any language whatsoever. "Abhorred Slaue," Miranda up-braids him:

> I pittied thee,
> Took pains to make thee speak, taught thee each houre
> One thing or other: when thou didst not (Sauage)
> Know thine owne meaning; but wouldst gabble, like
> A thing most brutish, I endow'd thy purposes
> With words that made them knowne. . . .
>
> (1.1.353, 355–360)

Before being taken in hand by Europeans, Caliban lived within an acoustic horizon totally filled with nonverbal sounds. To a degree he still lives in such a world in *The Tempest*—and the audience gets to hear it. "*A tempestuous noise of Thunder and Lightning heard*": the play begins by destroying language. All the physical attributes of sound that make speech possible—pitch, volume, timbre—are dissolved in a loud, inflectionless confusion of pitches across the full band of frequencies. The first human voices to emerge from the cacophony are shouts, eruptions of [o:]:

MASTER
    Bote-swaine.
BOATSWAIN
    Heere Master: What cheere? (1.1.1)

Out of the aural confusion Alonso, Sebastian, Antonio, Ferdinando, Gonzalo, and others attempt to establish a normal scene of speak-ing—until "*A confused noyse within*" explodes their efforts and returns the play to the sounded chaos with which it began. It is amid echoes of this noise, still ringing in the audience's ears, that Miranda and Prospero speak calmly and privately in the second scene. Caliban bursts into this conversation, a reminder of the barbaric noises just beyond earshot.

Caliban's relationship to language is unstable. Just what register of language does he speak? Stephen Orgel has called attention to the confusion among typesetters of the 1623 folio, as to whether Caliban's speeches should be set as verse or as prose (Orgel 1995). Certainly his speeches run the gamut from abrasive curses to rapturous poetry. "*A noyse of Thunder*" underscores the curses he pronounces at his en-trance in 1.2, carrying fardels of wood. In the course of the speech he conjures up sounds in the ambient world. Sometimes, he says, Pros-pero will punish him by setting spirits upon him "like Apes, that moe and chatter at me." Other times "am I / All wound with Adders, who with clouen tongues / Doe hisse me into madness" (2.2.9, 12–14). Squawking jays and tittering marmosets are added to these animal sounds in Caliban's offer to enslave himself to Trinculo and Stephano (2.2.168–169). The audience is actually assaulted with such threatening sounds in numerous instances of thunder, in the offstage barks that emerge amid Ariel's song "Come unto these yellow sands" (1.2.377ff), in the *"noyse of Hunters"* and offstage animal cries ("Harke, they rore") Ariel conjures to torment Stephano and Trinculo in 4.1. Amid these animal noises Ariel's song "Where the bee sucks" (5.1.88ff) insinuates the buzzing of bees and the whooping of owls.

For all its assaultive noises, the acoustic world of the island is also replete with charming, seductive sounds. "Be not affeard," Caliban reassures Stephano and Trinculo.

> the Isle is full of noyses,
> Sounds, and sweet aires, that giue delight and hurt not:
> Sometimes a thousand twangling Instruments
> Will hum about mine eares; and sometime voices,
> That if I then had wak'd after long sleepe,
> Will make me sleepe againe. . . .
>
> (3.2.138–143)

Though falling well short of a thousand, the twangling instruments of the Blackfriars broken consort are nonetheless called upon to supply for the audience, at several points in the play, soothing sounds of the sort Caliban claims to have heard on the island many times over. Temporally and spatially, philosophically and phenomenologically, music in the play reaches a climax in the cosmic harmonies of the wedding masque Prospero conjures for Ferdinand and Miranda in 4.1.

Within the broad acoustic horizons of *The Tempest*, between noise and music, Caliban stands dead center. Only he speaks gutteral curses one moment and mellifluous verse the next; only he runs the diapason from thunder to sweet airs. Among Robert Johnson's music for the original production no setting survives for Caliban's song "No more dams I'le make for fish." The rhythm of the words, however, suggests something out of an antimasque—or out of Africa. Its rhythmic irregularities are extreme. At first, in its alternation of four strong beats with three strong beats, it sounds a bit like an inverted ballad:

```
/   x   /   x    /   x  /
No more dams I'le make for fish
x   /   x  / x  x  x / x
Nor fetch in firing, at requiring,
\   x   /   x   \   x   /
Nor scrape trenchering, nor wash dish. . . .
```

The refrain or "burden"—a wonderful pun in this context—shifts to another kind of rhythm entirely:

> / x / x x / x /
> Ban' ban' Cacalyban
> /   x x / x / x x /
> Has a new Master, get a new Man. (2.2.179–184)

The strange apostrophes in the folio printing of the refrain's first line ("'ban' ban'," not the "'ban 'ban" of most modern editions) can only indicate silent but palpable beats. Multiple rhythmic patterns, the syncopation of beats felt in silence, accents falling in unexpected places: these are the essence of a distinctively African sense of rhythm that later manifested itself in jazz (Bate 1995: 155–162). The extra syllable in "Cacalyban" and the dactylic rhythm in the refrain's second line ("has a new . . .," "get a new . . .") only add to the effect of language being turned into music, of music being turned into dance, of sense being turned into sensation. Paul Zumthor's description of jazz fits perfectly Caliban's song—and Caliban's place in the acoustic design of *The Tempest:*

> Strongly outlined are accentuated tones, syncopes, and double-beat rhythms for which the percussion instrument provides the base, upon which the clarinet, trombone, and saxophone are set apart in jazz and toward which voice counterpoint moves. Speech looses the monotony that syntactic regularity engenders; discourse is constructed polymetrically. . . .

As a result,

> A massive unity is forged in the depths of a consciousness. The function of the text loses all clarity; its paucity and its mediocrity often cancel its impact: only music and dance remain. (1990: 150–151)

Caliban remembers a sound-world before language, and he constantly puts the audience in aural touch with that world's existence. As a result, the language of *The Tempest* threatens to devolve back into [o:]. The *"strange hollow and confused noyse"* that interrupts the dance of nymphs and reapers in the wedding masque (S.D. before 5.1.144) is "strange" in its spatial origins, "hollow" in its semantic emptiness, "confused" in its taking sounds that would be comprehensible if dispersed in time and crashing them together in a single moment. In that moment *The Tempest* comes closest to collapsing once again into the aural chaos with which the play began. Prospero's commanding speeches in Act V would seem to mitigate such a threat if not destroy it entirely. Caliban's last words in the play, after all, are a promise of reformation after his plot to gain his freedom: "Ile be wise hereafter, / And seeke for grace" (5.1.298–299). Prospero himself, however, asks for noise, asks for [o:] in the epilogue he speaks to the audience. Don't leave me stranded, he pleads in almost perfectly regular tetrameter:

> But release me from my bands
> With the helpe of your good hands:
> Gentle breath of yours, my Sailes
> Must fill, or else my proiect failes. . . .
> (Ep. 9–12)

The last thing scripted to be heard by the audience in the Blackfriars Theater is the noise of clapping hands and shouts of approval—accompanied, perhaps, by further music from the theater's consort of recorder, cittern, and viols. In that final combination of tempest and temper the audience hears a recapitulation of the play's acoustic design.

RETURNING, RETUNING

In their dealings with the Massachusetts Indians, John Eliot and his fellow missionaries devised for themselves a role that Europeans and North Americans have continued to play vis-à-vis peoples in other parts of the world across the four hundred years between then and now. Far from *forcing* the Indians to become Christians (as Catholic missionaries were reported to be doing in New Spain), Eliot and his brethren took the trouble to learn Algonkian, to speak to their potential converts in the listeners' own language. They saw themselves as establishing "discourse" with the Indians, even if they began with a "set speech" in the form of a sermon. Their "discourse," however, was not really a dialogue: its purpose, according to Eliot, was to "skrue by variety of meanes something or other of God into them" (1980: 4)—that is to say, "to implant firmly (a notion) by means of gradual insinuation; to contrive to insert" (OED, "screw" 10b). Eliot learned the Indians' language, but he did so primarily to speak, not to hear. He came to Waaubon's wigwam with his own agenda. He sat within his own culture's acoustemology. In so doing, he illustrates the kind of listening Michel Serres execrates in *Genesis*. If noise is an inescapable factor in human existence, then, in any encounter, there will be noise in both the listening subject and the speaking object. Really to listen requires that the listening observer "make less noise than the noise transmitted by the object observed. If he gives off more noise, it obliterates the object, covers or hides it. An immense mouth, minuscule ears, how many are thus built, animals in their misrecognition. Cognition is subtraction of the noise received and of the noise made by the subject" (1995: 61).

Most human cultures, Roland Barthes maintains, recognize only two modes of listening: "the arrogant listening of a superior, the servile listening of an inferior (or of their substitutes)." Clearly this is just the kind of listening, on both sides, that is going on in *The Day-Breaking, If Not the Sun-Rising of the Gospell with the Indians in New-England*. For his part, Barthes advocates a new kind of listening, a "free listening . . . a listening which circulates, which permutates, which disaggregates, by its mobility, the fixed network of the roles of speech: it is not possible to imagine a free society, if we agree in advance to preserve within it the old modes of listening: those of the believer, the disciple, and the patient" (1985: 259). For Barthes the inspiration for such listening is psychoanalysis, with its attention to the unseen, the unspoken, the unverbalized depths within the embodied speaker. The Native American example argues, however, that "free listening" is possible quite without Freud and Lacan. Indeed, the very physical circumstances of listening, the penetration into one's body of sounds made by another person, facilitate "free listening." All that is required is the will to open one's ears.

In one of the experiments catalogued in *Sylva Sylvarum*, Bacon observes that acuity of listening increases as the possibilities for vision decrease: "*Sounds* are *meliorated* by the *Intension* of the *Sense;* where the *Common Sense* is collected most, to the *particular Sense* of *Hearing*, and the *Sight* suspended: And therfore, *Sounds* are sweeter, (as well as greater,) in the *Night*, than in the *Day*" (1626: no. 235). Knowledge based on vision suppresses knowledge based on listening. The knowledge John Eliot, Roger Williams, and other missionaries wished to "skrue" into their Native American neighbors was of just this sort. It was knowledge codified in a printed book. As such, it was *objective* knowledge: something "out there," something any reasonable person could see. Eliot's listeners were themselves objectified in the printed accounts that circulated knowledge of them in Europe. In the accounts of Eliot and other Englishmen, Native Americans become collectibles, the living equivalents of the Native American artifacts Gerschow and other European travelers noted in their tours of Whitehall Palace. Print objectifies; listening subjectifies. Had Eliot listened—had he *freely* listened—and not just looked, he would have carried away a knowledge of the Native American that would not be so easy to place simply by writing it down. Free listening requires an acknowledgment of the physical presence of the speaker, the embodiedness of the sounds one is hearing, and it imparts a different kind of knowledge than seeing does.

The keepers of traditional Native American culture know the difference. In a session at the Shakespeare Association of America meeting in Albuquerque in 1994, storytellers from several of the pueblos of northern New Mexico shared with the academic audience some of the stories they knew, along with some of their experiences as storytellers. While the audience was free to take notes, the speakers insisted that no tape recordings be made. Why should that be so, I asked at the end. Because, one of the speakers replied, what has happened here has happened among us. We have heard one another's voices. Because you yourself were here, he continued, you can *tell* someone else about what has happened, but a tape recording would not give that person the kind of knowledge that only you can convey.

In the last analysis, free listening is an ecological concern. As *oikos* ("house") + *logos* ("knowledge"), ecology studies the conditions of inhabiting. Modern media and modern means of transportation have destroyed the physical barriers that once maintained the cultural differences of Aldersgate from Kenilworth, of England from Ireland, of Europe from North America. The horizon of hearing has shrunk at the same time that the voices clamoring within that horizon have increased exponentially. Anyone in the world who watches television or travels by jet plane finds herself in the position of Charles Ives in "Three Places in New England," surrounded by what sounds like cacophony. In those circumstances two courses of action present themselves most insistently: to speak very loudly oneself or to shut one's ears. Ecological viability lies in a third possibility: to look, to listen, and to know the difference.

$\mathcal{W}_{o\ r\ k\ s}\ \ C\ i\ t\ e\ d$

MANUSCRIPTS

Bodleian MS Mus.e.1–5. Compilation of part songs, made for and probably by John Sadler, 1585. Bodleian Library, Oxford.

Folger MS V.a.311. Thomas Fella, visual commonplace book, 1585–98. Folger Shakespeare Library, Washington, D.C.

Folger MS V.b.232. Thomas Trevelyon, visual commonplace book, 1608. Folger Shakespeare Library, Washington, D.C.

Folger MS V.a.345. Verse miscellany in several hands, associated with Christ Church, Oxford, 1630 and later. Folger Shakespeare Library, Washington, D.C.

Folger MS V.a.381. Commonplace book compiled c. 1600–c. 1650. Folger Shakespeare Library, Washington, D.C.

PRINTED MUSIC

Brett, Philip, ed. 1967. *Consort Songs.* Musica Britannica, vol. 22. London: The Royal Musical Association.

Bronson, Bertrand Harris, ed. 1976. *The Singing Tradition of Child's Popular Ballads.* Princeton: Princeton University Press.

Campion, Thomas. 1601. *A Booke of Ayres, Set foorth to be song to the Lute, Orpherian, and Base Violl.* London: Philip Rosseter.

Morley, Thomas. 1962. *Madrigalls to Fovre Voyces . . . The First Booke.* Edited by Edmund H. Fellowes. Revised by Thurston Dart. London: Stainer & Bell.

Ravenscroft, Thomas. 1611. *Melismata: Mvsicall Phansies. Fitting The Court, Citie, and Covntrey.* London: Thomas Adams.

Sharp, Cecil J., and Maud Karpeles. 1968. *Eighty English Folk Songs from the Southern Appalachians.* London: Faber, 1968.

PRINTED BROADSIDES

[Anonymous.] 1569. "A newe Ballade intytuled, Good Fellowes must go learne to daunce." London: William Griffith.

[Anonymous.] c. 1590. "A ditty delightfull of mother watkins ale." [London: publisher unspecified.]

[Anonymous.] 1598. "Luke Huttons lamentation: which he wrote the day before his death, being condemned to be hanged at Yorke this last assises for his robberies and trespasses committed." London: Thomas Millington.

[Anonymous.] 1603. "A lamentable Dittie composed upon the death of Robert Devereux late Earle of Essex, who was beheaded in the Tower of London, upon Ashwednesday in the morning. 1601." London: Margaret Allde.

[Anonymous.] c. 1620. "The Spanish Tragedy, Containing the lamentable Murders of *Horatio* and *Bellimpiria,* With the pitifull Death of old *Hieronimo*." London: H. Gosson.

[Anonymous.] [1658–1659]. "The Lamentable and Tragical History of *Titus Andronicus*." London: I. Clarke, W. Thackery, and T. Passinger,.

B———, R———. [1570]. "A new balade entituled as foloweth, / To such as write in Metres, I write / Of small matters an exhortation, / By reading of which, men may delite / In such as be worthy commendation." London: Alexander Lacie.

Birch, William. [1563]. "The complaint of a sinner, vexed with paine, / Desyring the ioye, that euer shall remayne. After W. E. moralized." London: Richard Applow.

Brice, Thomas. [1561–1562]. "Against filthy writing / and such like delighting." London: Edmond Halley.

Churchyard, Thomas. [1566]. "Churchyardes farewell." London: Edward Russell.

Deloney, Thomas. 1860. "The Queenes visiting of the Campe at Tilsburie with her entertainment there." 1588. In *Three Old Ballads on the Overthrow of the Spanish Armada,* edited by J. O. Halliwell-Phillips. London: printed for the editor.

Fulwood, William. [1562]. A Supplication to Eldertonne, for Leaches vnlewdnes: Desiring him to pardone, his manifest vnrudeness." London: John Alde.

I———, T———. 1588. "A Ioyful Song of the Royall receiuing of the Queenes most excellent Maiestie into her highnesse Campe at Tilburie in Essex: on Thursday and Fryday the eight and ninth of August, 1588." London: Richard Jones.

Peele, Steven. [1570]. "A letter to Rome, to declare to ye Pope, Iohn Felton his freend is hangd in a rope. . . ." London: Henry Kirkham.

Preston, Thomas. 1570. "A Lamentation from Rome, how the Pope doth bewayle, That the Rebelles in England can not preuayle." London: William Griffith.

[Smyth, Sir Thomas.] [1540a]. "A balade agaynst malycyous Sclanderers." London: John Gough.

[Smyth, Sir Thomas.] [1540b]. "An Envoye from Thomas Smyth upon thaunswer of one.W.G. Lurkyng in Lo[ ]rells Denne / for feare men shulde hym see." [London: printer unspecified.]

Tarlton, Richard. 1570. "A very Lamentable and woful discours of the fierce fluds whiche lately flowed in Bedford shire, in Lincoln shire, and iu [*sic*] many other places, with great losses of sheep and other Cattel. The v. of October. Anno Domini 1570." London: John Allde.

PRINTED TEXTS

A———, R———. 1615. *The Valiant Welshman.* London: Robert Lownes.

[Anonymous.] 1868. *The English Courtier, and the Cu[n]trey-gentleman.* In *Inedited Tracts Illustrating the Manners, Opinions, and Occupations of Englishmen during the Sixteenth and Seventeenth Centuries,* edited by W. C. Hazlitt. London: Roxburghe Society.

Albertus Magnus (attr.). 1977. *Quaestiones Alberti de Modis Significandi.* Ed. L. G. Kelly. Amsterdam: John Benjamins.

Ames, Susie M., ed. 1954. *County Court Records of Accomack-Northampton, Virginia, 1632–1640.* Washington, D.C.: American Historical Association.

Ames, Susie M., ed. 1973. *County Court Records of Accomack-Northampton, Virginia, 1640–1645.* Charlottesville: University of Virginia Press.

Anderson, Benedict. 1991. *Imagined Communities: Reflections on the Origin and Spread of Nationalism.* Rev. ed. London: Verso.

Arber, Edward, ed. 1875. *A Transcript of the Registers of the Company of Stationers of London: 1554–1640 A.D.* 5 vols. London: privately printed.

Aristotle (attr.). 1597. *The Problemes of Aristotle, with other Philosophers and Phisitions.* London: Arnold Hatfield.

Aristotle (abridged). [1637]. *A Briefe of the Art of Rhetorique. Containing in substance all that ARISTOTLE hath written in his Three Bookes of that subject, Except onely what is not applicable to the English Tongue.* London: Andrew Crook.

Aristotle. 1941. *The Basic Works.* Edited by Richard McKeon. New York: Random House, 1941. [Cited as *Works.*]

Appelbaum, David. 1990. *Voice.* Albany: State University of New York Press.

Aubrey, John. 1972. *Remains of Gentilism and Judaism.* In *Three Prose Works,* edited by John Buchanan-Brown. Fontwell, Sussex: Centaur Press.

Axtell, James. 1981. *The European and the Indian: Essays in the Ethnohistory of Colonial North America.* New York: Oxford University Press.

Bacon, Sir Francis. 1626. *Sylva Sylvarum: Or A Naturall Historie* and *New Atlantis.* Edited by William Rawley. London: William Lee.

Bacon, Sir Francis. 1640. *Of the Advancement and Proficience of Learning or the Partitions of Sciences.* Translated by Gilbert Wats. London: Robert Young and Edward Forrest.

Bacon, Sir Francis. 1648. *The Remaines . . . being Essayes and severall Letters to severall great Personages, and other pieces of various and high concernment not heretofore published.* London: B. Alsop.

Baillie, Hugh Murray. 1967. "Etiquette and the Planning of the State Apartments in Baroque Palaces." *Archaeologia* 101: 169–199.

Baker, David. 1997. *Between Nations: Shakespeare, Spenser, Marvell, and the Question of Britain.* Stanford, Calif.: Stanford University Press.

Bakhtin, M. M. 1981. *The Dialogic Imagination: Four Essays.* Translated by Michael Holquist. Austin: University of Texas Press.

Baldwin, William. 1584. [*Beware the Cat*]. London: Edward Allde.

Bales, Peter. 1590. *The Writing Schoolemaster: Conteining three Bookes in one.* London: Thomas Orwin.

Barber, Charles. 1997. *Early Modern English.* Rev. ed. Edinburgh: Edinburgh University Press.

Barthes, Roland. 1985. "Listening." In *The Responsibility of Forms,* translated by Richard Howard. Berkeley: University of California Press.

Baskervill, Charles Read. 1929. *The Elizabethan Jig and Related Song Drama.* Chicago: University of Chicago Press.

Bate, Jonathan. 1995. "Caliban and Ariel Write Back." *Shakespeare Survey* 48 (1995): 155–162.

Beadle, Richard, ed. 1982. *The York Plays.* London: Edward Arnold.

Beaumont, Francis, and John Fletcher. 1966–1996. *The Dramatic Works in the Beaumont and Fletcher Canon.* Edited by Fredson Bowers and others. 10 volumes. Cambridge: Cambridge University Press.

Beier, A. L. 1968. "Engine of manufacture: the trades of London." In *London 1500–1700: The Making of the Metropolis,* edited by A. L. Beier and Roger Finlay. London: Longman.

del Bene, Bartolomeo. 1609. *Civitas Veri sive Morum.* Paris: Drouart.

Benjamin, Walter. 1968. *Illuminations.* Edited by Hannah Arendt. Translated by Harry Zohn. New York: Schocken.

Bergeron, David. 1971. *English Civic Pageantry, 1558–1642.* London: Edward Arnold.

Berry, Cicely. 1988. *The Actor and His Text.* New York: Scribners.

Berry, Herbert. 1987. *Shakespeare's Playhouses.* New York: AMS.

Bevington, David. 1962. *From* Mankind *to* Marlowe. Cambridge, Mass.: Harvard University Press.

[Bible.] 1956. *Biblia Sacra juxta Vulgatum Clementinam.* Paris: Desclée & Co.

[Bible.] 1969. *The Geneva Bible.* Facsimile rpt. Edited by Lloyd E. Berry. Madison: University of Wisconsin Press.

Blatherwick, Simon. 1997. "The Archaeological Evaluation of the Globe Playhouse." In *Shakespeare's Globe Rebuilt,* edited by J. R. Mulryne and Margaret Shewring. Cambridge: Cambridge University Press.

Block, K. S., ed. 1922. *Ludus Coventriae or The Plaie called Corpus Christi.* Early English Text Society no. 120. London: Oxford University Press.

Boorman, Stanley. 1986. "Early Music Printing: Working for a Specialized Market." In *Print and Culture in the Renaissance: Essays on the Advent of Printing in Europe,* edited by Sylvia S. Wagonheim. Newark: University of Delaware Press.

Booth, Mark W. 1981. *The Experience of Song*. New Haven: Yale University Press.

Bossy, John. 1983. "The Mass as a Social Institution, 1200–1700." *Past and Present* no. 100: 29–61.

Boyer, Paul, and Stephen Nissenbaum, eds. 1977. *The Salem Witchcraft Papers*. 3 volumes. New York: Da Capo.

Bradshaw, Henry. 1889. *Collected Papers*. Cambridge: Cambridge University Press.

Brand, John. 1870. *Popular Antiquities of Great Braitin*. 3 volumes. London: J. R. Smith.

Brathwait, Richard. 1631. *Whimzies: Or, A New Cast of Characters*. London: Ambrose Rithirdon.

Bray, Alan. 1982. *Homosexuality in Renaissance England*. London: Gay Men's Press.

Brereton, John. 1983. *A Briefe and Trve Relation of the Discoverie of the North Part of Virginia*. In *The English New England Voyages 1602–1608*, edited by David B. Quinn and Alison M. Quinn. London: Hakluyt Society.

Breton, Nicholas. 1605. *A Poste with a Packet of madde Letters. The second part*. London: John Browne and John Smethwicke.

Breton, Nicholas. 1868. *The Court and the Country, or A briefe Discourse Dialogue-wise set downe betweene a Courtier and a Country-man*. In *Inedited Tracts Illustrating the Manners, Opinions, and Occupations of Englishmen during the Sixteenth and Seventeenth Centuries*, edited by W. C. Hazlitt. London: Roxburghe Society.

Brink, Jean R. 1990. "Constructing the *View of the Present State of Ireland*." *Spenser Studies* 11: 203–228.

Brinsley, John. 1612. *Ludus Literarius: Or, The Grammar Schoole*. London: Thomas Man.

Brinsley, John. 1622. *A Consolation for Our Grammar Schooles*. London: Thomas Man.

Bristol, Michael. 1996. *Big-time Shakespeare*. London: Routledge.

Brody, Alan. 1969. *The English Mummers and their Plays*. Philadelphia: University of Pennsylvania Press.

Brome, Richard. 1873. *The Dramatic Works*. 3 volumes. London: J. Pearson.

Brown, David. 1622. *The New Invention, Intituled, Calligraphia: Or, The Arte of Faire Writing*. St. Andrews: Edward Raban.

Buchan, David. 1972. *The Ballad and the Folk*. London: Routledge and Kegan Paul.

Burke, Peter. 1978. *Popular Culture in Early Modern Europe*. New York: Harper and Row.

Bursill-Hall, G. L. 1972. *Speculative Grammars of the Middle Ages*. The Hague: Mouton.

Busino, Orazio. 1995. "The Diary of Horatio Busino, Chaplain of Pietro Contarini, Venetian Ambassador in England." In *The Journals of Two Travelers in Elizabethan and Early Stuart England*. London: Caliban Books.

Butlin, R. A. 1979. "The enclosure of open fields and extinction of common rights in England *circa* 1600–1750: a review." In *Change in the Countryside: Essays on Rural England, 1500–1900*, edited by H. S. A. Fox and R. A. Butlin. London: Institute of British Geographers.

Calloway, Colin G. 1997. *New Worlds for All: Indians, Europeans, and the Remaking of Early America*. Baltimore: Johns Hopkins University Press.

Calloway, Colin G., ed. 1994. *The World Turned Upside Down: Indian Voices from Early America*. Boston: Bedford.

Campion, Edmund. 1633. *The Historie of Ireland*. Dublin: Society of Stationers.

Castiglione, Baldesar. 1900. *The Book of the Courtier*. Translated by Thomas Hoby. London: David Nutt.

Catterall, Helen Tunnicliff. 1926–1937. *Judicial Cases concerning American Slavery and the Negro*. 5 volumes. Washington, D.C.: Carnegie Institution.

de Certeau, Michel. 1984. *The Practice of Everyday Life*. Translated by Steven Randall. Berkeley: University of California Press.

Chalfont, Fran Cernocki. 1978. *Ben Jonson's London: A Jacobean Placename Dictionary*. Athens: University of Georgia Press.

Chamberlain, John. 1939. *The Letters*. Edited by Norman Egbert McClure. 2 volumes. Philadelphia: American Philosophical Society.

Chambers, E. K. 1923. *The Elizabethan Stage*. 4 volumes. Oxford: Clarendon Press.

Chambers, E. K. 1933. *The English Folk Play*. Oxford: Clarendon Press.

Chan, Mary. 1980. *Music in the Theatre of Ben Jonson*. Oxford: Clarendon Press.

Chappell, William. 1855–1859. *The Ballad Literature and Popular Music of the Olden Time*. 2 volumes. London: Cramer, Bede, and Chappell.

Charlton, John. 1964. *The Banqueting House Whitehall*. London: HMSO.

Child, Francis James, ed. 1957. *The English and Scottish Popular Ballads*. 5 volumes. Rpt. New York: Folklore Press, 1957.

[Church of England.] 1559. *The boke of common praier and administration of the Sacramentes, and other rites and ceremonies in the Churche of Englande*. London: R. Jugge and J. Cawode.

Churchyard, Thomas. 1560. *The Contention bettwyxte Churchyeard and Camell, upon David Dycers Dreame sett out in suche order, that it is bothe wyttye and profytable for all degryes. Rede this littell communication betwene Churchyarde: Camell: and others mo Newlye Imprinted and sett furthe for thy profyt gentill Reader*. London: Mitchell Loblee.

Cicero. 1939. *Orator*. Translated by H. M. Hubbell. Loeb Library. Cambridge, Mass.: Harvard University Press.

Cicero. 1967. *De Oratore*. Translated by E. W. Sutton and H. Rackham. Loeb Library. 2 volumes. Cambridge, Mass.: Harvard University Press.

Cicero (attr.). 1968. *De ratione dicendi (Rhetorica ad Herennium)*. Translated by

Harry Caplan. Loeb Library. Cambridge, Mass.: Harvard University Press.

Clark, Andrew, ed. 1907. *The Shirburn Ballads*. Oxford: Clarendon Press.

Clayton, Thomas. 1969. *The "Shakespearean" Addition in the Booke of Sir Thomas Moore*. Dubuque, IA: William C. Brown.

Cogswell, Thomas. 1989. *The Blessed Revolution: English politics and the coming of war, 1621–1624*. Cambridge: Cambridge University Press.

Collmann, Herbert L. 1912. *Ballads and Broadsides Chiefly of the Elizabethan Period*. London: Roxburghe Club.

Cooper, Geoffrey, and Christopher Wortham, eds. 1980. *The Summoning of Everyman*. Nedlands: University of Western Australia Press.

Coote, Edmund. 1596. *The English Schoole-Maister*. London: Ralph Jackson and Robert Dexter.

Cornwallis, Sir William. 1600. *Essayes*. London: Edmund Mattes.

Corradi-Fiumara, Gemma. 1990. *The Other Side of Language: A Philosophy of Listening*. Translated by Charles Lambert. London: Routledge.

Crane, Mary. 1993. *Framing Authority: Sayings, Self, and Society in Sixteenth-Century England*. Princeton: Princeton University Press.

Crawford, J. L. L. 1890. *Bibliotheca Lindesiana: A Catalogue of a Collection of English Ballads of the XVIIth and XVIIIth Centuries*. Aberdeen: Aberdeen University Press.

Creeth, Edmund. 1966. *Tudor Plays: An Anthology of Early English Drama*. New York: Anchor.

Cressy, David. 1980. *Literacy and the Social Order: Reading and Writing in Tudor and Stuart England*. Cambridge: Cambridge University Press.

Crooke, Helkiah. 1616. *Microcosmographia. A Description of the Body of Man*. London: Jaggard.

Cruttenden, Alan. 1992. "Intonation and the Comma." *Visible Language* 25, no. 1: 55–73.

Curry, E. Thayer. 1940. "The Pitch Characteristics of the Adolescent Male Voice." *Speech Monographs* 7: 48–62.

Daniel, Samuel. 1965. *Poems and A Defence of Ryme*. Edited by Arthur Colby Sprague. Chicago: University of Chicago Press.

Dasent, John Roche, gen. ed. 1894. *Acts of the Privy Council of England*. N. S. 8 and 9. London: Her Majesty's Stationery Office.

Davies, Sir John. 1612. *A Discoverie of the Trve Cavses why Ireland was neuer entirely Subdued, nor brought vnder Obedience of the Crowne of England, vntill the Beginning of his Maiesties happie Raigne*. London: John Jaggard.

Davis, Norman, ed. 1970. *Non-Cycle Plays and Fragments*. Early English Text Society. London: Oxford University Press.

Dekker, Thomas. 1612. *O per se O. Or a new Cryer of Lanthorne and Candle-light*. London: John Bushie, 1612.

Dekker, Thomas. 1616. *The Belman of London. Bringing to Light the Most Notori-*

ous Villanies that are now practised in the Kingdome . . . The Fourth impression, with new additions. London: Nathaniell Butters.

Dekker, Thomas. 1953–1961. *The Dramatic Works.* Edited by Fredson Bowers. 4 volumes. Cambridge: Cambridge University Press.

Dekker, Thomas. 1963. *The Non-Dramatic Works.* Edited by Alexander B. Grosart. 5 volumes. New York: Russell & Russell.

Della Casa, Giovanni. 1576. *Galateo . . . Or rather, a treatise of the ma[n]ners and behaviours, it behoueth a man to vse and eschewe, in his familiar conuersation.* Translated by Robert Peterson. London: Ralph Newberry.

Deloney, Thomas. 1619. *The Pleasant History of Iohn Winchcomb . . . called Iack of Newberie.* London: Humphrey Lownes.

Deloney, Thomas. 1912. *The Works.* Edited by Francis Oscar Mann. Oxford: Clarendon Press.

Derricke, John. 1581. *The Image of Irelande.* London: John Day.

Derrida, Jacques. 1973. *Speech and Phenomena and Other Essays on Husserl's Theory of Signs.* Translated by David B. Allison. Evanston: Northwestern University Press.

Derrida, Jacques. 1976. *Of Grammatology.* Translated by Gayatri Chakravorty Spivak. Baltimore: Johns Hopkins University Press.

Dobson, R. B., and J. Taylor. 1976. *Rymes of Robin Hood.* London: Heinemann.

Donawerth, Jane. 1984. *Shakespeare and the Sixteenth-Century Study of Language.* Urbana: University of Illinois Press.

Donne, John. 1953–1962. *The Sermons.* Ed. George R. Potter and Evelyn Simpson. 10 volumes. Berkeley: University of California Press.

Donne, John. 1965. *The Elegies and The Songs and Sonnets.* Edited by Helen Gardner. Oxford: Clarendon Press.

Dorson, Richard M. 1968. *The British Folklorists: A History.* London: Routledge & Kegan Paul.

Dow, George Francis, ed. 1912. *Records and Files of the Quarterly Courts of Essex County, Massachusetts.* Volume 2. Salem, Mass.: The Essex Institute.

Drayton, Michael. 1961. *Poly-Olbion.* In *The Works of Michael Drayton,* edited by J. William Hebel. Volume 4. Oxford: Blackwell.

Dugdale, George. 1950. *Whitehall Through the Centuries.* London: Phoenix House.

Dugdale, William. 1656. *The Antiquities of Warwickshire.* London: Thomas Warren.

Earle, John. 1628. *Micro-cosmographie. Or, A Peece of the World Discovered; In Essayes and Characters.* London: Edward Blount.

Earle, John. 1629. *Micro-cosmographie. Or, A Peece of the World Discovered; in Essayes and Characters. The fift Edition much enlarged.* London: Robert Allot.

Egan, M. Davis. 1988. *Architectural Acoustics.* New York: McGraw-Hill.

Eisenstein, Elizabeth L. 1979. *The Printing Press as an Agent of Change: Communications and cultural transformations in early-modern Europe.* 2 volumes. Cambridge: Cambridge University Press.

Eliot, John. 1834. *The Day-Breaking, If Not the Sun-Rising of the Gospell with the Indians in New-England.* In Massachusetts Historical Society *Collections,* 3rd series, 4 (1834).

Eliot, John. 1980. *Indian Dialogues,* For Their Instruction in that great Service of Christ, in calling home their Country-men to the Knowledge of God, And of Themselves, and of Iesus Christ. Edited by Henry W. Bowden and James P. Ronda as *John Eliot's Indian Dialogues: A Study in Cultural Interaction.* Westport, Conn.: Greenwood Press.

England, George, ed. 1897. *The Towneley Plays.* Early English Text Society no. 71. London: Oxford University Press.

Erasmus, Desiderius. 1969–1995. *Opera Omnia.* Edited by J. H. Waszink *et al.* 22 volumes. Amsterdam: North-Holland.

Erasmus, Desiderius. 1978. *De recta pronuntatione.* Translated by Maurice Pope as *The right way of speaking.* In *Collected Works,* volume 4, edited by J. K. Sowards. Toronto: University of Toronto Press.

Erasmus, Desiderius. 1985. *De conscribendis epistolis.* Translated by Charles Fantazzi as *On the writing of letters.* In *Collected Works,* volume 25, edited by J. K. Sowards. Toronto: University of Toronto Press.

Fairbanks, Grant. 1960. *Voice and Articulation Drillbook.* 2nd ed. New York: Harper.

Faret, Nicolas. 1632. *The Honest Man: Or, the Art to please in Court.* Translated by E. G[rimestone]. London: Thomas Harper.

Farley, Henry. 1621. *St. Paules-Church Her Bill for the Parliament.* London: [R. Milbourne].

Feld, Steven. 1996. "Waterfalls of Song: An Acoustemology of Place Resounding in Bosavi, Papua New Guinea." In *Senses of Place,* edited by Steven Feld and Keith H. Basso. Santa Fe: School of American Research Press.

Fellowes, E. H. 1967. *English Madrigal Verse 1588–1632.* Rev., Frederick W. Sternfeld and David Greer. Oxford: Clarendon Press.

Fineman, Joel. 1991. "The Sound of O in *Othello.*" In *The Subjectivity Effect in Western Literary Tradition.* Cambridge: MIT Press.

Florio, John. 1578. *Florio His first Fruites: which yeelde familiar speech, merie Prouerbes, wittie Sentences, and golden sayings.* London: Thomas Woodcocke.

Foakes, R. A. 1985. *Illustrations of the English Stage 1580–1642.* Stanford: Stanford University Press.

Fry, Dennis. 1977. *Homo Loquens: Man as a Talking Animal.* Cambridge: Cambridge University Press.

Fry, Dennis. 1979. *The Physics of Speech.* Cambridge: Cambridge University Press.

Furnivall, F. J., ed. 1896. *The Digby Plays.* Early English Text Society. London: Oxford University Press.

Gainsford, Thomas. 1616. *The Rich Cabinet Furnished with varietie of Excellent discriptions, exquisite Charracters, witty discourses, and delightfull Histories, Devine and Morrall.* London: Roger Jackson.

Galpin, F. W. 1965. *Old English Instruments of Music*. 4th ed. Rev., Thurston Dart. London: Methuen.

Garry, Jane. 1983. "The Literary History of the English Morris Dance." *Folklore* 94, no. 2: 219–228.

Gascoigne, George. 1907–1910. *The Complete Works*. Edited by John W. Cunliffe. 2 volumes. Cambridge: Cambridge University Press.

Gerschow, Frederic. 1892. "Diary of the Journey of Philip Juilius, Duke of Stettin-Pomerania, through England in the Year 1602." Edited by Gottfried von Bülow. *Transactions of the Royal Historical Society*, n. s. 6: 1–67.

Gillies, John. 1994. *Shakespeare and the Geography of Difference*. Cambridge: Cambridge University Press.

Giraldus Cambrensis. 1892. *Historical Works*. Edited by Thomas Wright. London: Bell, 1892.

Girouard, Mark. 1978. *Life in the English Country House: A Social and Architectural History*. New Haven: Yale University Press.

Goldberg, Jonathan. 1990. *Writing Matter from the Hands of the English Renaissance*. Stanford: Stanford University Press.

Goody, Jack. 1977. *The Domestication of the Savage Mind*. Cambridge: Cambridge University Press.

Gouk, Penelope. 1991. "Some English Theories of Hearing in the Seventeenth Century." In *The Second Sense: Studies in Hearing and Musical Judgment from Antiquity to the Seventeenth Century*, ed. Charles Burnett, Michael Fend, and Penelope Gouk. London: Warburg Institute.

de Grazia, Margreta. 1991. "Shakespeare in Quotation Marks." In *The Appropriation of Shakespeare*, edited by Jean I. Marsden. New York: St. Martins.

Guazzo, Stephano. 1967. *The Civile Conversation*. Translated by George Pettie and Bartholomew Young. Edited by Edward Sullivan. 2 vols. New York: AMS.

Guilpin, Everard. 1974. *Skialetheia or A Shadowe of Truth, in Certaine Epigrams and Satyres*. Edited by D. Allen Carroll. Chapel Hill: University of North Carolina Press.

Gumbrecht, Hans Ulrich. 1994. "A Farewell to Intrepretation." In *Materialities of Communication*, edited by Hans Ulrich Gumbrecht and K. Ludwig Pfeiffer, translated by William Whobrey. Stanford: Stanford University Press.

Gumbrecht, Hans Ulrich, and K. Ludwig Pfeiffer, eds. 1994. *Materialities of Communication*. Translated by William Whobrey. Stanford, Calif.: Stanford University Press.

Gurr, Andrew. 1992. *The Shakespearean Stage 1574–1642*. 3rd edition. Cambridge: Cambridge University Press.

Gurr, Andrew. 1996. *The Shakespearian Playing Companies*. New York: Oxford University Press.

Hakluyt, Richard, trans. 1893. *The Trve Pictures and Fashions of the People in*

That Part of America Now Called Virginia. . . . In Thomas Hariot, *Narrative of the First English Plantation of Virginia.* Rpt. London: Quaritch.

Hales, John W., and Frederick J. Furnivall, eds. 1867–1868. *Ballads and Romances.* 3 volumes. London: Trubner.

Halliwell-Phillips, J. O., ed. 1849. "Some account of the popular tracts which composed the Library of Captain Cox." In *The Shakespeare Society's Papers.* Volume 4. London: The Shakespeare Society.

Hancock, Ian. 1986. "The Domestic Hypothesis, Diffusion, and Componentiality: An Account of Atlantic Anglophone Creole Origins." In *Substrata Versus Universals in Creole Genesis,* edited by Pieter Muysken and Norval Smith. Amsterdam: John Benjamins.

Handel, Stephen. 1989. *Listening: An Introduction to the Perception of Auditory Events.* Cambridge, Mass.: MIT Press.

Hariot, Thomas. 1588. *A Briefe and True Report of the New Found Land of Virginia.* London: [no printer specified].

Harsnett, Samuel. 1603. *A Declaration of egregious Popish Impostures. . . .* London: James Robert.

Hart, John. 1955. *An Orthographie, conteyning the due order and reason, howe to write or paint th*[*'*]*image of mannes voice, most like to the life or nature.* In *Works on English Orthography and Pronunciation,* edited by Bror Danielsson. 2 volumes. Stockholm: Almquist & Wiksell.

Hartog, François. 1988. *The Mirror of Herodotus: The Representation of the Other in the Writing of History.* Translated by Janet Lloyd. Berkeley: University of California Press.

Harvey, Richard. 1590?. *Plaine Percevall the Peace-Maker of England.* London: G. Seton.

Hattaway, Michael. 1982. *Elizabethan Popular Theatre: Plays in Performance.* London: Routledge and Kegan Paul.

Hecht, Anthony. 1986. *Obbligati: Essays in Criticism.* Boston: Athenaeum.

Heidegger, Martin. 1962. *Being and Time.* Translated by John Macquarrie and Edward Robinson. New York: Harper & Row.

Henslowe, Philip. 1907. *Henslowe Papers.* Edited by W. W. Greg. London: A. H. Bullen.

Henslowe, Philip. 1961. *Henslowe's Diary.* Edited by R. A. Foakes and R. T. Rickert. Cambridge: Cambridge University Press.

Hentzner, Paul. 1612. *Itinerarium Germaniae, Galliae, Angliae, Italiae. . . .* Nürnberg: Abraham Wagenmann.

Hentzner, Paul. 1901. *Travels in England during the Reign of Queen Elizabeth.* Translated by Richard Bentley. Edited by Henry Morley. London: Cassell.

Herbert, Thomas. 1634. *Discription of the Persian Monarchy Now beinge: The Orientall Indyes, Iles & other part's of the Greater Asia, and Africa.* London: William Stansby and Jacob Bloome.

Hercolani, Guilantonio. 1574. *Lo Scrittor' Utile, et brieve Segretario.* [Bologna.]

Heywood, Thomas. 1614. *The Rape of Lucrece*. London: Nathaniel Butter.

Heywood, Thomas, and Richard Brome. 1634. *The Late Lancashire Witches*. London: Benjamin Fisher.

Historical Manuscripts Commission (HMC). 1925. *Report on the Manuscripts of Lord de L'Isle and Dudley*. Volume 1. London: HM Stationery Office.

Higginbotham, A. Leon, Jr. 1978. *In the Matter of Color: Race and the American Legal Process: The Colonial Period*. New York: Oxford University Press.

Highley, Christopher. 1997. *Shakespeare, Spenser, and the Crisis in Ireland*. Cambridge: Cambridge University Press.

Holinshed, Raphael. 1577. *The Chronicles of England, Scotlande, and Irelande*. 2 volumes. London: Lucas Harrison.

Holloway, John, ed. 1971. *The Euing Collection of English Broadside Ballads*. Glasgow: University of Glasgow Press.

Holyday, Barten. 1942. *Technogamia*. Edited by M. Jean Cavanaugh. Washington, D.C.: Catholic University of America Press.

Hooker, Richard. 1593–1597. *Of the Lawes of Ecclesiasticall Politie*. London: John Windet.

Hosley, Richard. 1969. "A Reconstruction of the Second Blackfriars." In *The Elizabethan Theatre 1*, edited by D. Galloway. Toronto: Macmillan.

Howard, Skiles. 1998. *The Politics of Courtly Dancing in Early Modern England*. Amherst: University of Massachusetts Press.

Howell, James. 1650. *Epistolae Ho-Elianae, Familiar Letters*. 2 volumes. London: Humphrey Moseley.

Huarte de San Juan, Juan. 1594. *The Examination of mens Wits*. Translated by Richard Carew. London: Richard Watkins.

Hudson, Nicholas. 1994. *Writing and European Thought 1600–1830*. Cambridge: Cambridge University Press.

Hughes, W. J. 1924. *Wales and Welsh in English Literature from Shakespeare to Scott*. Wrexham: Hughes and Son.

Hulme, Peter. 1994. *Colonial Encounters: Europe and the Native Caribbean*. London: Routledge.

Husserl, Edmund. 1960. *Cartesian Meditations*. Translated by Dorion Cairns. The Hague: Mouton.

Hutton, Ronald. 1994. *The Rise and Fall of Merry England: The Ritual Year 1400–1700*. Oxford: Oxford University Press.

Hutton, Ronald. 1996. *The Stations of the Sun: A History of the Ritual Year in Britain*. Oxford: Oxford University Press.

Hymes, Dell H. 1972. "Models of the Interaction of Language and Social Life." In *Directions in Sociolinguistics: The Ethnography of Communication*, edited by John J. Gumperz and Dell Hymes. New York: Holt, Rinehart and Winston.

Hymes, Dell H. 1974. "Sociolinguistics and the Ethnography of Speaking." In *Language, Culture, and Society: A Book of Readings*, edited by Ben G. Blount. Cambridge, Mass.: Winthrop.

Iamblichus. 1926. *Life of Pythagoras*. Translated by Thomas Taylor. London: Watkins.

Ihde, Don. 1976. *Listening and Voice: A Phenomenology of Sound*. Athens: University of Ohio Press.

Ihde, Don. 1993. *Postphenomenology: Essays in the Postmodern Context*. Evanston, IL: Northwestern University Press.

Ingram, Martin. 1984. "Ridings, Rough Music and the 'Reform of Popular Culture' in Early Modern England." *Past and Present* 105: 77–113.

Ingram, R. W., ed. 1981. *Records of Early English Drama: Coventry*. Toronto: University of Toronto Press.

Iselin, Pierre. 1995. "Music and Difference: Elizabethan Stage Music and Its Reception." In *French Essays on Shakespeare and His Contemporaries*, edited by Jean-Marie Maguin and Michèle Willems. Newark: University of Delaware Press.

Iser, Wolfgang. 1978. *The Act of Reading: A Theory of Aesthetic Response*. Baltimore: Johns Hopkins University Press.

Jackman, W. T. 1962. *The Development of Transportation in Modern England*. London: Cass.

Johnson, Mark. 1987. *The Body in the Mind: The Bodily Basis of Meaning, Imagination, and Reason*. Chicago: University of Chicago Press.

Johnson, Richard. 1612. *A Crowne-Garland of Goulden Roses*. London: J. Wright.

Johnson, Robert. 1612. *The New Life of Virginea*. London: Felix Kyngston.

Jones, James H. 1961. "Commonplace and Memorization in the Oral Tradition of the English and Scottish Popular Ballads." *Journal of American Folklore* 74 (1961): 97–112.

Jonson, Ben. 1925–1963. [Complete works.] Edited by C. H. Herford, Percy Simpson, and Evelyn Simpson. 11 volumes. Oxford: Clarendon Press.

Jordan, Winthrop D. 1968. *White over Black: American Attitudes Toward the Negro, 1550–1812*. Chapel Hill: University of North Carolina Press.

Josselyn, John. 1988. *An Account of Two Voyages to New-England*. Edited by Paul J. Lindholt as *John Josselyn, Colonial Traveler*. Hanover: University Press of New England.

Kemp, William. 1884. *Kempes nine daies wonder . . . written by himselfe to satisfie his friends*. Edited by Edmund Goldsmid. Edinburgh: privately printed.

Kennedy, Jean. 1971. *Isle of Devils: Bermuda under the Somers Island Company 1609–1685*. Hamilton: Collins.

Kintgen, Eugene R. 1990. "Reconstructing Elizabethan Reading." *Studies in English Literature 1500–1900* 30, no. 1: 1–18.

Knight, Stephen. 1994. *Robin Hood: A Complete Study of the English Outlaw*. Oxford: Blackwell.

Knutson, Roslyn. 1994. "A Caliban in St. Mildred Poultry." In *Shakespeare and Cultural Traditions*, edited by Tetsuo Kishi, Roger Pringle, and Stanley Wells. Newark: University of Delaware Press.

Kupperman, Karen. 1980. *Settling with the Indians: The meeting of English and Indian cultures in America, 1580–1640*. Totowa, NJ: Rowman & Littlefield.

Kyd, Thomas. 1615. *The Spanish Tragedie*. London: I. White and T. Langley.

Langham [Laneham], Robert. 1983. *A Letter*. Edited by R. J. P. Kuin. Leiden: Brill.

Leggatt, Alexander. 1992. *Jacobean Public Theatre*. London: Routledge.

Lemnius, Levinus. 1576. *The Touchstone of Complexions*. Translated by Thomas Newton. London: Thomas Marsh.

Lester, G. A., ed. 1981. *Three Late Medieval Morality Plays*. New York: Norton.

Levine, Laura. 1994. *Men in Women's Clothing: Anti-theatricality and Effeminization, 1579–1642*. Cambridge: Cambridge University Press.

Lieberman, Philip, and Sheila E. Blumstein. 1988. *Speech Physiology, Speech Perception, and Acoustic Phonetics*. Cambridge: Cambridge University Press.

Livingston, Carole Rose. 1991. *British Broadside Ballads of the Sixteenth Century*. Volume 1: *A Catalogue of the Extant Sheets and an Essay*. New York: Garland.

Lockridge, Kenneth A. 1974. *Literacy in Colonial New England*. New York: Norton.

London County Council (LCC). 1930. *Survey of London*. Gen. eds., Montagu H. Cox and Philip Norman. Volume 13 (Westminster). London: Batsford.

Long, John H. 1961–1971. *Shakespeare's Use of Music*. 3 volumes. Gainesville: University of Florida Press.

Love, Harold. 1993. *Scribal Publication in Seventeenth-Century England*. Oxford: Clarendon Press.

Luhmann, Niklas. 1989. *Ecological Communication*. Translated by John Bednarz, Jr. Chicago: University of Chicago Press.

Luhmann, Niklas. 1990. *Essays on Self-Reference*. New York: Columbia University Press.

Luhmann, Niklas. 1991. "How Can the Mind Participate in Communication?" In *Materialities of Communication*, edited by Hans Ulrich Gumbrecht and K. Ludwig Pfeiffer, translated by William Whobrey. Stanford: Stanford University Press.

Lumiansky, R. M., and David Mills, eds. 1974. *The Chester Mystery Cycle*. London: Oxford University Press.

Lyotard, Jean-François. 1991. "Can Thought Go on Without a Body?" In *Materialities of Communication*, edited by Hans Ulrich Gumbrecht and K. Ludwig Pfeiffer, translated by William Whobrey. Stanford: Stanford University Press.

M———, R———. 1891. *Micrologia*. In *Character Writings of the Seventeenth Century*, edited by Henry Morley. London: Routledge.

Machyn, Henry. 1848. *Diary of a Resident in London (1550–1563)*. London: Camden Society.

Magnusson, A. Lynne. 1992. "The Rhetoric of Politeness and *Henry VIII.*" *Shakespeare Quarterly* 43, no. 4: 393–409.

Magnusson, A. Lynne. 1998. *Eloquent Relations: Shakespeare's Social Discourse and Elizabethan Epistolary Rhetoric.* Cambridge: Cambridge University Press.

Mallette, Richard. 1998. *Spenser and the Discourses of Reformation England.* Lincoln: University of Nebraska Press.

Manningham, John. 1976. *Diary.* Edited by Robert Parker Sorlein. Hannover, NH: University Press of New England.

Marcus, Leah. 1978. *The Politics of Mirth.* Chicago: University of Chicago Press.

Marcus, Leah. 1995. "Elizabeth and Parliament: Speech, Manuscript, and Print." Paper delivered at annual meeting of Modern Language Association of America, December.

Margeson, J. M. R. 1967. *The Origins of English Tragedy.* Oxford: Clarendon Press.

Markham, Gervase. 1617. *Hobsons Horse-load of Letters: Or, a President for Epistles.* London: Richard Hawkins.

Marlowe, Christopher. 1981. *The Complete Works.* Edited by Fredson Bowers. 2nd edition. 2 volumes. Cambridge: Cambridge University Press.

Marston, John. 1939. *The Plays.* Edited by H. Harvey Wood. 3 volumes. Edinburgh: Oliver and Boyd.

Martin, Henri-Jean. 1994. *The History and Power of Writing.* Translated by Lydia D. Cochrane. Chicago: University of Chicago Press.

Martyr, Peter. 1555. *The Decades of the newe worlde or west India, Conteyning the nauigations and conquestes of the Spanyardes . . . .* Translated by Richard Eden. London: William Powell.

Marx, Karl. 1972. *The Marx-Engels Reader.* Edited by Robert C. Tucker. New York: Norton.

Merleau-Ponty, Maurice. 1989. *Phenomenology of Perception.* Translated by Colin Smith. London: Routledge.

Meteren, Emanuel van. 1865. *History of the Netherlands* (1559). Translated by William Brenchley Rye. In *England as Seen by Foreigners in the Days of James the First.* London: John Russell Smith.

Mignolo, Walter D. 1995. *The Darker Side of the Renaissance: Literacy, Territoriality, and Colonization.* Ann Arbor: University of Michigan Press.

Morley, Henry, ed. 1891. *Character Writings of the Seventeenth Century.* London: Routledge.

Morton, Thomas. 1637. *The New English Canaan . . . Containing an Abstract of New England.* Amsterdam: Jacob Frederick Stam.

Moryson, Fynes. 1617. *An Itinerary . . . containing his ten yeeres of trauell.* London: John Beale.

Moryson, Fynes. 1904. [Unpublished notes for *Itinerary.*] In *Illustrations of*

Irish History and Topography, Mainly of the Seventeenth Century, edited by C. Litton Falkiner. London: Longmans.

Moryson, Fynes. 1967. [Unpublished notes for *Itinerary.*] In *Shakespeare's Europe . . . being unpublished chapters of Fynes Moryson's Itinerary,* edited by Charles Hughes. New York: Benjamin Bloom.

Moxon, Joseph. 1958. *Mechanick Exercises on the Whole Art of Printing.* Edited by Herbert Davis and Henry Carter. London: Oxford University Press.

Mulcaster, Richard. 1582. *The First Part of the Elementarie Which Treateth Chefelie of the right writing of our English tung.* London: Thomas Vautroullier.

Mulryne, R. L., and Margaret Shewring. 1995. *Making Space for Theatre: British Architecture and Theatre since 1958.* Stratford-upon-Avon: Mulryne and Shewring Ltd.

Munday, Anthony. 1923. *John a Kent and John a Cumber.* Oxford: Malone Society.

Munrow, David. 1976. *Instruments of the Middle Ages and Renaissance.* London: Oxford University Press.

Murphy, Cullen. 1998. "Jamestown Revisited." *Preservation* 50, no. 4 (1998): 40–51.

Murray, David. 1991. *Forked Tongues: Speech, writing, and representation in North American Indian texts.* Bloomington: Indiana University Press.

Mysak, Edward D. 1959. "Pitch and Duration Charactersitics of Older Males." *Journal of Speech and Hearing Research* 2: 46–54.

Nashe, Thomas. 1904–1910. *Works.* Edited by F. B. McKerrow. 3 volumes. London: A. H. Bullen.

Naylor, Edward W. 1931. *Shakespeare and Music.* 2nd ed. London: Dent.

Neill, Michael. 1994. "Broken English and Broken Irish: Nation, Language, and the Optic of Power." *Shakespeare Quarterly* 45, no. 1: 1–32.

Nelson, Alan. 1982. "'King I Sit': Problems in Medieval *Titulus* Verse." *Mediaevalia* 8: 189–210.

Nelson, Alan. 1994. *Early Cambridge Theatres: College, University, and Town Stages, 1464–1720.* Cambridge: Cambridge University Press.

Nelson, Alan. 1995. [Untitled: on the Earl of Oxford's contemporary reputation as a sodomite.] Paper delivered at the Folger Shakespeare Library, February.

Nichols, John. 1823. *The Progresses and Public Processions of Queen Elizabeth.* 3 volumes. London: Nichols.

Norden, John. 1618. *The Surveiors Dialogue.* London: Thomas Snodham.

Norden, John. 1593. *Speculum Britanniae. The first parte[.] An historicall and chorographicall discription of Middlesex. . . .* London: [Eliot's Court Press].

Ogilby, John. 1670. *Africa: Being an Accurate Description of the Regions of AEgypt, Barbary, Lybia, and Billedulgerid, The Land of Negroes, Guinee, AEthiopia, and the Abyssines. . . .* London: Thomas Johnson.

Ogilby, John. 1671. *America: Being the Latest, and Most Accurate Description of the New World.* London: printed by the author.

Ong, Walter J. 1944. "The Historical Backgrounds of Elizabethn and Jacobean Punctuation Theory." *PMLA* 59: 349–360.

Ong, Walter J. 1965. "Oral Residue in Tudor Prose Style." *PMLA* 80, no. 3: 145–154.

Ong, Walter J. 1968. "Tudor Writings on Rhetoric." *Studies in the Renaissance* 15: 39–69.

Ong, Walter J. 1982. *Orality and Literacy: The Technologizing of the Word.* London: Methuen.

Orgel, Stephen. 1995. Unpublished paper delivered at the University of Maryland, November.

Orgel, Stephen. 1996. *Impersonations: The Performance of Gender in Shakespeare's England.* Cambridge: Cambridge University Press.

Orlin, Lena. 1994. "The Elizabethan Long Gallery and the Progress of Privacy." Paper delivered at annual meeting of Modern Language Association of America, December.

Orrell, John. 1990. "Beyond the Rose: Design Problems for the Globe Restoration." In *New Issues in the Reconstruction of Shakespeare's Theatre,* edited by Franklin J. Hildy. New York: Peter Lang.

Orrell, John. 1997. "Designing the Globe: Reading the Documents." In *Shakespeare's Globe Rebuilt,* edited by J. R. Mulryne and Margaret Shewring. Cambridge: Cambridge University Press.

Orwin, C. S. 1967. *The Open Fields.* 3rd edition. Oxford: Clarendon Press.

Osborn, James M. 1958. "New Shakespearian Discovery: Benedick's Song in 'Much Ado.'" *The Times* (London), 17 November: 11.

Osborn, James M., ed. 1960. *The Quenes Maiesties Passage through the Citie of London to Westminster on the Day before her Coronacion.* New Haven: Yale University Press.

Overbury, Sir Thomas [and others]. 1616. *His Wife. With Addition of . . . New Newes, and diuers more Characters. . . .* London: Lawrence L'Isle.

Palingenius, Marcellus. 1576. *The Zodiake of life.* Translated by Barnaby Googe. London: Ralph Newbury.

Paré, Ambroise. 1634. *Workes.* Translated by Thomas Johnson. London: Thomas Cotes and R. Young.

Parkes, M. B. 1993. *Pause and Effect: An Introduction to the History of Punctuation in the West.* Berkeley: University of California Press.

Peckham, Sir George. 1940. *A true reporte of the late discoveries.* In *The Voyages and Colonising Enterprises of Sir Humphrey Gilbert,* edited by David B. Quinn. Volume 2. London: The Hakluyt Society.

Pepys, Samuel, comp. 1987. *Catalogue of the Pepys Library at Magdalene College, Cambridge: The Pepys Ballads: Facsimiles.* Edited by Robert Latham. 5 volumes. Cambridge: D. S. Brewer.

Percy, Thomas. 1756. *Reliques of Ancient English Poetry.* 3 volumes. London: J. Dodsley.

Perin, Etienne. 1809. "A Description of England and Scotland." Translator unknown. In *The Antiquarian Repertory*, edited by Francis Grose and Thomas Astle. Volume 4. London: Edward Jeffrey.

Perkins, William. 1608–1609. *Workes*. 3 volumes. Cambridge: John Legate.

Plato. 1937. *The Dialogues*. Translated by B. Jowett. 2 volumes. New York: Random House. [Cited as *Dialogues*.]

Platter, Thomas. 1937. *Travels in England*. Translated by Clare Williams. London: Jonathan Cape.

Potter, Lois. 1999. "Humor Out of Breath: Francis Gentleman and the *Henry IV* Plays." In *Shakespeare: Text and Theater: Essays in Honor of Jay Halio*, edited by Arthur F. Kinney. Newark: University of Delaware Press.

La Primaudaye, Pierre de. 1618. *The French Academie Fully Discoursed and finished in foure Bookes*. Trans. Thomas Bowes, Richard Dolman, and W. P. London: Thomas Adams.

Prouty, C. T. 1942. *George Gascoigne, Elizabethan Courtier, Soldier and Poet*. New York: Columbia University Press.

Quintilian. 1921–1922. *The Institutio Oratoria*. Translated by H. E. Butler. 4 volumes. Loeb Library. London: Heinemann.

Rackin, Phyllis. 1987. "Androgyny, Mimesis, and the Marriage of the Boy Heroine on the English Renaissance Stage." *PMLA* 102: 29–41.

Rainolds, John. 1986. *Oxford Lectures on Aristotle's Rhetoric*. Ed. and trans., Lawrence D. Green. Newark: University of Delaware Press.

Raleigh, Sir Walter. 1632. *Instructions to his Sonne, and to Posterity*. London: Benjamin Fisher.

Rammsla, J. W. Neumayr von. 1865. [Journal of journey made by Johann Ernst, Duke of Saxe-Weimar, through France, England, and Netherlands in 1613–14.] Translated by William Brenchley Rye. In *England as Seen by Foreigners in the Days of James the First*. London: John Russell Smith.

Rathgeb, Jacob. 1865. [Journal of journey made to England by Friedrich Duke of Württemberg-Mömpelgard, 1592 (published 1602)]. Translated by William Brenchley Rye. In *England as Seen by Foreigners in the Days of James the First*. London: John Russell Smith.

Rhodes, Neil. 1992. *The Power of Eloquence and English Renaissance Literature*. New York: St. Martins.

Rice, George P. Jr. 1951. *The Public Speaking of Queen Elizabeth*. New York: Columbia University Press.

Rich, Barnabe. [1609]. *A Short Survey of Ireland. Trvely Discovering Who It Is That Hath so armed the hearts of that people with disobedience to their Prince*. London: B. Sutton and W. Parenger.

Rich, Barnabe. 1610. *A New Descrition of Ireland: Wherein is described the disposition of the Irish whereunto they are inclined*. London: Thomas Adams.

Rich, Barnabe. 1619. *The Irish Hvbbvb, Or, The English Hve and Crie . . . No lesse smarting then tickling*. London: John Marriot.

Rickford, John R. 1997. "Prior Creolization of AAVE [African-American Vernacular English]? Sociohistorical and Textual Evidence from the 17th and 18th Centuries." *Journal of Sociolinguistics* 1, no. 3: 315–336.

Roberts, B. K. 1973. "Field Systems of the West Midlands." In *Studies of Field Systems in the British Isles,* edited by Alan R. H. Baker and Robin A. Butlin. Cambridge: Cambridge University Press.

Robinson, Clement. 1584. *A Handefull of pleasant delites.* London: Richard Jones.

Rollins, Hyder E. 1919. "The Black-Letter Broadside Ballad." *PMLA* 34: 258–339.

Rollins, Hyder E. 1924. *An Analytical Index to the Ballad-Entries (1557–1709) in the Registers of the Company of Stationers of London.* Chapel Hill: University of North Carolina Press.

Rutter, Carol Chillington, ed. 1984. *Documents of the Rose Playhouse.* Manchester: Manchester University Press.

S———, E———. 1949. *A vewe of the present state of Irelande.* In Edmund Spenser, *Works.* Volume 10. Balitmore: Johns Hopkins University Press.

St. Clare Byrne, M. 1954. *Elizabethan Life in Town and Country.* London: Methuen.

Salmon, Vivian. 1979. "Pre-Cartesian Linguistics." In *The Study of Language in Seventeenth-century England.* Amsterdam: John Benjamins.

Salmon, Vivian. 1985. "Missionary Linquistics in Seventeenth-Century Ireland and a North American Analogy." *Historigraphia Linguistica* 12, no. 3: 321–349.

Salmon, Vivian. 1992. "Thomas Harriott and the English Origins of Algonkian Linguistics." *Historigraphica Linguistica* 19, no. 1: 23–56.

Saltonstall, Wye. 1631. *Picturae Loquentes.* London: Thomas Slater.

Saltonstall, Wye. 1946. *Picturae Loquentes.* Enlarged edition (1635). In *Picturae Loquentes,* edited by C. H. Wilkinson. Oxford: Basil Blackwell.

Salzman, L. F., ed. 1951. *The Victoria County History of Warwickshire.* Volume 6. London: University of London Press.

Schafer, R. Murray. 1993. *The Soundscape: Our Sonic Environment and the Tuning of the World.* Rochester, Vt.: Inner Traditions International.

Schreiber, W. L. 1928. *Handbuch der Holz- under Metallschnitte des XV. Jahrhunderts.* Volume 6. Leipzig.

Schwartzstät, Georg von. 1950. "England in 1609." Translated by G. P. V. Akrigg. *Huntington Library Quarterly* 14: 75–94.

Scott, David. 1977. "William Patten and the Authorship of 'Robert Laneham's Letter' (1575)." *ELR* 7: 297–306.

Seng, Peter J., ed. 1978. *Tudor Songs and Ballads from MS Cotton Vespasian A-25.* Cambridge, Mass.: Harvard University Press.

Serres, Michel. 1995. *Genesis.* Translated by Geneviève James and James Nielson. Ann Arbor: University of Michigan Press.

Shakespeare, William. 1968. *The First Folio of Shakespeare.* Edited by Charlton Hinman. New York: Norton.

Shakespeare, William. 1981. *Shakespeare's Plays in Quarto*. Edited by Michael J. B. Allen and Kenneth Muir. Berkeley: University of California Press.

Shakespeare, William. 1988. *The Complete Works*. Edited by Stanley Wells and Gary Taylor. Oxford: Clarendon Press.

Shapiro, Michael. 1977. *Children of the Revels: The Boy Companies of Shakespeare's Time and Their Plays*. New York: Columbia University Press.

Sheale, Richard, comp. 1860. *Songs and Ballads with Other Short Poems, Chiefly of the Reign of Philip and Mary*. Edited by Thomas Wright. London: J. B. Nichols.

Shesgreen, Sean. 1990. *The Criers and Hawkers of London*. Stanford: Stanford University Press.

Sidney, Sir Philip. 1595. *The Defence of Poesie*. London: William Ponsonby.

Simpson, Claude M. 1966. *The British Broadside Ballad and Its Music*. New Brunswick: Rutgers University Press.

Simpson, J. A., and E. S. C. Weiner, eds. 1989. *Oxford English Dictionary*. 20 volumes. Oxford: Clarendon Press.

Sisson, C. J. 1936. *Lost Plays of Shakespeare's Age*. Cambridge: Cambridge University Press.

Smith, B. J. 1971. *Acoustics*. London: Longman.

Smith, Bruce R. 1979. "The Contest of Apollo and Marsyas: Ideas about Music in the Middle Ages." In *By Things Seen: Reference and Recognition in Medieval Thought*, edited by David L. Jeffrey. Ottawa: University of Ottawa Press.

Smith, Bruce R. 1988. *Ancient Scripts and Modern Experience on the English Stage 1500 to 1700*. Princeton: Princeton University Press.

Smith, Bruce R. 1991a. *Homosexual Desire in Shakespeare's England: A Cultural Poetics*. Chicago: University of Chicago Press.

Smith, Bruce R. 1991b. "Reading Lists of Plays in Early Modern England." *Shakespeare Quarterly* 42: 127–144.

Smith, Bruce R. 1996. "L[o]cating the Sexual Subject." In *Alternative Shakespeares, Volume 2*, edited by Terry Hawkes. London: Routledge.

Smith, Irwin. 1964. *Shakespeare's Blackfriars Playhouse: Its History and Its Design*. New York: New York University Press.

Smith, John. 1986. *The Complete Works*. Edited by Philip L. Barbour. 3 volumes. Chapel Hill: University of North Carolina Press.

Snyder, Jon R. 1989. *Writing the Scene of Speaking: Theories of Dialogue in the Late Italian Renaissance*. Stanford: Stanford University Press.

Somerset, J. Alan B., ed. 1994. *Records of Early English Drama: Shropshire*. 2 volumes. Toronto: University of Toronto Press.

Southern, Richard. 1975. *Medieval Theater in the Round*. 2nd ed. London: Faber.

Stanyhurst, Richard. 1587. "A Treatise Conteining a plaine and perfect description of Ireland." In *Chronicles*, compiled by Raphael Holinshed, London: John Harrison, George Bishop, Ralph Newbury, Henry Denham, and Thomas Woodcock.

Sternfeld, F. W. 1963. *Music in Shakespearean Tragedy*. London: Routledge and Kegan Paul.

Stevens, John. 1961. *Music and Poetry in the Early Tudor Court*. London: Methuen.

Stewart, Garrett. 1990. *Reading Voices: Literature and the Phonotext*. Berkeley: University of California Press.

Stokes, James. ed. 1996. *Records of Early English Drama: Somerset*. 2 volumes. Toronto: University of Toronto Press.

Stow, John. 1971. *A Survey of London*. 2 vols. Edited by Charles Lethbridge Kingsford. Oxford: Clarendon Press.

Strachey, William. 1953. "The Historie of travell into Virginia Britania . . . in part gathered, and obteyened, from the industrious and faithful Obseruations, and Commentaries of the first Planters and elder Discouerers; and in parte obserued, by *William Strachey* gent, three yeeres thether imployed. . . ." Edited by Louis B. Wright and Virginia Freund. London: Hakluyt Society.

Stubbes, Phillip. 1877–1879. *Anatomy of Abuses*. Edited by Frederick J. Furnivall. London: Trubner.

Styan, J. L. 1967. *Shakespeare's Stagecraft*. Cambridge: Cambridge University Press.

Tabourot, Jehan (*alias* Thoinot Arbeau). 1925. *Orchesography*. Translated by Cyril W. Beaumont. London: C. W. Beaumont.

Thirsk, Joan. 1984. *The Rural Economy of England*. London: Hambledon.

Thomas of Erfurt. 1972. *Grammatica Speculativa*. Ed. and trans., G. L. Bursill-Hall. London: Longman.

Thomas, Keith. 1986. "The Meaning of Literacy in Early Modern England." In *The Written Word: Literacy in Transition*, edited by Gerd Baumann. Oxford: Clarendon Press.

Thomas, Keith. 1977. "The Place of Laughter in Tudor and Stuart England." *TLS*, 21 January: 77–81.

Thornton, John. 1992. *Africa and Africans in the Making of the Atlantic World, 1400–1680*. Cambridge: Cambridge University Press.

Tite, Sir William. 1845. *A Garland for the New Royal Exchange*. London: J. D. White.

Tory, Geoffroy. 1529. *Champ Fleury*. Paris.

Tory, Geoffroy. 1927. *Champ Fleury*. Translated by George B. Ives. New York: Grolier Club.

Trousdale, Marion. 1982. *Shakespeare and the Rhetoricians*. Chapel Hill: University of North Carolina Press.

Truax, Barry. 1984. *Acoustic Communication*. Norwood, NJ: Ablex.

Tusser, Thomas. 1557. *A Hundreth Good Pointes of Husbandrie*. London: Richard Tottel.

Tuvill, Daniel. 1629. *Vade Mecum. A Manuall of Essayes Morall, Theologicall. Interwouen with moderne Obseruations, Historicall, Politcall*. London: J. Spencer.

Udall, Nicholas (attr.). 1952. *Respublica*. Edited by W. W. Greg. Early English Text Society o. s. 226. London: Oxford University Press.

Underdown, David. 1985. *Revel, Riot, and Rebellion: Popular Politics and Culture in England, 1603–1660*. New York: Oxford University Press.

Valdśstejna, Zdenek Brtnickyz. 1981. *The Diary of Baron Waldstein*. Translated by G. W. Groos. London: Thames and Hudson.

Vansina, Jan. 1985. *Oral Tradition as History*. Madison: University of Wisconsin Press.

Vigne, Louise. 1975. *The Five Senses: Studies in a Literary Tradition*. Lund (SW): Royal Society of Letters.

Vitruvius. 1567. *De Architectura*. Edited by Daniele Barbaro. Venice: Francesco de Francheschi Senese and Giovanni Chrieger Alemano.

Vitruvius. 1931. *On Architecture*. Translated by Frank Granger. Loeb Library. London: Heinemann.

Waage, Frederick O. 1977. "Social Themes in Urban Broadsides of Renaissance England." *Journal of Popular Culture* 11: 731–742.

Wall, Wendy. 1993. *The Imprint of Gender: Authorship and Publication in the English Renaissance*. Ithaca: Cornell University Press.

Walters, John L. 1996. "On Phonography." *TLS*, 26 April: 10.

Wedel, Lupold von. 1895. "Journey Through England and Scotland Made by Lupold von Wedel in the Years 1584 and 1585." Translated by Gottfried von Bülow. *Transactions of the Royal Historical Society*, n. s. 9: 223–270.

Weimann, Robert. 1996. *Authority and Representation in Early Modern Discourse*. Edited by David Hillman. Baltimore: Johns Hopkins University Press, 1996.

Wells, Stanley, and Gary Taylor. 1987. *William Shakespeare: A Textual Companion*. Oxford: Clarendon Press.

Wheatley, Henry, and Peter Cunningham. 1891. *London Past and Present*. London: John Murray.

Wickham, Glynne. 1972. *Early English Stages 1300 to 1660*. Volume 2. 2. London: Routledge.

Wickham, Glynne. 1987. *The Medieval Theatre*. 3rd ed. Cambridge: Cambridge University Press.

Wiles, David. 1981. *The Early Plays of Robin Hood*. Cambridge: D. S. Brewer.

Wiles, David. 1987. *Shakespeare's Clown: Actor and Text in the Elizabethan Playhouse*. Cambridge: Cambridge University Press.

Wilkinson, Henry. 1933. *The Adventurers of Bermuda*. Oxford: Oxford University Press.

Williams, Gordon. 1994. *A Dictionary of Sexual Language and Imagery in Shakespearean and Stuart Literature*. 3 volumes. London: Athlone Press.

Williams, Raymond. 1973. "Base and Superstructure in Marxist Cultural Theory." *New Left Review* 82: 3–16.

Williams, Roger. 1643. *A Key into the Language of America: Or, An help to the*

Language of the Natives in that part of America, called New-England. London: Gregory Dexter.

Wilson, K. J. *Incomplete Fictions: The Formation of the English Renaissance Dialogue*. 1985. Washington, D.C.: Catholic University of America Press.

Wilson, Thomas. 1557. *The Arte of Rhetorique*. London: [Richard Grafton].

Wood, William. 1634. *New Englands Prospect. A true, liuely, and experimentall description of that part of* America *commonly called* NEW ENGLAND. London: Thomas Cotes.

Woodfield, Ian. 1995. *English Musicians in the Age of Exploration*. Stuyvesant, NY: Pendragon Press.

Wright, Thomas. 1630. *The Passions of the Mind in Generall*. London: Miles Flesher for Robert Dawlman.

Wrightson, Keith. 1982. *English Society, 1580–1680*. New Brunswick: Rutgers University Press.

Würzbach, Natascha. 1990. *The Rise of the English Street Ballad 1550–1650*. Translated by Gayna Walls. Cambridge: Cambridge University Press.

Yates, Frances A. 1966. *The Art of Memory*. London: Routledge and Kegan Paul.

Zemlin, Willard R. 1964. *Speech and Hearing Science: Anatomy and Physiology*. Champagne, IL: Stipes.

Zumthor, Paul. 1990. *Oral Poetry: An Introduction*. Minneapolis: University of Minnesota Press.

Zumthor, Paul. 1994. "Body and Performance." In *Materialities of Communication*, edited by Hans Ulrich Gumbrecht and K. Ludwig Pfeiffer, translated by William Whobrey. Stanford: Stanford University Press.

I n d e x

Page references to figures and musical quotations appear in italics.